The Victorian Eye

The Victorian Eye

A Political History of Light and Vision in Britain, 1800–1910

CHRIS OTTER

The University of Chicago Press   *Chicago and London*

CHRIS OTTER is assistant professor in the Department of History at
Ohio State University.

The University of Chicago Press, Chicago 60637
The University of Chicago Press, Ltd., London
© 2008 by The University of Chicago
All rights reserved. Published 2008
Printed in the United States of America

17 16 15 14 13 12 11 10 09 08   1 2 3 4 5
ISBN-13: 978-0-226-64076-1 (cloth)
ISBN-13: 978-0-226-64077-8 (paper)
ISBN-10: 0-226-64076-0 (cloth)
ISBN-10: 0-226-64077-9 (paper)

Library of Congress Cataloging-in-Publication Data

Otter, Chris.
   The Victorian eye : a political history of light and vision in
Britain, 1800–1910 / Chris Otter.
     p. cm.
   Includes bibliographical references and index.
   ISBN-13: 978-0-226-64076-1 (cloth : alk. paper)
   ISBN-10: 0-226-64076-0 (cloth : alk. paper)
   ISBN-13: 978-0-226-64077-8 (pbk. : alk. paper)
   ISBN-10: 0-226-64077-9 (pbk. : alk. paper)
   1. Lighting—Great Britain—History—19th century.
2. Lighting—Great Britain—History—20th century.   3. Lighting—
Social aspects—Great Britain—History.   4. Lighting—Political
aspects—Great Britain—History.   5. Optical   engineering—Great
Britain—History—19th century.   6. Visual perception.   7. Great
Britain—Social life and customs—19th century.   I. Title.
TH7703.O88   2008
303.48′3—dc22   2007043184

Contents

Illustrations

Acknowledgments

This book began in Manchester in the late 1990s and was finished in the spring of 2007 in New York, thanks to a generous fellowship from the International Center for Advanced Studies at NYU. Throughout the many years of the book's slow gestation, I have received more help than I can possibly acknowledge here. Special thanks to Patrick Joyce, whose idea it was to "write a book about light" and whose intellectual vigor and curiosity continues to be a great inspiration to me. Special thanks also to James Vernon, who read the entire manuscript and provided the most detailed, astute commentary, and who also helped make the transition to American academe a real pleasure. Over the years, many, many friends and colleagues have offered generous insights and advice, in response to draft chapters and papers airing some of the book's many themes, or simply while chatting over coffee or beer. Among them are Jordanna Bailkin, Jane Burbank, Tom Bender, Neil Brenner, Herrick Chapman, Harry Cocks, Lisa Cody, Steven Connor, Fred Cooper, Tom Crook, Francis Dodsworth, Kate Flint, Elaine Freedgood, Graeme Gooday, Manu Goswami, Simon Gunn, Stephen Kotkin, Yanni Kotsonis, Andy Lakoff, Sharon Marcus, Matthew McCormack, Frank Mort, Lynda Nead, Molly Nolan, Susan Pedersen, John Pickstone, Nathan Roberts, John Shovlin, Maiken Umbach, Daniel Ussishkin, Carl Wennerlind, and Caitlin Zaloom.

At the University of Chicago Press, Doug Mitchell has provided enthusiasm and wisdom in equally large amounts. I would also like to thank Mark Reschke, Joe Brown, Robert

Hunt, and Timothy McGovern for their patience and assistance at various stages of this book's germination. An article serving as the basis for part of chapter 2 originally appeared as "Making Liberalism Durable: Vision and Civility in the Late Victorian City," *Social History* 27, no. 1 (2002): 1–15.

Finally, I would like to thank my family. Tina Sessa has been the sharpest reader, and greatest encourager, of my work for a long time. Without her, this book would not exist. During the final year of the book's completion, we were joined by our first son, Nicholas. This book is dedicated to my parents, Eva and Patrick Otter, who inspired me to read, think, and write in the first place.

Introduction:
Light, Vision, and Power

This is a book about light, vision, and power in nineteenth-century Britain. It argues that the ways in which streets, houses, and institutions were lit, and the ways in which people saw within them, have a political history. *Who* could see *what*, *whom*, *when*, *where*, and *how* was, and remains, an integral dimension of the everyday operation and experience of power. Yet the critical tools, concepts, and frameworks usually used to analyze this visual form of power are inadequate and misleading. The history of vision and power over the past couple of European centuries is invariably written as a history of either discipline or spectacle, or some combination of both. This book rejects such an approach. Instead, it argues that the nineteenth-century history of light and vision is best analyzed as part of the history of freedom, in its peculiarly and specifically British form.

At the beginning of the twentieth century, artificial light was routinely viewed as the supreme sign of "modernity" or "civilization." In 1902, for example, the chemist William Dibdin reflected on the previous century's advances in illumination and nocturnal perception in tones simultaneously reverential and pensive: "The necessities of modern civilization having to so large an extent turned night into day both in the working world as well as in that of the world of pleasure and social intercourse when the day's work is done, a state of things has arisen in which artificial illumination holds the very first place, as without it the whole scheme of present day society would at once fall to the ground."[1] This "modernity" of illumination systems, particularly electric ones, and their capacity to "turn night into day," has become integral to a narrative that (often tacitly) pervades cultural theory and history, not to mention society more broadly.[2] At its crudest, but also most powerful, the European past is dark and gloomy, and its historical present, formed over the nineteenth century, is glittering and radiant. Wolfgang Schivelbusch, for example, describes the appearance of electric light as a visual "apotheosis," an effulgence so shocking and radical that Parisian ladies were forced to unfurl parasols to protect their delicate retinas.[3] Electric light was the "culmination" of a century's relentless drive toward spectacular radiance, generating a "fairyland environment" or "celestial landscape."[4] Night, in turn, has been conquered, colonized, divested of mystery.[5] The future, meanwhile, will be only more brilliant and starless: "The urban landscape of the future will be characterised by an almost perpetual illumination which practically defies the natural order of day and night."[6]

Most of this scholarship is sophisticated and scrupulous, and I have no desire to caricature it. Nonetheless, it is clear that twentieth-century cultural historians have created a powerful, influential narrative that depicts "Western modernity" in terms of the relentless expansion of illumination. The production of illuminated, disenchanted modern space is, moreover, invariably seen as integral to two specifically visual historical processes: the rise of surveillance and the development of spectacle. In the former, illumination is the means through which society is permeated by a nefarious, anonymous, disciplinary gaze: light is a glittering trap. In the latter, illumination is seductive and dazzling, creating the stage on which the commodity makes its breathtaking appearance: light is deceptive and narcotic. The cultural history of light and vision thus becomes inseparable from two political histories, those of discipline and of capital. These two paradigms, the *disciplinary* and the *spectacular*, are embodied in two figures, one architectural and one human: the panopticon and the flâneur, both of which have developed a cultural and theoretical significance far beyond studies of illumination or visual culture. This book will complicate, critique, and unsettle the paradigms of discipline/panopticism and spectacle/flâneur, and the particular political histories that support them, by arguing that the visual dimensions of space were, in general, engineered with neither coercion nor seduction in mind. It aims to replace these rather procrustean paradigms with a suppler and broader range of terms that are both more empirically satisfying and more analytically useful and, thus, to recast the political history of light and vision as part of a material history of Western liberalism. The first thing to do, then, is to examine these hegemonic visual concepts and ask in what ways they are unsatisfactory.

Questioning Visual Concepts: Panopticism and the Flâneur

Over the past three decades, drawing on the foundational texts of Foucault and Benjamin, scholars have produced a rich, interdisciplinary body of work on the historical relations between vision and power.[7] The panopticon and the flâneur loom large in such analyses. No lexicon of contemporary cultural theory would be complete without them. Neither, however, is particularly useful when attempting to understand the politics of light and vision in nineteenth-century Britain, and they are of probably as equally limited use elsewhere.

Bentham's panopticon, devised between 1787 and 1791, was, according to Foucault, a cogent solution to several pressing contemporary problems of government, relating to crime, health, and morality: "A fear haunted the latter half of the eighteenth century: the fear of darkened spaces, of the pall of gloom which prevents the full visibility of things, men and truths. . . . A form of power whose main instance is that of opinion will refuse to tolerate areas of darkness. If Bentham's project aroused interest, this was because it provided a formula applicable to many domains, the formula of 'power through transparency,' subjection by 'illumination.'"[8] The architectural details of the panopticon will probably be familiar to most readers.[9] The inmate, according to Bentham, should always "*conceive* himself to be" inspected from the panopticon's central watchtower, even if the inspector was actually absent. He would be "awed to silence by an invisible eye" and rendered compliant and docile. The windows of each cell would be "as large as the strength of the building . . . will permit," while firm partitions prevented inmates from seeing each other. As dusk fell, lamps would "throw the light into the corresponding cells, [which] would extend to the night the security of the day."[10] Blinds or lanterns prevented the inspector from being detected in his lodge while allowing him to read or work there. One could never verify whether one was not being watched. The panopticon produced total asymmetry of vision: a gallery of illuminated inmates helplessly "subjected to a field of visibility," imagining themselves to be permanently watched by an omniscient, invisible, and possibly absent, inspector.[11] This was, Bentham declared, an "instrument of government," and he began his *Panopticon Letters* with a list of the moral and physical benefits of his "*simple idea in Architecture!*"[12]

This was asymmetrical "subjection by illumination," in which light is used as a direct, coercive instrument of power. Since the publication of Foucault's *Discipline and Punish* (1977), with its memorable analysis of Bentham's "simple idea," panopticism has become the dominant paradigm for understanding the visual operation of power in post-Enlightenment Europe. It is "the universal optical machine of human groupings," a model, or "diagram of power," used to structure not just prisons and schools but even whole cities and societies.[13] The panopticon was "a type that flourished for a century" and might, in the later twentieth century, be in the process of mutating into something perhaps even more pernicious, a telematic or computerized society of "control," typified by conduct so predictable that the massive, forbidding paraphernalia of panopticism has become obsolete.[14] Other scholars, however, see in the early-twenty-first-century world of CCTV (closed-circuit television), Internet surveillance, and nightsun helicopters nothing less than an intensification or a perfection of panopticism, through far more sophisticated, pervasive, and miniaturized, even nanopanoptic, techniques: "The Panopticon is 'present' nearly everywhere."[15] Nowhere is this statement more correct than in academe, where a cursory search of databases and books finds panopticism being used to explicate the politics of photography, physical appearance, the Internet, CCTV, cartography, children's playgrounds, consumer space, sport, incest, audit culture, travel, and the novels of Charles Dickens.[16]

The resulting narrative, again, will be familiar: the past two hundred years have witnessed the rise of malign, insidious surveillance.[17] Modern brightness is inescapable. Illumination, and the gazes it makes possible, traps us all, not just the prisoner in the cell. The tacit premise of much of this literature—that vision and power are symbiotic and have taken specific forms over the past two hundred years—is indisputable. But these forms have almost invariably not been panoptic. *Panopticism* has been emptied of meaning to the point where it simply refers to any configuration of vision and power, any technological or architectural arrangement designed to facilitate the observation of some humans by others. We have seen the retrospective panopticization of a Western society that was, historically, not panoptic. As Lauren Goodlad states, the contemporary obsession with panopticism has made us, ironically, historically myopic: we think, talk, and write "more about panopticism than [ask] why it was that nineteenth-century Britons declined to build any Panopticons."[18]

This last point is obvious but deceptive. There were several failed attempts to build panopticons: for example, at the Edinburgh Bridewell

(1791), where an already compromised semicircular arrangement was further thwarted by the imposition of a belt of workshops beyond the cells that occluded nocturnal visibility.[19] In 1854, the Panopticon of Science and Art opened in Leicester Square. The purpose here was to exhibit notable objects (self-acting lathes, plaster casts) rather than subjects, and the venture lasted only three years.[20] Many putative "panopticons" turn out, on closer inspection, not to be panoptic at all: for example, the "Jamaican Panopticon" discussed by Thomas Holt or the "panoptical" inspection of prostitutes described by Philippa Levine.[21] These visual regimes were asymmetrical and coercive, to be sure, but this asymmetry and coercion never took a panoptic form. My point might appear pedantic. It might be argued that I have conflated panopticism with physical panopticons and that the absence of the latter does not disprove the force and drive toward the former. Such is the reasoning of those who argue that panopticism should be interpreted as an "ideal analytic type" rather than a set of concrete structures.[22] But we must seriously doubt the extent to which panopticism can function without material systems to sustain it.

Perfect, transparent vision of society remained, and remains, elusive, undesirable, impossible, and probably meaningless: Western governments have usually been quite happy to tolerate broad areas of darkness, indeed, to actively create them. Even Bentham acknowledged the need for rudimentary screens to preserve the dignity of defecating panopticon inmates.[23] Little techniques of privacy infiltrated the apparatus, which, in its purest form, could never work, as Gauchet and Swain have argued: "Some of our contemporaries . . . act as if the project could work; as if, for example, the 'eye of power' that is positioned at the core of the panoptic machine, infinitely open and theoretically infallible in its exhaustive force, although lacking a gaze, could finally notice anything but the void."[24] This is not, again, to deny that vision and power have often operated in forms that were both asymmetrical and disciplinary. It is merely to suggest that we need a far more expansive vocabulary, and a much more flexible topographic framework, to capture multiple modalities or patterns of vision: Foucault certainly described others.[25] Indeed, several neologisms have recently been coined, suggesting the need for analytic plurality: the *synopticon*, the *polyopticon*, the *omnicon*, and the *oligopticon*.[26] We should also be aware, in passing, of modalities of vision that have less to do with power than with emotional and affective experience.[27] The historical understanding of vision needs to be thickened and nuanced, and escaping the panopticon is the best place to start.

The second hegemonic visual paradigm, that of the flâneur, owes more to Baudelaire and Benjamin than to Bentham and Foucault. In

"The Painter of Modern Life," Baudelaire described the flâneur as a "passionate spectator" who "set[s] up house in the heart of the multitude . . . a kaleidoscope gifted with consciousness," recording or capturing the fleeting and transient.[28] The flâneur is aloof, anonymous, and perpetually fascinated with the spectacle of urban life. Drifting slowly through the crowd, often nocturnally, sometimes intoxicated, he possesses a form of urban intelligence that those around him lack: "Preformed in the figure of the flâneur is that of the detective."[29] His skills are those of the physiognomist, as Benjamin noted: "The flâneur has made a study of the physiognomic appearance of people in order to discover their nationality and social station, character and destiny, from a perusal of their gait, build and play of features." Just as physiognomic knowledge circulated via texts and images, so the flâneur was a profoundly literary creature: "The social basis of flânerie is journalism."[30]

The physical locus of flânerie was, thus, the urban crowd and spectacle. The flâneur meandered through the arcades, streets, department stores, and hotels of the city center, thriving amid "urban brilliance and luxury."[31] The flâneur was drawn mothlike to the light of seduction, where "capitalism is illuminated and made brilliant."[32] Such urban centers were, of course, the glittering milieus where electric light first appeared and where the phenomenology of urban modernity—speed, distraction, alienation, fragmentation, illumination—has been most frequently located by scholars.[33] The perambulating flâneur, intoxicated with spectacle, was, thus, a product of the same forces of commodification that he theoretically resisted and critiqued.

In *The Spectator and the City*, Dana Brand argues that the flâneur's origins are less French than English, emerging from cultural spectacle and literary practice in seventeenth- and eighteenth-century London.[34] In the nineteenth century, journals like *Blackwood's Magazine* and Dickens's literature popularized this form of urban spectatorship. In a *Household Words* article entitled "The Secrets of the Gas" (1854), George Augustus Sala produced a paradigmatic description of this nocturnal, omniscient figure: "He who will bend himself to listen to, and avail himself, of the secrets of the gas, may walk through London streets proud in the consciousness of being an Inspector—in the great police force of philosophy—and of carrying a perpetual bull's-eye in his belt. . . . Not a bolt or bar, not a lock or fastening, not a houseless night-wanderer, not a homeless dog, shall escape that searching ray of light which the gas shall lend him, to see and to know."[35] Urban journalism, wry taxonomies of urban "types," and visual spectacles displaying collections of objects, images, or even people (exhibitions, art galleries, waxworks) flourished during the nineteenth century in most major European cities, not just London and Paris. In such conditions, Brand continued, the flâneur "becomes a dominant form."[36]

Rather like that of the panopticon, however, the flâneur's dominance is more evident in late-twentieth-century cultural studies texts than on the streets of the nineteenth-century city. Flânerie was an exclusive, metropolitan, elitist, narcissistic practice, limited to a select group of writers who seldom used the term *flâneur* to describe themselves. Benjamin himself thought that the flâneur's extinction began in the 1840s: he smoked hashish and rambled round Marseille in a valiant attempt to replicate nineteenth-century languor. Sala's omniscient stalker is a purely fantastic figure. In a move reminiscent of panopticism's unyoking from the panopticon, abstract flânerie has been disconnected from the physical flâneur: "If the flâneur has disappeared as a specific figure, the perceptive attitude that he embodied saturates modern experience, specifically, the society of mass consumption. In the flâneur, we recognise our own consumerist mode of being-in-the-world."[37] But the fact that crowds flocked to exhibitions, aquariums, and art galleries does not prove that their experience was in any way aloof, ironic, all knowing, or literary. It is unhelpful and reductive to characterize any isolated figure moving through urban space with his or her eyes open, or any journalist recording the minutiae of city life, as a flâneur. Then as well as now, a spectacle produces boredom and indifference among many, if not most, of its spectators. Furthermore, such practices were highly occasional and limited to urban centers. Flânerie, quite simply, cannot be seen as a representative visual practice in nineteenth-century Britain.[38]

Panopticism and flânerie have radically different histories and embody completely different power relations. The former implies a cruel, cold, fixed gaze, the latter a more playful, empowered, and mobile one. But they are, in fact, mutually reinforcing. Both, after all, are fantasies, one architectural, the other literary. And their fantasy is of total knowledge of a subject population, be it of a body of criminals or of an urban crowd. The flâneur moves everywhere and sees everything, while the prisoner of the panopticon is permanently seen and known. A fantasy of omniscience underlies both models. The flâneur has maximum freedom and knowledge, the panopticon inmate a minimum of both.

It is easy, too easy, to slide from the empirical to the abstract and to allow such abstract concepts to assume a life of their own. Before we know it, we are characterizing nineteenth-century visual culture in terms of "floods of light," "panopticism" and flânerie, even though hardly anybody wanted floodlights, nobody built panopticons, and flâneurs were almost entirely absent. Factories, asylums, and workhouses were built at an impressive pace, but none of these structures were panoptic, despite the occasionally deceptive appearance of towers or annular rooms. Cities swelled, and their public spaces thronged with crowds, but flânerie remained a marginal practice, a luxury few could afford and still fewer desired. Oil, gas, and electric illumination flourished, but they produced neither panoptic trap nor flâneuristic stage. What, then, were the forms taken by light and vision? For the most part, this book will answer this question positively rather than negatively, but it is helpful to stress a set of formal limits to the spread and intensity of illumination.

The Limits of Illumination

Here, I will emphasize three fundamental limitations to the development of illumination: technological, optical, and politicocultural.[39] The first, the technological, is the most obvious. Networks themselves routinely broke down or leaked, while money for expansion and repair was frequently lacking. Even functional systems produced illumination that was often derided as gloomy: following a flawed installation of electric light in Hull in 1882, for example, observers commented on the wretched quality of all contemporary light forms, despite their burgeoning number.[40] Most late-nineteenth-century Britons still relied largely on oil lamps and candles. In 1888, only thirteen hundred of Crewe's six thousand houses were supplied with gas.[41] In 1895, the *Engineer* scathingly observed: "The principal streets [of London] are lighted [*sic*] in a manner which astonishes the foreigner and incites the American to contemptuous scorn."[42] The point is basic. We must avoid talking abstractly about things like "the city" being "flooded with light." Instead, we must speak with more spatial specificity about particular cities, explore their idiosyncratic networks, and listen to what contemporaries had to say about them. Only then can tentative generalizations be drawn.

Nonetheless, it might be argued that engineers wanted to flood cities with light but that the technology was simply incapable of achieving this. The second limit to illumination, the optical, confounds this straightforward conclusion. As Schivelbusch's anecdote about parasols suggests, overillumination was often as much of a problem as dimness. In his popular nineteenth-century textbook *Diseases of the Eye*, Edward Nettleship, surgeon to the Royal London Ophthalmological Hospital, observed that astringent, piercing electric arc light could damage the optical apparatus: "Attacks, apparently identical with snow-blindness, but of shorter duration, sometimes occur in men engaged in trimming powerful electric lights."[43] Ocular physiology is routinely ignored in both technological and cultural histories of light, but we must take it very seriously since nineteenth- and early-twentieth-century engineers did. "The science of artificial lighting," noted the civil engineer William Webber in *Town Gas* (1907), entailed simultaneous attention "to physical, physiological, and economic considerations."[44] William Dibdin, too, laid great emphasis on the physiological dimension of illuminating engineering:

Now-a-days many people spend a considerable proportion of their lives, especially during the winter months, working by artificial light. Every time the direct rays from a light source impinge upon the retina, the iris, or "pupil" of the eye rapidly closes until the intensity of the rays passing through it is reduced to bearable limits. As soon as the direct rays cease to enter the eye the pupil expands in order that sufficient light from a less illuminated object can act upon the optic nerve, otherwise the object viewed would be invisible, or nearly so. The constant action of the pupil, or guardian angel of the eye, as it might be termed, combined with that on the optic nerve, and the crystalline lens, becomes most fatiguing, and in time unquestionably affects the power of vision.[45]

Dibdin drew attention to variable or erratic, rather than intense, illumination. Light should be both steady and tempered because the eye was an active producer of vision rather than a passive orifice through which light streamed en route to the mind. Even if the technology were available to genuinely flood a city with light, using it would be futile since vision would be rendered dysfunctional and unbearable.

This leads into the third limit to the spread of illumination. There was, baldly stated, much resistance to, and almost no support for, the idea of a totally illuminated society. Opposition did not come just from isolated aesthetes like William Morris and Robert Louis Stephenson, who saw in gas and electricity, respectively, something vulgar and disturbing.[46] There was a far more pervasive and less splenetic resistance that had little to do with aesthetics, as the *Electrician* made clear in 1880: "To light a whole city with a huge electrical sun is a great scientific achievement; but it is not the sort of light that anybody wants."[47] A city from which darkness had been expunged to allow generalized omniscience was a city devoid of that most cherished value, personal privacy, or the ability to altogether escape from the gaze of others, and this is why the idea was repugnant. Pace Foucault and Schivelbusch, in Western Europe, this idea has never been successful or popular, despite the seductive appeal of occasional utopian schemes.[48]

The moment one begins to write a history of light and vision, then, one finds oneself simultaneously engaging with technology, the eye, and politics. In particular, writing such a history draws us to the question of illumination's function, in terms of the visual and bodily capacities it was routinely used to produce, shape, and stimulate. Illumination was used in streets to facilitate the detection of moving objects or the discernment of street signs; in operating theaters to scrutinize the inner surface of bodily organs; in factories to allow workers to accurately match colors at night; in warfare to transmit signals or illuminate enemy troop movements; in housing for bedtime reading or comfortable nocturnal visits to the lavatory. These tacit perceptual practices were deeply embedded in habitual, daily routines. They are not, however, too tacit or banal for historical analysis. A central thesis of this book is that these visual practices, and the technologies securing them, have a political history that cannot be captured with the limited range of concepts provided by cultural theory. To repeat: who could see what, whom, when, where, and how was a profoundly important political question, but the answer never came in the form of a single architectural plan, text, or treatise. There were many, contested answers, something that makes the political history of light and vision the history of multiple, overlapping perceptual patterns and practices rather than singular paradigms. To contextualize this, I will here examine in more detail the prevalent political ideas of the day, which were liberal ones, and specifically address the relation between liberalism and technological proliferation. If we view the growth and management of gas and electricity networks as part of the development of a "technological state" and then examine how technologies were themselves believed to shape or encourage forms of conduct in ways that are irreducible to discipline and spectatorship, we will, one hopes, be in a better position to assess the politics of light and vision during the period.

Liberalism

Liberalism is a notoriously protean, slippery term, as Isaiah Berlin famously argued: "Like happiness and goodness, like nature and reality, the meaning of this term is so porous that there is little interpretation that it seems able to resist."[49] Yet this has, probably, been integral to its success. In Victorian Britain, liberalism was expansive and heterogeneous, drawing on numerous intellectual traditions, including utilitarianism, political economy, evangelicalism, and romanticism.[50] These disparate strands contributed to one of liberalism's central features: its restlessly critical and self-critical nature. This was, perhaps, most manifest in the general suspicion of state power that persisted across the century and united Smith and Burke with Mill, Spencer, and even L. T. Hobhouse. Whatever these thinkers' differences, they all argued that state power should be significantly limited, particularly toward economic activity.[51] They simply disagreed over precisely where to draw the line. A complementary feature of liberal thought was the emphasis on self-government, be this individual, municipal, or local. As Mill observed in *Considerations on Representative Government*: "It is but a small portion of the public business of a country which can be well done, or safely attempted, by the central authorities."[52] A society could be deemed civilized only to the extent to which its citizens were acting under their own volition.

Liberalism can, thus, be characterized, to adopt Berlin's parlance, rather "negatively," as a critique of overgovernment.[53] But liberals also had very strong "positive" and normative ideas.[54] These centered on the "liberal subject," the kind of human being targeted and presupposed by such a minimal state, a being simultaneously free and self-governing, on the one hand, and subjected and governed, on the other, which rather elides the classic distinction between negative and positive liberty.[55] "By the 1880s," notes Peter Mandler, "something recognisable as the 'liberal subject' was widespread in stabilising urban communities across Britain."[56] The rights and freedoms of this subject, stated Mill, were "accessible to all who are in the normal condition of a human being."[57] Mill's normal condition was rather exacting and far from universal: it involved being rational, sane, self-disciplined, independent, thrifty, sober, and energetic.[58] Some of these attributes translated into legal rights, like the vote, but others, operating in a more socially normative fashion, secured important forms of cultural capital, like respect. All these qualities were subsumed under the nebulous but central concept of character.[59] The creation of character was, critically, a deeply bodily enterprise, a process by which one took the physical attributes of oneself as an object to be worked on, improved, and disciplined. Thus, the ethical formation of the subject involved the cultivation of cleanliness, sexual moderation, sobriety, physical fitness, and good health. A society composed of such well-drilled "men of character" barely needed a state to govern it: "Men of character are not only the conscience of society, but in every well-governed State they are its best motive power."[60] Conversely, the state might cautiously be called on to provide institutions or resources to assist the positive process of character building.

Liberalism, as understood here, is not a cohesive body of ideas.[61] Adam Smith, for example, was explicit about the basically pragmatic nature of liberal political practice, attacking the "man of system" who aims to implement something like an "ideal plan of government."[62] I am not referring to it, as some have done, as a *doctrine*, and it cannot be reduced simply to the thoughts or philosophies of undeniably important writers like Mill.[63] Patently "liberal" ideas appear in innumerable discursive sites, from evangelical pamphlets to charitable programs, popular history to novels. They are also clearly found in the language and practice of doctors, engineers, builders, and sanitarians. The monotony, anonymity, and pervasiveness of such ideas mark liberalism as a *discourse*, albeit one riddled with contradictions and contestations.[64] It is also unhelpful to rigidly equate liberalism with the practice or ideology of a formally "Liberal" political party. Demonstrably "liberal" policies and ideas (commitment to freedom of trade and conscience, balanced budgets) were never the preserve of a single party in Victorian Britain.[65] In fact, one of the key nineteenth-century developments in government was the creation of administrative systems, like the civil service, that would be unaffected by election results. Government, we might argue, was being institutionally unhooked from politics.[66]

Building and Governing a Technological State

As historians have often noted, the rise of liberal critique of the British state coincided with substantial expansion of administration and bureaucracy. This occurred for many reasons, for example, the development of specific state projects like the 1834 Poor Law or the consolidation and extension of empire.[67] Government also expanded along with technological infrastructure: the rapid development of a national railway network is, perhaps, the most obvious example, but gas, water, sewerage, telegraphy, and, later, electricity all ultimately required some form of official government action, in the form of legislation, institutions, inspectorates, or even state ownership. This action did not necessarily take the form of centralization: it invariably operated through the locality, and attempts at genuine centralization, like Edwin Chadwick's General Board of Health, rarely succeeded.

Liberal writers seldom, if ever, argued that the state should do nothing about such matters. In *The Wealth of Nations*, Smith was adamant that the sovereign had a duty to build and maintain some public works, like roads, for example, through the use of tolls. He saw lighting as less significant and purely a matter for private interest: "Were the streets of London to be lighted and paved at the expence of the treasury, is there any probability that they would be so well lighted and paved as they are at present, or even at so small an expence?"[68] Eighty-five years later, such matters were taken altogether more seriously by Mill, who argued that local government should take responsibility for such things: "The different quarters of the same town have seldom or never any material diversities of local interest; they all require to have the same things done, the same expenses incurred. . . . Paving, lighting, water supply, drainage, port and market regulations, cannot without great waste and inconvenience be different for different quarters of the same town."[69] Formal bodies, in particular the newly reorganized municipal governments, should be equipped with the capacity to fund, build, and run technological systems. The realm of legitimate intervention had increased, not so much because liberalism was being eroded, as because of a demonstrable growth in large technical systems that required organization and regulation. The historian must view the development of gas and electric light in this context. A parliamentary select committee met in 1809 to consider the question of incorporating gas companies, and others sat thereafter to discuss issues like explosions or the quality of gas. In 1847, Parliament passed the Gasworks Clauses Act, revised on numerous occasions thereafter, which greatly expedited the process whereby a municipality or other local political unit could obtain permission to construct a gas network. The 1859 Sale of Gas Act defined the "cubic foot" as the legal unit in which gas was to be measured, bought, and sold, while gas meters were to be officially stamped and the range of legitimate inaccuracy fixed. A year later, the Metropolitan Gas Act laid down firm requirements for testing London's light, which had been subject to legal minima by a series of acts from 1850 on. The provision of electric light was legally regulated via acts of 1882 and 1888.

Illumination was, thus, a collective need, requiring legislation and judicious government. It was too precious to be left only to the vicissitudes of the market. Even private gas companies had their prices regulated and were forced to accept (very generous) legal maximum dividends. Was this socialism? Palpably not, declared Winston Churchill in 1906:

Collectively we have an Army and a Navy and a Civil Service; collectively we have a Post Office, and a police, and a Government; collectively we light our streets and supply ourselves with water; collectively we indulge increasingly in all the necessities of communication. But we do not make love collectively, and the ladies do not marry us collectively, and we do not eat collectively, and we do not die collectively, and it is not collectively that we face the sorrows and the hopes, the winnings and the losings of this world of accident and storm.[70]

The characteristic liberal urge to split state responsibility from that of the individual is preserved, but the realm of legitimately "collective" practices had greatly expanded in the century or so since the *The Wealth of Nations*. The bulk of these (gaslight, large-scale sanitary networks, telegraphy, the police service, the penny post) were simply nonexistent in Smith's day.

It has become commonplace in histories of the European state to posit a shift from the government of territory to that of population, one occurring over the course of the eighteenth century and the early nineteenth and demonstrable through the rise of police, statistical reason, and public health.[71] The later eighteenth century and the nineteenth clearly saw the emergence of a third stratum of government, that of technology. This did not, of course, mean that territory and population ceased to be government concerns, but it did mean that they were increasingly seen as, potentially, technologically governable. The state itself, its cities and its colonies, was often physically engineered into a state of governability, something visible in schemes of land drainage, bridge building, and canal construction in Ireland and Scotland in the early nineteenth century and in the huge imperial railway projects later in the same century.[72] Infrastructure and engineering were, thus, integral to the development of the British state.[73] Infrastructure was often explicitly viewed as the provider of a liberty that was both positive and collective: it enhanced the capacity of potentially large groups of individuals. Here is Samuel Smiles in his *Lives of the Engineers*: "Freedom itself cannot exist without free communication,—every limitation of movement on the part of the members of society amounting to a positive abridgement of their personal liberty. Hence roads, canals, and railways, by providing the greatest possible facilities for locomotion and information, are essential for the freedom of all classes, of the poorest as well as the richest."[74]

To repeat: the management of these systems constituted a substantial part of the Victorian "growth of government."[75] Parliament was simply too small and too full of politicians to be able to consider in detail the physics of road surfaces, the chemistry of foodstuffs, or the biology of sewage. An increasing volume of government business was delegated to "experts," individuals with specialist knowledge.[76] In his 1842 *Sanitary Report*, Edwin Chadwick urged that the "most important branches of administration" should be lifted "out of the influence of petty and sinister interests, and of doing so by securing the appointment of officers of superior scientific attainments."[77] Such experts (chemists, engineers) held regular meetings, founded countless journals, and developed professional identities across the period, while municipal councils began appointing city engineers, medical officers of health, surveyors, and public analysts.

In the nineteenth century, engineering as a profession first emerged and then flourished. The Institute of Civil Engineers was established in 1818; by the 1830s, its members were heavily involved in railway construction. It had 220 members in 1830 and 3,000 by 1880.[78] Professional differentiation continued with the founding of separate institutes for mechanical, telegraph, and electrical engineers (in 1847, 1871, 1889, respectively). There were seventeen distinct national organizations by 1914 and some forty thousand practicing engineers. Chadwick famously saw engineering as more significant than medicine in the war on fever, while Smiles depicted engineers as archetypal self-helpers, devoted purely to the common good, and utterly disconnected from anything governmental: "Government has done next to nothing to promote engineering works. These have been the result of liberality, public spirit, and commercial enterprise of merchants, traders, and manufacturers."[79] In 1903, Henry Armstrong argued: "Modern society would be impossible without this class of workers; but their value has yet to be fully appreciated."[80] Like Dibdin, Armstrong made the straightforward point that society did not exist apart from the technical networks that knitted it together. Whatever their formal relation to the state, engineers were clearly responsible for building the networks that governed many aspects of daily life.

Governing through Technology

Thus far, the term *government* has been used somewhat elastically. It has been used to refer to formal political and administrative structures, themselves ranging from central legislative and executive organs to municipal institutions and their various departments. But I have also suggested that technological infrastructures and systems can themselves govern. Chadwick made this point explicitly: "The course of the present enquiry shows how strongly circumstances that are governable *govern* the habits of the population, and in some instances appear almost to breed the species of the population."[81] Government, here, is something quite clearly performed both by formal institutions and their experts and by the networks they manage. This latter sense of government that is delegated and diffuse, carried out by routine administration, and increasingly reliant on technology has certain affinities toward what Foucault called *governmentality*, a neologism intended to capture modes of governance irreducible to state, sovereign, or law.[82]

It is important not to ignore traditional forms of government, but a focus on the heterogeneity of government agencies, and their palpably material nature, is clearly vital to any attempt to discuss the interrelation of power and technology. The governmentality literature, however, frequently displays ambivalence toward infrastructure and technology.[83] Thus, the Foucauldian expression *technology of power* is often invoked with only the most cursory consideration of the precise technologies that have enabled "modern" forms of power to operate.[84] Liberalism becomes a technology of power largely disembedded from the physical systems that sustain it. This drift toward abstraction has already been noted in discussions of illumination, panopticism, and flânerie.

Foucault, however, urged that any analysis of power should remain close to material systems: "Every discussion of politics as the art of government of men necessarily includes a chapter or a series of chapters on urbanism, on collective facilities, on hygiene, and on private architecture."[85] Technology, then, was both governed (by laws, central and local institutions, engineers, inspectors, and, increasingly, as we shall see, itself) and governor. Engels reflected on this when evoking Saint-Simon's observation that the European state was increasingly "replacing the government of persons by the administration of things."[86] These things, however, were not simply bearers of "rationality" or "power" but themselves performative actors. The concept of "material agency," developed within the field of science studies, is helpful here. Humans, according to Bruno Latour and others, do not have a monopoly on action in the world, and their agency has itself been magnified, translated, and stabilized by the calculated co-option of nonhuman systems that perform actions that are simply beyond human capacity.[87]

The idea that nonhuman systems shape the actions of humans is hardly novel. The premise obviously underpinned Bentham's panopticon. The panopticon is usually characterized as being a visual mechanism that reformed through the calculated play of light and shade. But it was much more than this: it envisaged the total control of the environment within which the human body was situated. Each cell would be warmed "upon the principles of those in hot-houses," there would be individual privies, with earthen pipes, and Bentham even toyed with the idea of piped water for each cell.[88] Even in this most visual of environmental technologies, the emphasis was on the entire body, its warmth, salubrity, and fitness. The health of the inmate's body was as vital to reformation as that of the soul. One was far more amenable to improvement if one was fit and free of infection. Light was but one element of a totally controlled space. It was never a totally discrete environmental element, dissociated from others.

Nineteenth-century engineers, sanitarians, and doctors reiterated ad infinitum this idea that environments could themselves be instruments of improvement. They routinely argued, and more commonly simply assumed, that, if one built houses, networks, and other structures in particular ways, one could encourage, promote, or stimulate forms of being or conduct (health, independence, sobriety) that can be referred to in terms of "liberal subjectivity." Here is the brother-in-law of Charles Dickens, the engineer Henry Austin, reflecting on the failure of model tenements to inculcate character in the 1845 *Health of Towns Report*:

The independence of the tenants has not been preserved. . . . The privies, wash-houses, water, and other necessaries, have been in common, and the inmates, being thus constantly thrown together, continual disturbance has been the invariable consequence, and hence the necessity of a control being exercised, which those possessing the means of providing other accommodation, and so far being independent, will not brook; such institutions, to be successful, must be removed in their character, as far as practicable, from any appearance of charity or dependence. The object should be to render the tenant's position an independent and responsible one. One man's habits or interest should interfere as little as possible with his neighbour's. All things necessary for his comfort being provided, he should be made to feel that the possession of it depends entirely on his own good conduct. With such inducements for improvement, he will soon discover that he has a responsible part to act, and become a better character.[89]

Independence, lack of interference, conduct, character: this might not have been political philosophy, but it was demonstrably the language of liberalism. Such ideas suffused writing on illumination. Factory owners eulogized electric light's ability to generate states of productive attention, while the introduction of even modest illumination into public spaces was invariably promoted as an aid to public order. Electric light was also advanced as an agent of salubrity: the chief electrical engineer to the Post Office Department, William Preece, reported in 1890 that its introduction increased health and, consequently, enhanced productivity.[90] Productivity, health, morality, and improvement all required a firm material base. This did not, of course, mechanically guarantee the production of character. Hobhouse grumbled that it was erroneous to think "that the art of governing men is as mechanical a matter as that of laying drain-pipes."[91]

Technology was one governmental tool among many: it would be utterly misleading to refer to this as *technocracy*.[92] Nonetheless, the belief that infrastructure could promote liberal subjectivity itself stimulated changes in liberal principles. If such structures were so vital to individual and national progress, then it became imperative to ensure their construction. New liberals like the idealist philosopher D. G. Ritchie, writing in 1889, saw this as an argument for greater state power: "What the State can do, and what it ought to do, is to provide all its members so far as possible with such an environment as will enable them to live as good lives as possible—good in every sense of the term. 'Compulsion,' 'interference,' 'liberty,' are ambiguous words and give us little help in determining such matters."[93] There was, then, a looping effect between the two forms of government: the move toward council housing or municipal ownership of utilities was a clear response to the prevailing idea that certain environmental and technological systems were both collectively necessary and not adequately secured by the market. New liberalism, and the emergence of more recognizably social government, was as much a product of material change as of ideological mutation.

The argument here is that freedom, whether conceived by J. S. Mill or by sanitarians and engineers, was routinely conceived to be at least partially securable through technology. I would like to close with a few more precise observations on this question. "To govern," observed Foucault, "is to structure the possible field of action of others." In other words, it is to make specific kinds of agency or capacity possible: it is a positive act.[94] We are speaking here of capacities: the ability to be clean, to read at night, to move at speed, to fight infection. How could we call these capacities and the mode of generating them *liberal*? Such bodily capacities were, as we have seen, the physiological foundation of a liberal subjectivity that was never idealist in essence. They were also stimulated and secured in a fundamentally noncoercive way: it was, and remains, far easier and more acceptable to pierce houses with wires and pipes than to admit government officials, although inspectors would increasingly enter the home as the century progressed. As networks were built with greater durability and became operative over greater distances, they became less obtrusive, more part of the background of routine existence.

Such networks also governed through norms rather than laws, something that is often seen as characteristic of liberal modes of governance.[95] A prison sentence or a fine did not await the woman who willfully refused to wash, but social opprobrium did: she was not "in the normal condition of a human being." Smashing your gas meter, of course, was another matter, which reminds us that laws retained their traditional force. To provide the tools to be decent, healthy, sober, and self-governing was to create an apparatus within which the self could be worked on, and through, as an autonomous agent. It could also do this without force or threat, as Bentham had noted when he described liberty as "not any thing that is produced by positive Law. It exists without Law, and not by means of Law."[96] In this sense, the notion of "liberal governmentality" maintains some of its usefulness, provided it is always anchored in specific spaces and not turned into an explanatory abstraction that is operative everywhere.[97] This was liberty in a positive and normative sense, as Albert Borgmann has noted: "Without modern technology, the liberal programme of freedom, equality and self-realisation is unrealisable."[98] We might say that technology was a necessary condition for liberalism's operation, but it was far from a sufficient one.

The profusion of gas and electric light will be examined within this context of technological government as well as in the more cultural or phenomenological context of vision and perception. This new explanatory framework, provided by liberalism, technology, and perception, is far more historical and empirical than the existing one of spectacle, panopticism, and flânerie. It enables me to stay close to historical actors, their words, their eyes, and their physical environment while integrating a theoretical and analytic approach that clearly has its origins in the late twentieth century. I will close this introduction with a brief sketch of the chapters to follow.

Outline of the Book

Chapter 1 provides a historical overview of three dimensions of the history of nineteenth-century perception. First, it examines the development of ophthalmology and the attempt to calculate and normalize perception as well as schemes to protect and nurture vision in institutions like schools. Second, it explores some very concrete connections between liberal subjectivity and visual practices like attention to detail, recognition, and reading. Finally, I look at how these practices produced social cohesion among the respectable and facilitated distinction from, and exclusion of, those individuals incapable of visual control. Such perceptual differentiation, I argue, was inseparable from the built form of the city: while the suburb, the library, and the boulevard were sites where visual command could be exercised, the slum, the court, and the alley were spaces where this command crumbled and gave way to something more tactile and intimate.

These final, spatial points are developed further in chapters 2 and 3, which explore certain prevalent, recurring and vital visual trends or patterns in nineteenth-century Britain. Chapter 2 looks at "oligoptic" visual practice, whereby small groups of individuals freely monitored each other, and the role of nonpanoptic supervision as techniques of organizing such liberal self-observation. More specifically, it looks at four material strategies designed to embed such visual conditions: street widening, smoke abatement, glazing, and soundproof paving. In all these cases, I show how careful attention to material systems reveals as much mess, breakdown, and confusion as system and pattern. Chapter 3 examines inspection, or "inspectability," which, I argue, thrived as a viable way of producing knowledge about Victorian cities and society precisely because it was distinguished from omniscient surveillance or furtive spying. I explore the various ways in which inspection grew in scope and detail without fundamentally undermining the sacred liberal tenets of subjective freedom or privacy. I also analyze a set of private visual practices that were tied to self-government rather than the government of others. For example, private inspection of one's own body (as well as one's soul) was made possible through the partitioning of domestic space.

The final three chapters analyze illumination technologies. Chapter 4 studies the development of gaslight and the networks that secured it. As physical systems delivering gas to its point of illumination proliferated, questions of management became pressing. Mains and gasworks had to be inspected and maintained, while the standards of both gas and the light it produced needed calculating. The science of photometry had emerged in the eighteenth century but became particularly important when light levels, for both public and private use, required measuring. This was an extremely complicated task, and I trace the institutionalization of municipal photometry and various attempts to establish minimal light levels for streetlights. Chapter 5 moves from networks and measurement to perception itself, in the context of the emergence of functional electric light systems from the late 1870s. It looks at particular perceptual capacities and subjective practices that artificial illumination systems, both gas or electric, were frequently expected to secure or stimulate. I also explore the "total environment" of illuminants, by examining the numerous public health dimensions of artificial illumination. Chapter 6 offers a detailed study of early electricity infrastructure, from mains and streetlamps to domestic meters and switches. This enables me to explore the attempt to construct autonomous systems that secured perceptual aptitudes that were themselves vital elements of liberal subjectivity. I conclude with a case study of the illumination of the City of London in the final two decades of the century.

In the conclusion, I make some general remarks on the main themes of the book: the particular forms of British liberalism and its relation to both material systems and perception and the complicated, nonlinear nature of technological change. Most important, I suggest that reductive visual paradigms should be replaced by a multiplicity of overlapping, intersecting, and contrasting perceptual "patterns" that recur throughout the nineteenth century and capture visual experience in all its everyday richness and complexity far better than monolithic abstractions like the panopticon ever could.

ONE

The Victorian Eye: The Physiology, Sociology, and Spatiality of Vision, 1800–1900

The question of eyesight has in late years assumed such an importance in social and commercial economy as was not dreamt of twenty years ago. HARRY CRITCHLEY, *HYGIENE IN SCHOOL* (1906)

The eye made the Victorians particularly verbose. It was the only organ for which a royal doctor was deemed necessary, a Ruskinian portal to the soul, and a sublimely complex interface between body and world (figure 1.1). Clumsy rhapsody proliferated: "It has been designated 'the queen of the senses,' 'the index of the mind,' 'the window of the soul;' nay, it has even been esteemed 'in itself a soul;' and 'He who spake as never man spake' has declared that 'the light of the body is the eye,' at which we cannot marvel when we contemplate the inestimable pleasures and advantages it confers upon mankind."[1] The eye's anatomical salience demonstrated the superiority of human civilization over that of animals and quadrupeds in particular.[2] The idea that the eye might be the product of utterly blind evolutionary processes struck even Charles Darwin as fundamentally improbable: "To suppose that the eye, with all its inimitable contrivances for adjusting the focus to different distances, for admitting different amounts of light, and for the correction of spherical and chromatic aberration, could have been

Figure 1.1 Horizontal view of the human eye, showing aqueous humor, crystalline lens, vitreous humor, and optic nerve. From MacKenzie, *Practical Treatise* (1855).

formed by natural selection, seems, I freely confess, absurd in the highest possible degree."[3]

Such ingenuous ocularphilia should not surprise us: after all, it is precisely at this historical juncture ("modernity") and in this geographic location ("the West") that the "hegemony of vision" is often said to have emerged.[4] Over the past couple of hundred years, this thesis goes, vision has become unquestionably the most venerated, potent, and socially significant sense. This visual hegemony has not, however, been universally lauded. As Martin Jay has authoritatively documented, most "great" European thinkers of the twentieth century have condemned this dominance of the visual. To take one example among many: Georg Simmel famously argued that the modern city, and particularly its crowds and speed, reduced lived experience to something fragmented, alienated, and isolated. This anomic existence was largely experienced through the eye:

Social life in the big city as compared with the towns shows a great preponderance of occasions to *see* rather than *hear* people. . . . Before the appearance of omnibuses, railroads, and streetcars in the nineteenth century, men were not in a situation where for periods of minutes or hours they could or must look at each other without talking to one another.... The greater perplexity which characterises the person who only sees, as contrasted with the one who only hears, brings us to the problems of the emotions in modern life: the lack of orientation in collective life, the sense of utter lonesomeness, and the feeling that the individual is surrounded on all sides by closed doors.[5]

Explicating historically specific modes of perception, in other words, has been a fruitful way of historicizing, and critiquing, hegemonic modes of nineteenth- and twentieth-century experience and subjectivity. Alienation, objectification, coercion, gendered and racialized identities, all have been approached through the historical analysis of perception—and vision in particular.[6] Such analyses, however, often become formulaic. Vision is identified as having become hegemonic, and then certain malign forms of visual subjectification are outlined and vilified: the coercive disciplinary gaze, the rapacious sexual gaze, the stupefied narcosis of the consumer gaze. This formulation is present in the most sophisticated scholarship. "The senses of smell, taste, and touch," concludes Henri Lefebvre in *The Production of Space*, "have been almost completely annexed and absorbed by sight."[7] "Interaction in the big city," concurs Christoph Asendorf, "is characterised by consistent visualisation and the retreat of verbal and tactile components; it does not admit of contacts other than visual ones devoid of touch."[8] To write a convincing history of perception, linked carefully to broader modes of subjectivity, the historian must avoid such polemic and reduction, attending instead to the multiple, intersecting, and conflicting forms of visual perception operative in the "modern" period, as well as addressing the interaction between visuality and other modes of sensory experience.

Before moving on, I want to clarify the meaning of the rather ugly word *visuality*. Obviously, the term incorporates our everyday understanding of *vision* itself: the act of looking and seeing, which is stubbornly delimited by the body. The body does not, however, absolutely determine its own perceptual capacities: physiology frames but underdetermines visuality. Vision is also only one of the five senses of which the Western sensorium is routinely deemed to consist, and we cannot ignore the historically shifting relations between these senses.[9] Historians must also consider all writings relating to the senses, from philosophical discourse to the countless imperatives to use one's senses in particular ways that, through conduct manuals and more tacit sets of rules, have permeated society. Maintaining silence in a library and averting one's eyes from a passing funeral procession have become, implicitly or otherwise, part of consecrated perceptual practice. This separation between vision itself and the discourses relating to the use of vision is probably at best heuristic. Ophthalmologic science also produced normative and prescriptive discourse about the eyes, discourse through which individuals came to think about, care for, and use their eyes in historically fresh ways. Finally, visuality includes physical, environmental, and spatial factors: air, light, humidity, and heat all affect the actions and balance of the sensory organs, while architectural arrangements inevitably carry with them unarticulated perceptual injunctions. Technologies accentuate some perceptual experiences while occluding others. The term *visuality* captures the simultaneously physiological, practical, discursive, and technospatial nature of vision.

I will now explore this visuality in more detail, charting a course between Victorian eulogy and Foucauldian gloom, beginning with the eye itself, then examining some connections between vision and liberalism, before placing the visual subject in a broader social and spatial context.

Ophthalmologic Science and the Nineteenth-Century Physiology of Vision

Jonathan Crary has produced one of the most compelling theses about the historicity of perception. He argues that a specifically "modern" understanding of perception emerged between 1810 and 1840 with the work of Goethe, Maine de Biran, and others, which "effectively broke with a classical regime of visuality and grounded the truth of vision in the density and materiality of the body."[10] Crary is correct to emphasize the historical development of a more thoroughly physiological comprehension of vision; nevertheless, his adherence to a model of epistemological rupture dividing classical from modern is certainly problematic, and the implication that the classical (i.e., pre-nineteenth-century) comprehension of perception was devoid of physiological grounding is unsustainable.[11] Lockean psychology, for example, was ostentatious in its sensationalism. The senses channeled all impressions to the mind, where they were associated into ideas, including fundamental conceptions like space: "We get the *Idea* of Space . . . by our Sight and Touch; by either of which we receive into our Minds the *Ideas* of Extension or Distance."[12] Such resolute anti-innatism gave the senses a foundational role in the formation of interiority and selfhood and marked a significant shift from Cartesianism.[13]

Nonetheless, Locke did not address the physiology of sensation. He was content to ignore the body when considering the processes through which the human subject gained knowledge about the world. The second paragraph of his *Essay concerning Human Understanding* includes the following disclaimer: "I shall not at present meddle with the Physical Consideration of the Mind; or trouble my self to examine, wherein its essence consists, or by what Motions of our Spirits, or Alterations of our Bodies, we come to have any Sensation by our Organs, or any *Ideas* in our Understandings; and whether those *Ideas* do in their Formation, any, or all of them, depend on Matter, or no."[14] To illustrate how perception worked, Locke used the model of the camera obscura: a dark chamber, with a hole punctured on one side, through which light streamed, forming an image on the back wall. It was an ancient idea, revived by Renaissance thinkers, who demonstrated its homology to the eyeball.[15] According to one commentator, it "affected the scientific imagination so greatly that by the seventeenth century it had become *the* model for the eye."[16] Certainly, Locke was convinced, arguing famously: "For, methinks, the *Understanding* is not much unlike a Closet wholly shut from light, with only some little openings left, to let in external visible Resemblances, or *Ideas* of things without; would the pictures coming into such a dark Room but stay there, and lie so orderly as to be found upon occasion, it would very much resemble the Understanding of a Man, in reference to all Objects of sight, and the *Ideas* of them."[17]

This model of knowledge as coextensive with sensation left no place for bodily mediation. Yet Locke never argued that there was no physiological mediation: he simply chose to ignore the issue. Such strategic bracketing was, moreover, not shared by all eighteenth-century thinkers, who made the understanding of sensation fleshier, particularly through study of the nerves. This was evident in the work of the philosopher David Hartley, who mixed Lockean sensationalism with Newton's corpuscular ideas to argue that the nerves were fibrous channels transmitting discrete quanta of physical data from world to mind. Knowledge thus became more fragile and body dependent, the psychological implications of which haunted Hume.[18] By the later nineteenth century, for some writers, like the psychologist Alexander Bain, will and thought could seem the captives of capricious nervous economies. This physiological model, however, was never totally dominant but developed alongside and in opposition to alternative, less materialistic ideas about perception, an opposition as political as it was scientific.[19]

Well before the nineteenth century, then, analysis of perceptual organs was dismantling the homologies and identities on which the camera obscura model was based. The commonsense philosopher Thomas Reid argued in 1764:

There is not the least probability that there is any picture or image of the object either in the optic nerve or brain.... Nor is there any probability, that the mind perceives the pictures upon the *retina*.... We acknowledge, therefore, that the *retina* is not the last and most immediate instrument of the mind in vision. There are other material organs, whose operation is necessary to seeing, even after the pictures on the *retina* are formed. If ever we come to know the structure and use of the choroid membrane, the optic nerve, and the brain, and what impressions are made upon them by means of the pictures on the *retina*, some more links of the chain may be brought within our view, and a more general law of vision discovered.[20]

As yet, however, the sensory systems were physiologically indistinct. The nerves mediated perception, but the tactile fibers, say, did not differ in any material or structural sense from the auditory. This was enshrined in the theory of the "sensorium commune," a unified nervous plenum devoid of anatomical specificity.[21] The senses themselves had been anatomically separated from the motor functions in the mid-eighteenth century by the anatomist Albrecht von Haller, whose work was furthered in the early nineteenth century by Charles Bell and François Magendie.[22] The qualitative differentiation of the five sensory systems was demon-strated by the physiologist Johannes Müller in 1826.[23] His idea of "spe-cific nerve energies" led Müller to attempt to correlate perceptual with cerebral geography. The apparatus of vision was now physiologically dis-tinct and localized. This functional, immanent model of discrete sensory systems often assumed an evolutionary character by the later nineteenth century: "As we go higher up the scale of animals, in order to give that wider and more accurate knowledge of the various properties of mat-ter necessary for the complex relations of the higher animals, sensory nerve-fibres are differentiated into several kinds, so that each may give clear knowledge of a different property."[24]

This separation of the senses produced, or reinforced, a correspond-ing cleavage between sensation itself (a physiological process) and per-ception (a psychological one). This distinction was advanced by the commonsense philosophers, nineteenth-century experimentalists like Wheatstone and Flourens, and developed in coherent form by Herman von Helmholtz in the *Handbuch der physiologischen Optik* (Handbook of physiological optics), published between 1856 and 1866. Our everyday usage of the words *sensation* and *perception* (to which I adhere through-out this book) is a historical product of a physiological approach to sensory action unwilling to forgo the final, stubbornest remnants of Cartesianism.[25] Sensation was, thus, easily conceived as a physiological, and perception as a psychological (or moral), category. This differentiation also generated critique: the physiological paradigm was neither internally consistent nor free from external challenge.[26] Some physiologists argued for fuzzier, more dynamic, and less rigid interconnections between the various senses and their corresponding mental states. As early as 1802, the physiologist Pierre Cabanis noted: "As each sense is able to come into operation only by virtue of the earlier operation of all the general systems of organs, and able to continue only by virtue of their simultaneous operation, it always necessarily feels the effects of their behaviour and shares to a greater or lesser degree their most ordinary defects."[27] More radically, some philosophers, artists, and poets rejected the whole project of functional disaggregation, a line of revolt reaching its apogee in Merleau-Ponty's phenomenology: "The senses translate each other without any need of an interpreter, and are mutually comprehensible without the intervention of any idea."[28] The history of perception is altogether more dialectical and contested than Crary's epistemic model allows.

What were the defining characteristics of this new physiology of vision? The eye itself was analyzed, measured, and tested more than in any previous historical period.[29] Here is the polymath Thomas Young, writing in 1807: "The Eye is an irregular spheroid, not very widely differing from a sphere; it is principally composed of transparent substances, of various refractive densities, calculated to collect the rays of light, which diverge from each point of an object, to a focus on its posterior surface, which is capable of transmitting to the mind the impression of the colour and intensity of the light, together with a distinction of the situation of the focal point, as determined by the angular place of the object."[30] This eye is dense, stratified, delicate: not a neat aperture through which light silently passes en route to the mind, but a variegated zone of mediation. The biologist Gottfried Treviranus demonstrated in 1834 that the retina was composed of two distinct forms of sensor, the cone (or bulb) and the rod (figure 1.2). Cones detected color, rods light: thus emerged the anatomist Max Schultze's concept of the *duplex retina*, a field of reception operative at two levels, one bright and chromatic, the other duller and monochromatic, one functional by sunlight, the other functional by moonlight.[31] The sensors were scattered unevenly across the retinal surface, with the cones clustered at the heart, around the *macula lutea*, or yellow spot (figures 1.3 and 1.4). At the very center of this was the *fovea centralis*, a minute depression, half a millimeter across, entirely composed

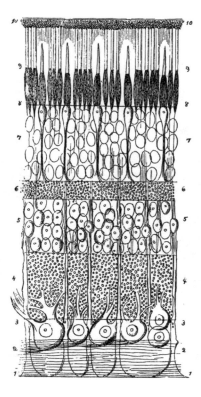

Figure 1.2 Cross section of the retina, with layer of rods and cones at the top (indicated by "9"). From Carter, *Practical Treatise* (1876).

of cones and, consequently, the location of acutest color vision. Rods predominated on the outer periphery of the retina, giving night vision greater width but less focus than that by day.[32] Retinal discernment of color was, finally, affected by levels of brightness. At twilight, the perception of greens and blues is enhanced relative to reds, a reversal of daylight perception: this is the so-called Purkinje shift, named after the Czech physiologist who wrote on the subject in 1825.[33]

The front of the eye was equally complex. In his influential 1864 textbook, *On the Accommodation and Refraction of the Eye*, the Dutch physician F. C. Donders observed: "The lens, however, is no homogeneous mass, but consists of layers of refractive power, increasing towards the centre. In the lens itself, therefore, innumerable refractions take place from layer to layer, which cannot, nevertheless, be separately traced."[34] No simple geometry could capture the trajectory of light rays through this dense mass. The layers further within the eye also clouded vision. The eye, pace Darwin and Ruskin, was imperfect. Entoptic phenomena (visible entities within the eyeball) had been noted centuries earlier by

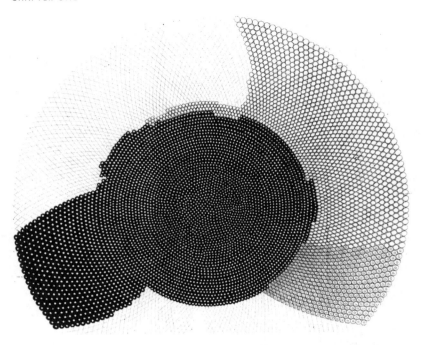

Figure 1.3 Mosaic of cones in the fovea centralis and area, including the middle of the macula lutea, magnified four hundred times. From Schultze, *Zur Anatomie und Physiogie der Retina* (1866).

Benvenutus Grassus, who in 1474 attributed the squiggles, circles, and dots floating in eyes to excessive black bile.[35] Donders dwelt on the internal spectra thrown by "tears, mucus, fat globules and bubbles of air, moving on the cornea," making the eye literally self-distorting, while John Tyndall observed "snake-like lines, beads, and rings."[36] The familiar *muscae volitantes* were refractions caused by a "ropy substance in the aqueous humour," while other entities (blood vessels, corneal scars, morbid retinal fragments, "the pigmented remains of iritic adhesions") threw ripples, nebulae, and spangles across the visual field.[37] Like dust silently accumulating in the inner whorls of the ear, here was palpable proof of the bodily thickness of perception, the confusing, chiasmatic hinterland between demonstrably inner and outer worlds.

The eye was, within limits, self-regulating. Most significant here was the ability of the muscles within the iris to adjust the pupil's size in accordance with the amount of light striking the eye. This relation was shown to be calculable—and applicable to the planning of illumination

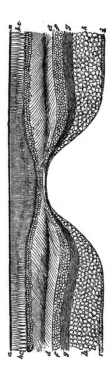

Figure 1.4 Diagrammatic section of the macula lutea, or yellow spot. From Huxley, *Lessons in Elementary Physiology* (1917).

systems.[38] In 1895, the electrical engineer André Blondel noted that "an object may.... appear one to sixteen times more luminous... according to the opening of the pupil," something illuminating engineers could not afford to ignore unless they wanted to waste candlepower.[39] The eye, effused William Preece, "adapt[s]... itself with wonderful rapidity to change of light... owing to the remarkable power possessed by the iris, to expand and contract with the variation of light."[40] Illumination systems should always respect the eyes: as we shall see, the self-regulating nature of the eye would be considered by many illuminating engineers.

The emphasis was also shifting to eyes: in contrast to the monocular architectonics of the camera obscura, nineteenth-century physiology treated the optical apparatus as a binocular system. Vision was composite, made from the interaction of two distinct images transmitted to the brain from either side of the nasal ridge. This gave vision its depth and tangibility: it "necessarily forces upon us a definite bodily form, and almost makes us touch and feel—which is precisely the characteristic of bodily vision."[41] With one eye, the sense of distance was confused, dimensions dissolved, and the rest of the body called in to compensate:

"A person deprived of the sight of one eye sees therefore all objects, near and remote, as a person with both eyes sees remote only, but that vivid effect arising from binocular vision of near objects is not perceived by the former; to supply this deficiency he has recourse unconsciously to other means of acquiring more accurate information. The motion of the head is the principal means he employs."[42] The act of seeing was demonstrably muscular, from the large recti and oblique muscles allowing the eye to swivel in its orbit, to the tiny, intricate ligaments within the eye itself, like the iris muscles and the zonule of Zinn.[43] "The delicate muscular adaptations which effect the accommodation of the eye," noted the psychiatrist Henry Maudsley, "seem really to give to the mind the ideas of distance and magnitude . . . the muscular adaptations . . . imparting the suitable intuitions."[44]

Vision was also ineluctably temporal. The essential jerkiness caused by restless, twitching eyes gave rise to the term *saccadic motion*, devised by the ophthalmologist Emile Javal in 1878.[45] Vision did not occur instantaneously: it was a temporal, as well as a spatial, composite. Helmholtz's 1850 calculation of the speed of nerve transmission—as substantially slower than the speed of light—clearly influenced this.[46] The electrophysiologist William Steavenson described this transmission as "a molecular disturbance propagated along the nerve in the form of a wave the length of eighteen millimetres, and possessing a velocity of twenty-eight metres per second."[47] The present was no longer a point. It thickened and distended, forever merging with futures and pasts. This was most evident in the afterimage, which fascinated Goethe and, when correctly understood, formed the basis for cinematic technology and perception. As David Brewster, the inventor of the kaleidoscope, concluded, arresting perceptual flux was physiologically impossible: "A visible object cannot, in all its parts, be seen single *at the same instant of time*."[48]

A final, glacial, temporal dimension was ocular aging. As the body grows old, the aqueous humor loses transparency, the cornea and conjunctiva become dull, and the eye becomes more rigid, making accommodation of near objects more difficult. This inability to clearly see close objects was manifested by the need to hold books at arm's length or beneath a lamp and was acknowledged to become palpable, for normal eyes, around the age of forty-five. The term *presbyopia* was coined to name this process. As Donders explained, it was "the normal condition of the normally constructed eye at a more advanced period of life," "no more an anomaly than are grey hairs or wrinkling of the skin." In his choice of language, Donders demonstrated that perception had become normalized. The technical term for normal vision was *emmetropia*, which

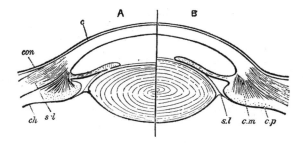

Figure 1.5 The mechanism of accommodation: "A" shows the lens adjusted for distant vision, and "B" shows it adjusted for near vision. From Huxley, *Lessons in Elementary Physiology* (1917).

occurred when "the principal focus of the media of the eye at rest falls on the anterior surface of the most external layer of the retina."[49] Donders admitted that the normal eye was not an ordinary eye. It was a necessary simplification, an eye reduced "to a single refracting surface, bounded anteriorly by air, posteriorly by aqueous and vitreous eye, and this *reduced eye*, where the greatest accuracy is not required, *may be the basis of a number of considerations and calculations*."[50] He removed the crystalline lens to simplify the refractive geometry of his normal eye. Elsewhere, the normal eye was described as "purely an arbitrary term" yet one absolutely necessary to measure vision and render it amenable to statistical analysis.[51] Future ophthalmologists should learn what normal vision was before exploring optical defects, according to the ophthalmologist Otto Haab in 1910: "*I cannot too strongly advise the beginner to study normal eyes as often and as thoroughly as possible.*"[52]

Normalization, of course, implies corresponding abnormality and error. There were two basic kinds of visual error: that of accommodation and that of refraction. Accommodation was the ability of the lens to change its shape, via the action of the ciliary muscles, to adjust focal length for near and far objects (figure 1.5). It was noticed by William Porterfield in 1738, a discovery implicitly critiquing the rigid camera obscura model of vision.[53] Donders noted: "*The change consists in an alteration of form of the lens; above all, its anterior surface becomes more convex and approaches the cornea.*"[54] Presbyopia was caused by the slow decline in accommodative capacity with age. Refraction, by contrast, was anatomical rather than muscular, a result of the spatial correspondence between lens and retina. Myopia, long recognized, was caused by an overly long eyeball, which brought rays to a focus before they reached the retina. Myopes found close inspection of small objects easy, but their capacity to distinguish detail at a distance was greatly diminished: the myope "cannot read the inscriptions on doors and houses, nor recognise persons across the street; if he go into a large room, in which there are many persons, he cannot readily distinguish those he knows."[55] Hypermetropia, or farsightedness, was distinguished from presbyopia in 1864 by Donders. Its cause was a small, or flat, eyeball: it was an *"imperfectly developed eye."*[56]

The growth in comprehension of eye defects, errors, and disorders was greatly facilitated by institutional and technical developments, including the eye hospital. British eye hospitals were founded at Moorfields (in London) in 1805, Bristol in 1810, Bath in 1811, and Manchester and Dublin in 1814, with others following over the next two decades.[57] These clinical spaces aggregated individuals with ocular defects and allowed doctors to inspect, compare, and classify them. Of Moorfields, the anatomist William Lawrence noted: "You may see more of diseases of the eye in this institution in three months than in the largest hospital in fifty years."[58] Other forms of ocular institution thrived, for example, those for the blind: circulating Braille libraries, societies, small hospitals. There were twenty-seven institutions for educating the blind in England in 1876, providing rudimentary education, "mostly confined to the teaching of some manual trade and reading raised type."[59]

Eye treatment was becoming more thoroughly medicalized and professionalized. This professionalization can be traced through the establishment of journals: the *Optician* was founded in 1891, followed by the *Dioptric and Ophthalmometric Review* in 1896. The three specialized terms employed in these titles corresponded to the basic professional division of opticians (who made spectacles), dioptricians (who used written or other subjective tests), and ophthalmologists (who examined vision with the ophthalmoscope). The policing of the often porous boundaries between these specialties was undertaken first by the British Optical Association (BOA), founded in 1895, and then by a set of mandatory examinations for potential practitioners. Nonetheless, of twenty thousand individuals claiming to be opticians in 1903, only around six hundred had BOA certification.

Texts devoted to the identification and treatment of ocular disorders proliferated and grew to unwieldy proportions: the eye surgeon William MacKenzie's *Practical Treatise on the Diseases of the Eye* ran to well over a thousand pages by its fourth edition (1854). MacKenzie's text eclectically mingled ancient and modern forms of treatment. For strabismus (squinting), he recommended purging, prismatic lenses, or electricity, while numerous diseases, including ophthalmia and iritis, might still be treated with leeches.[60] But it was in eye surgery that many of the most radical nineteenth-century developments were made. Such surgery is not, of course, a modern invention: cataract operations, at least, were performed in ancient times. But the scope and frequency of eye operations grew enormously during the nineteenth century. "The eye and its appendages," observed the optimistic ophthalmic surgeon Robert Carter in 1876, "are extremely tolerant of discreet surgical interference, and are made the subjects of a great variety of operations.[61] Cataracts, in multiple forms (figure 1.6), remained central objects of surgical rectification, and numerous operative procedures (displacement, excision, and division) were developed to rectify them. Other forms of eye surgery included the treatment of strabismus. According to MacKenzie, the first successful strabismus surgery, involving the careful division of the eye's internal rectus muscle, came in 1839 (figure 1.7). Various defects of the eyelids, including their adhesion to the ocular globe following accidents and their eversion, became surgically treatable. Surgeons grew adept at creating artificial pupils in eyes clouded and damaged by disease or injury.[62] Attempts at transplants, using the eyes of rabbits or parts thereof, were also made toward the end of the century.[63] Eye surgery remained, however, highly dangerous by today's standards. Edward Nettleship reported that 5 percent of eyes were still lost during cataract operations in the later nineteenth century.[64]

If eyes were lost—through, for example, a gunshot wound—artificial eyes were fitted with increasing regularity. Usually made of enamel and decorated with a crimson network of imitation veins, such eyes were often convincing replacements: "Not only is the casual observer deceived, but even the professional man who is conversant with ophthalmic practice may not detect the substitute."[65] MacKenzie recommended that a small eye should be fitted first, then slowly replaced over time by a succession of larger ones, "till at length the lids shall appear to have reached nearly their natural degree of expansion."[66] The artificial eye thus supported the lids and, if its wearer was fortunate, might revolve in its orbit. Its use remained laborious and uncomfortable: the eye was always to be removed at night and wiped clear of accumulated mucus, while the eyelids should be washed with tepid water. The surface of the globe coarsened over time and required replacing: "Whatever care is taken of it, the external surface of an artificial eye becomes roughened, in course of time, by the chemical action of the tears; and the rough surface irritates the lining membrane of the lids, and produces an obstinate form of conjunctivitis, which commonly leads to contraction.... A year is generally about the longest period during which it can be worn with safety."[67]

More routine and less invasive were the various testing techniques used to measure eyesight. Most significant was the ophthalmoscope, through which clinical perception extended to the eye's inner surfaces (figure 1.8). In 1846, William Cumming, a London student, demonstrated that it was

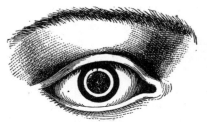

Laminar Cataract, with transparent periphery.

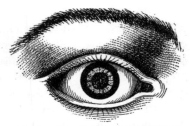

Laminar Cataract, with opaque striation of periphery.

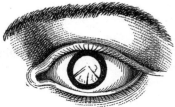

Striæ of Commencing Cortical Cataract (senile).

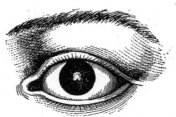

The same Eye, with the pupil undilated and the gaze directed to the front.

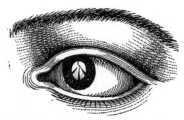

The same Eye, with the pupil undilated and the gaze directed laterally.

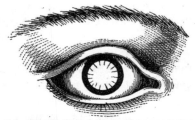

Striæ of Senile Cortical Cataract in an early stage and limited to the periphery of the lens.

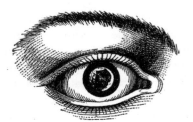

Nuclear Cataract (senile).

Cataractous Striæ in both anterior and posterior cortex.

Figure 1.6 Various kinds of cataract compared. From Carter, *Practical Treatise* (1876).

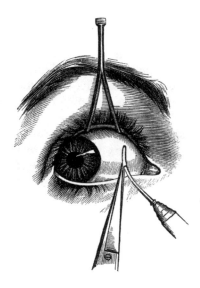

Figure 1.7 Operation for strabismus (squinting), showing retractor, hook, and scissors. From Walton, *Treatise on Operative Ophthalmic Surgery* (1853).

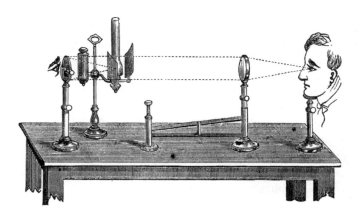

Figure 1.8 Table ophthalmoscope. The observer's eye is to the left, looking through the aperture in the mirror, which reflects light into the observed eye at the far right. From Carter, *Practical Treatise* (1876).

possible to see through the pupil, into the eye itself, stimulating the invention of numerous ophthalmologic devices.[68] The basic premise of ophthalmoscopy was straightforward: artificial light was shone into the eye through the pupil, via a mirror, while the observer, peering through an aperture in the mirror, observed the image of the bottom of the eye, or its "fundus" (the mirror being necessary to prevent the observer from obstructing the rays entering the eye).[69] A convex lens magnified the surface of the fundus: this could be positioned close to the eye for detail (the "direct method") or further away, at a distance of around eighteen inches, to illuminate a larger portion of the fundus (the "indirect method").[70] The device was difficult to use, its operation requiring the skillful, practiced coordination of lamp, mirror, lens, and two bodies, but it was remarkably useful, enabling the detection of everything from basic conditions like myopia to retinal disease and cataracts. In 1854, a writer in the *Medical Times and Gazette* declared: "The discovery. . . . of a mode of examining directly the interior of the eye in the living subject is by far the most important improvement made in ophthalmology in modern times."[71] Haab concurred: "The fundus of the observed eye is converted into a luminous object which we can see like any other object in the outside world."[72] The ophthalmoscope remains the basic device used for routine optical inspection: one leading brand goes by the name of the Welch Allyn PanOptic Ophthalmoscope.[73]

The ophthalmoscope was not the only way of measuring ocular capacity. One Professor Eduard von Jaeger, of Vienna, developed the "first well considered" test for "acuteness of vision," involving a series of prints of various sizes, numbered from 1 to 20, with crosses and asterisks used for the illiterate. This test was improved by the Dutch ophthalmologist Hermann Snellen, who factored in the visual angle and defined normative letters composed of squares. Standard test types allowed visual acuity to be measured and generated the concept of twenty-twenty vision (the ability to read type 20 at twenty feet), which, according to Carter, was "taken as the normal standard."[74] Astigmatism, caused by an elliptical corneal surface, was detectable by a sequence of letters composed of parallel lines at various angles (figure 1.9).[75] By 1900, an eye examination had become a complex composite, involving devices and charts as well as more manual, physical inspection of the eyeball.

Normalizing and Protecting Perception

Basic rectification of myopia, hypermetropia, presbyopia, and astigmatism came through the more widespread use of spectacles. Convex lenses date back at least to the medieval period, and, through Keplerian and Newtonian optics, principles of refraction had been scientifically grasped. The Duke of Wellington argued that the spectacles made for him by John Dolland gave him a decisive edge over the French in battle. Less pugilistically, the capacity to correct defective vision with spectacles

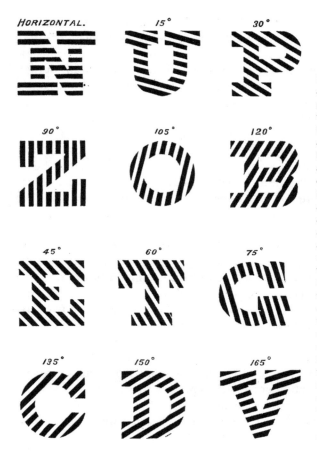

Figure 1.9 Pray's astigmatism test. The patient is situated ten to twelve feet from the letters. "He is then asked whether the black and white lines are equally conspicuous in all the letters; and if he answers in the affirmative there is no astigmatism. If he is astigmatic, he will name some letter in which the lines are more sharply defined than in any of the rest, and will say on inquiry that they are least defined in the companion letter, below or above the first, which has its stripes in the opposite direction." From Carter, *Practical Treatise* (1876), 490 (quote).

was often perceived as a great social blessing, those without spectacles finding "themselves completely shut out from the occupations to which, in a busy society, they are called."[76] Social practice, Brewster argued, would collapse without spectacles:

They enable us to see the faces of our friends in the same apartment or across a table, to enjoy the beautiful in nature or in art, and to count the stars in the firmament when we can hardly see with distinctness a few inches before us, and are obliged to bring close to the eye every object which we examine. Those only can understand how miserable must have been the condition of the aged and the shortsighted before the invention of spectacles, who have themselves long experienced the great blessings which they confer.[77]

Those needing, but lacking, this vital apparatus, Donders concurred, were often recognizable: they might have "a peculiar freeness" of bearing and an "awkward gait," and, worse, by not basing judgments on tangible visual evidence, they plugged gaps in their knowledge by wanton exercise of the imagination.[78] Losing one's spectacles was socially disastrous. Should such a calamity occur, Brewster recommended rather cumbersome remedies, including creating an "extempore lens with wine or varnish on a plane of glass, or by crossing at right angles two cylindrical bottles filled with water, and looking through the portion that is crossed."[79]

Spectacles appeared in manifold shapes and forms, reflecting the diversity of nineteenth-century visual practice. There were predictable gender dimensions here: in 1902, a special edition of the *Dioptric and Ophthalmometric Review* focused on ladies' glasses and eye care.[80] Glasses were increasingly tailored for individual tasks like reading or drawing: these "should be worn well down the nose."[81] Pantoscopic (now called *bifocal*) glasses were devised for "those who are addressing assemblies, when the attention of the eyes is divided between the manuscript or notes and the hearers."[82] Travelers' spectacles might have gauze around the sides to prevent dust and stones injuring the eyeball, while special magnifying devices facilitated watchmaking and engraving. The glare from sun, snow, and electric arc lights could be subdued with sunglasses or veils (for women).[83] Trials with different colors of glass suggested: "Dark-grey glasses afforded the best protection against strong light; blue glasses are also useful. Glasses of other colours seem to be useless."[84]

Spectacles were routinely depicted as one of the contemporary world's most useful and necessary devices, mainly because this world was itself regularly viewed as being more ocularly demanding than previous epochs and perennially on the brink of destroying vision altogether. The destruction of vision presaged the destruction of all the senses and, indeed, of "modern man" himself. Addressing the San Francisco Medical Society in 1893, W. F. Southard summed up these fears: "The eye is the nurse and foster mother of all the other senses and the patron of all the arts and sciences, and the modern man is looking minutely into a myriad of things and taxing his eyesight accordingly, and many are the hopes that have failed, and bitter has been the disappointment, when eyesight has given way under stress and ceaseless burden of the varied avocations and professions of modern life."[85]

The stress and burden were most manifest in the rise of myopia. Surveys and measurements revealed that myopia grew in proportion to "civilization" or "modernization." It was, consequently, a Western phenomenon. Shortsightedness, intoned the American eye surgeon George

Harlan, was "one of the penalties of advancing civilisation. . . . Its greatest prevalence is in Germany. . . . It is comparatively rare among seamen and farmers, and no one ever heard of a short-sighted Indian." It was particularly associated with the rapidly moving, detailed visual field produced by urbanization: "One of the disadvantages of city life is the constant occupation of the eyes with close objects and the absence of anything like a long, free range."[86] Simmel's sociological anxieties might have a physiological basis: fin de siècle fears of nervous exhaustion and fatigue had a significant ocular dimension.[87] Correctional techniques, from surgery to spectacles, were necessary but not sufficient. More significant was the protection of normal vision, particularly during the formative years of childhood. This involved the inculcation of sound individual visual habits and more collective strategies addressing institutional environments, both designed to nurture society's most precious organ.[88]

Reading and learning were obviously integral to education and to the West's sense of its own "civilized" nature, but bad reading habits were, in turn, destroying civilized eyes. Accumulated injunctions suggest that reading had become physiologically perilous and that people needed to be taught to read healthily: one should not read lying down, with the book too close to the eyes, while riding in carriages or trains, or until too late an hour. The arrangement of book and light should not be left to chance. "Always turn your back to the source of light when you are reading, so that the light may fall on to the book, instead of coming into your eyes," John Browning, the first president of the British Optical Association, instructed. "Always lean well back when reading, and hold the book up [figure 1.10]. Do not lean forward and face the light [figure 1.11]."[89] Harlan recommended a reading angle of between forty and forty-five degrees and advised readers to take regular breaks. The modern reader, then, should be acutely self-conscious of the bodily act of reading and his or her own delicate ocular economy. When one felt one's eyesight beginning to fail, one should visit the ophthalmologist and acquire spectacles: "As soon as it is found that the figure 3 cannot be readily distinguished from 5 in the popular railway guide by artificial light, spectacles should at once be obtained."[90] Eyesight's failure was noted through the declining discernment of letters and numbers. Once acquired, spectacles should be cared for: "They are essential to the proper exercise of vision by a large proportion of the inhabitants of civilised countries; and, this being so, they should be treated as carefully and as respectfully as the eyes themselves."[91] Browning recommended a soft cambric handkerchief or wash leather for cleaning.

Figure 1.10 Correct position for reading by lamplight. From Browning, *How to Use Our Eyes* (1883).

Figure 1.11 Incorrect position for reading by lamplight. From Browning, *How to Use Our Eyes* (1883).

Hygiene should extend from the hands, ears, and armpits to the eye itself. The eyelids were a particularly moist and inviting haven for every kind of tiny object circulating in the atmosphere: bits of twig, stone chippings, cigar ash. Insects found them attractive, and oculists regaled readers with possibly apocryphal stories about larvae hatching there.[92] "It is a good plan," suggested Browning, "to sluice the eyes well every morning with cold water."[93] Workers in dusty, sooty atmospheres should exercise particularly scrupulous ocular hygiene. Communal washing facilities with shared towels were vilified as ideal environments for the spread of contagious eye diseases like ophthalmia. Temperance, or at least moderation, was also upheld as a critical strategy of ocular self-protection. Alcohol produced short-term effects like double vision and loss of command over accommodation, while its more permanent effects included corneal abscesses. A drunkard's eyes, noted the doctor Robert MacNish, were "red and watery," while "the delicacy of the retina is probably affected. . . . The tunica adnata which covers the cornea must lose its original clearness and transparency."[94] Smoking, meanwhile, was blamed for amblyopia, a progressive lessening of visual acuity: "The patients are, almost without exception, males, and at or beyond middle life. With very rare exceptions they are smokers, and have smoked for many years, and a large number are also intemperate in alcohol. . . . It is now generally agreed that tobacco has a large share in the causation, and in the opinion of an increasing number of observers it is the sole excitant."[95]

Children were central objects of ocular concern. Carter recommended the cleansing of newborn's eyes and urged that babies' eyes be protected from excessive light.[96] The most innocent forms of play might have serious long-term ocular ramifications: "Holding the child's toy near its eyes, or amusing it by suddenly presenting some favourite object close to its face, may excite squinting."[97] Such matters, of course, remained within the ambit of individual, or at least parental, control. This was not the case, however, at school, where numerous environmental and practical tactics were used to normalize visual development, which could then stand synecdochically for normal development itself. "The aim of those having care of children should be to promote the just development towards the standard," noted the hygienist Edward Hope and the ophthalmologist Edgar Browne in their *Manual of School Hygiene* (1901). "The perfect eye is to be regarded not merely as a good thing in itself but as a sign of well-conducted and symmetrical growth of the whole body." The ultimate aim was to equip children with eyes fit to repel the stresses of the world beyond the school, "to render [vision] . . . strong and healthy and fitted to withstand the strain that may be thrown upon it in after-school life."[98]

Rather than producing ocularly normal subjects, however, schools were routinely castigated as dark and badly equipped, machines for generating myopes: "Apart from certain cases of disease and faulty development, the vast preponderance of all short sight is acquired, and is

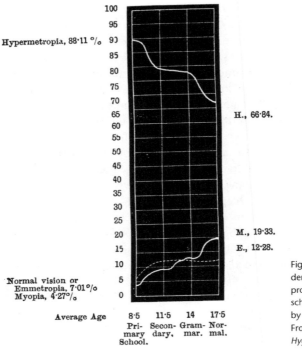

Hypermetropia, 88·11 %

H., 66·84.

M., 19·33.
E., 12·28.

Normal vision or
Emmetropia, 7·01%
Myopia, 4·27%

Average Age   8·5   11·5   14   17·5
               Pri-   Secon- Gram- Nor-
               mary   dary, mar. mal.
               School.

Figure 1.12 Graph demonstrating the production of myopia in school, which is indicated by the line labeled "M." From Newsholme, *School Hygiene* (1887).

a purely artificial condition, induced by the misuse of the eyes during the period of growth."[99] Statistical measurements showed that myopia steadily increased as children progressed through school (figure 1.12). In *School Hygiene* (1887), the epidemiologist Arthur Newsholme declared: "School-life has, under conditions which commonly prevail, a most deleterious influence on eyesight."[100] The causes of this were material and practical. Badly designed desks, small windows, and defective illumination contributed to an atmosphere of willfully contrived dinginess. This was exacerbated by habitual practices that could permanently damage the eyes: gratuitous detentions, pointless writing exercises, needlework drill, excessive homework, and even wall maps with small letters, which "require the microscope of an expert to decipher."[101]

The reform of visual practice was slow and uneven but discernible. Some schools, for example, began deploying eye tests. "Every child on entering a school," urged Hope and Browne, "should be examined as to his capability of reading 20 type at 20 feet easily."[102] Such tests particularly helped identify the hypermetropic (farsighted) child, often dismissed simply as a slow learner or an idler. Architects and designers produced new desk designs for schools.[103] Much attention was paid to the physical form of books themselves. Paper should be white, opaque, and unglossed: "The *smooth glazed* surface, now greatly in fashion, is highly objectionable. It gives rise to dazzling, and in certain positions the reflexions, especially from artificial lights, render the print almost invisible."[104] Hope and Browne recommended short paragraphs, a maximum line length of three inches, the abandonment of hyphenation, and reasonable spaces between letters. When fonts became ground down, Newsholme urged, printers must replace them: "Letter-press derived from a *worn-out fount* [sic] gives an imperfect impression of the letters. The loops of *a* and *e*, of *b d p g* are apt to form a black spot; long letters become broken, and fine up-strokes are imperceptible."[105] Through a combination of techniques, one could provide writing that a normal child's eye could read at the normal distance (ten to twelve inches) provided by a normal desk and, thus, environmentally normalize ocular development.

In ideal circumstances, then, the young adult would leave school with a pair of reasonably sound eyes, ready for the world of work, which presented a set of more varied optical challenges. As Southard noted: "There is no profession, no business or trade of any kind which is not, in some degree, dependent upon accuracy of vision for success."[106] The growing necessity for demonstrable and measurable levels of perceptual capacity was evident in the increasing number of occupations (train driving, the army) routinely using eye tests.[107] Each business or trade required particular visual capacities that, over time, might leave permanent, tangible traces in the eyes, face, and body. "Shortsight [sic] is most frequent in artisans who require to have their work brought near the eye, and in literary men who are devoted to reading," noted Brewster, "while shepherds and sailors, and labourers in the field, have their sight lengthened by their profession."[108] Working underground produced the nystagmus of miners, while tailors and needlewomen often developed ophthalmia or asthenopia.[109] Magnifying glasses were used with such frequency by watchmakers that they "affect[ed] . . . the expression of the face," while flickering illuminants aggravated optic nerves.[110]

The protection of workplace perception took multiple forms. This protection was most materially and crudely needed among those working with hot, sharp, or corrosive substances. Factories sometimes had a particular worker who was adept at removing fragments of metal, often with a magnet or a penknife. The wearing of protective goggles or glasses for jobs like stone breaking or ironworking was becoming more common by the early twentieth century. Their use tended to be discernible along generational lines, with younger workers far more likely to wear them than stubborn older ones.[111] More generally, the development of illuminating engineering would subject work space to varying degrees of photometric and chromatic standardization, in the name of both efficiency and ocular protection.

To repeat: visuality is always a synthetic act, involving bodily, discursive, and material factors. A schoolchild scrutinizing a blackboard, a train driver obeying a signal, or an ophthalmologist examining a myopic eye: none of these acts of seeing was simply an effect of crushing, extrinsic normalization, technological discipline, or pure, unmediated human will. These acts of perception were, instead, the particular, relatively durable fusion of extrinsic agents (light, desk, sign, scope), forms of discourse (rules, tests, laws, norms), and physiological entities (eye, retina, optic nerve, brain, hands). Vision, as Merleau-Ponty argued, was essentially chiasmatic, or simultaneously, necessarily, subjective and objective, and its government too should be regarded as a composite, multiple practice.[112]

Visuality and Liberal Subjectivity

Ophthalmologic science and the social concern for protecting vision developed at roughly the same time period as liberalism. This was, I think, more than simple historical coincidence. The huge increase in the use of spectacles and eye tests corresponded to a world of increasing visual information, in the form of language, numbers, and other signs, although perhaps not to the overwhelming, exclusive extent claimed by Simmel. The self-governing liberal subject was formed within this world of ramifying signification and was expected to possess many different visual capacities: attention, observation, recognition, introspection, discernment, literacy. The liberal subject, particular forms of visual practice, and visual technologies like the ophthalmoscope all appeared reasonably contemporaneously.

This assembly, however, was forged at least partly on the body's own terms, which does suggest that the eyes and vision, and liberalism as it developed in the nineteenth century, had certain mutual affinities. What can this mean? In *The Absent Body*, Drew Leder develops the concept of the "phenomenological vector" to examine the relation between our bodies and the kinds of societies we form, or "practical vectors established by the body's structure." "Our fundamental anatomy and physiology delimit and suggest the modes of usage to which different corporeal regions will be put," he concludes.[113] This is not reductive determinism or essentialism: rather, the body "suggests" manifold possible use vectors, forms kinds of practice or connection. Can this approach be applied to vision in the nineteenth century? Some quite strong connections do exist between vision and liberalism, connections that help explain why the Victorians became so concerned with eyesight. There is, obviously, nothing natural or inevitable about the liberal uses of vision, but there is nothing absolutely constructed either. I will suggest four phenomenological dimensions to this politicoperceptual congruence.

First is the dimension of *volition*. Self-determination has been integral to liberal ideas of self-government, and, in using the eyes, the human being has more choice than in using the ears or the nose and possibly even the hands. We can choose what to look at, by bodily and ocular motion, and can negate vision altogether by closing our eyelids. No other sensory organ provides such options, as Brewster noted: "By means of six muscles attached to it, it can direct itself, without moving the head, to almost every point of a hemisphere; but when the motion of the body is combined with that of the eyeball, it can command almost a continuous picture—a panorama of everything around it."[114] This issue of controlling one's vision perhaps explains why occult and extrasensory modes of perception were regarded with such mistrust: they threatened the ability to determine one's own sensations, to fill the mind with the data one wanted, to form oneself from within rather than be controlled by nefarious outside forces.[115]

Second is the function of *distance*. This is vital, as Hans Jonas states: "Sight is the only sense in which the advantage lies not in proximity but in distance: the best view is by no means the closest view; to get the proper view we take the proper distance, which may vary for different objects and different purposes, but is always realised as a positive and not a defective feature in the phenomenal presence of the object."[116] To view something properly involves disembedding oneself from the viewed world. Vision is, thus, the sense of phenomenological individuality par excellence: unlike smell and taste, vision involves not so much incorporating or merging into the world as setting oneself against and apart from it. Again, this accorded with liberalism's individuating tendencies. As Huxley put it: "if a man had no other sense than that of smell, and musk were the only odorous body, he could have no sense of *outness*—no power of distinguishing between the external world and himself."[117]

Third is the related question of *objective judgment*. Vision, Jonas suggests, gives the (technically erroneous) impression of having no cause: unlike, say, the barking of a dog, objects present themselves for observation without any suggestion of activity. Similarly, the eye's physiological activities remain frequently buried beneath consciousness: "In *sight*, the impression is so completely projected outward—seems so absolutely objective—and the consciousness of anything taking place in the eye is so completely lost, that it is only by careful analysis that we can be convinced of its essential subjectiveness."[118] The liberal subject was a self-judging being, rational and objective. Vision provided the phenomenological structure for such potential, providing the impression that judgment was being formed wholly from within the body rather than from a complex, compound perceptual interaction with the world. This is the structure necessary for empiricism, induction, and those approaches to reality generically termed *scientific*. It also has a potentially egalitarian, even Protestant dimension, in that this object world is essentially unveiled to all and not in need of deciphering by privileged hermeneuts. It is simply there. Empiricism was as much a feature of democratic as of scientific thought, as Tocqueville stated: "Equality stimulates each man to want to judge everything for himself and gives him a taste in everything for the tangible and real."[119]

Fourth is *thought*. The act of thinking has often been associated with the kind of distance and objectification associated with vision, which has been readily co-opted to serve as a sensory analogue for thinking itself. For Descartes, seeing and thinking were closely interconnected: thought is expressed through (now altogether hackneyed) visual analogies and metaphors.[120] The curious invisibility of seeing is suggestive of the experience of thought. Paranoid, Sartrean grumbles about the invisibility of the other's gaze spring to mind.[121] And this structure, through which visual activity and control recede from immediate consciousness, creates an effect of concentration, of pure communion with an entirely separate world. If ocular health was good, noted Harlan, there should "be nothing to remind us that we have eyes."[122] Despite the rise of physiological optics, vision can seem the most unmediated, and freest, way of interacting with, and thinking about, a world we can potentially control.

These visual aspects of liberal subjectivity are meant to be suggestive and heuristic rather than in any sense definitive. They appeared in numerous eighteenth-century contexts, with reference, in particular, to morality. We find several references to self-observation, perhaps the most famous and often-cited of which was made by Smith in *The Theory of Moral Sentiments*:

I divide myself, as it were, into two persons; and that I, the examiner and the judge, represents a different character from that other I, the person whose conduct is examined into and judged of. The first is the spectator, whose sentiments with regard to my own conduct I endeavour to enter into, by placing myself in his situation, and by considering how it would appear to me, when seen from that particular point of view. The second is the agent, the person whom I properly call myself, and of whose conduct, under the character of a spectator, I was endeavouring to form some opinion. The first is the judge; the second the person judged of.[123]

This is the classic formulation of the "specular subject," for whom all moral judgments are formed following scrupulous observation of oneself and others: spectatorship is a kind of social adhesive. One's own self can be "judged of" by disaggregating a viewing from a viewed subject. The guilty moral subject is specularly fissured, gazing ever inward, casting a critical eye over its blackest thoughts and impulses. "We are so afraid of being fools, and above all of looking like fools, that we are always watching ourselves even in our most violent thoughts," noted Benjamin Constant of such narcissistic introspection.[124] Gentlemanly character, argued Smiles in *Self-Help*, involved self-inspection: behavior was appraised "as he sees it himself; having regard for the approval of his inward monitor."[125] This was, of course, an imaginary act, but it extended to the more concrete sphere of social interaction, where "moral sense" and "sympathy" were perceived as the sensory equipment of visual communication. Here is the Earl of Shaftesbury: "No sooner are actions viewed, no sooner the human affections and passions discerned (and they are most of them as soon discerned as felt) than straight an inward eye distinguishes, and sees the fair and shapely, the amiable and admirable, apart from the deformed, the foul, the odious, or the despicable."[126] The man of feeling was a man of seeing, and he saw by moving through society. Through such public perception, one was also invited to consider oneself as an object available for the critical inspection of others.

This dynamic, unstable fusion of social observation and introspection, of interaction and autonomy, which became central to much nineteenth-century public and private practice, suggests a more complex form of visuality than that provided by models of urban spectatorship or panopticism. Excessive focus, in particular, on a Foucauldian "faceless gaze" occludes understanding of more prevalent modalities of vision that were at once visible, embodied, and voluntary. It also ignores the ways in which vision was an active subjective technique: the knowledge gained from judicious observation was useful largely for the subject alone. Norbert Elias, in his schematic way, captures this when noting how, following the medieval period, there was a greater "oblig[ation] to observe," an "increased tendency of people to observe themselves and others."[127]

The historian should be careful not to make the absurd assumption that, before the modern period, people never bothered to notice or observe anything. Rather, the emphasis should be on the slow social diffusion of attentive modes of world reading and particularly their intensification in urban contexts, where a greater number of encounters were mediated through the eye and the threat of misrecognition and misunderstanding proliferated. The idea of the obligation to observe or notice is critical. It allows us to connect the civilizing process to the history of perception. Crary's analysis of attention is again relevant. "Attention," he suggests, comes to refer to the subjective capacity to distinguish significant from insignificant signification, "an imprecise way of designating the relative capacity of a subject to selectively isolate certain contents of a sensory field at the expense of others in the interests of maintaining an orderly and productive world."[128] The city might have been a text, but nobody would ever read it from cover to cover. Attention creates islands of detail, foregrounds and backgrounds, dividing social language from streams of noise or nonsense. Focus, scrutiny, discernment: all are little techniques of sifting and managing the flux of sensory data.

"Subjectivity, as the innermost core of the private," argued Habermas in *The Structural Transformation of the Public Sphere*, is "always already oriented to an audience."[129] The public sphere was a nonpanoptic space where one attentively looked and was available to be attentively seen. To enter public space was to present oneself before what Gabriel Tarde called the *social retina*, which is composed of the rods and cones of fellow humans, usually strangers.[130] The penalties for incivility no longer had real eschatological consequence, and they were less physical than emotional, psychological, or normalizing: shame, rejection, humiliation. The tiniest, pettiest smirk can shatter confidence. Rose describes such strategies as *"government through the calculated administration of shame,"* taking shame in its sociological sense as defined by Elias as "a kind of anxiety which is automatically reproduced in the individual on certain occasions by force of habit."[131] This kind of visual system is bidirectional, tacit, normative, and pervasive in modern society. The avoidance of shame was entirely presupposed in Smiles's evocation of its opposite, setting an example: "Even the humblest person, who *sets before his fellows an example* of industry, sobriety, and upright honesty of purpose in life, has a present as well as a future influence upon the well-being of his country: for his life and character pass unconsciously into the lives of others, and propagate good example for all time to come."[132]

Society was visually experienced as a large group of strangers, every one of whom was potentially individually legible, and, hence, the whole was atomizable into its parts. Appearances, as Sennett has out-lined, became vital because "in the high Victorian era people believed their clothes and their speech disclosed their personalities; they feared that these signs were equally beyond their power to mould, but would instead be manifest to others in involuntary tricks of speech, body ges-ture, or even how they adorned themselves."[133] The fundamentally im-manent conception of self and character should be noted: individual personality erupted from deep within, left its tangible traces on the skin, hair, and raiment, and could be read from without. Thus, there were two dimensions to public performance. First, one should be attentive, watch-ful of details, aware of signs. Spectacles, obviously, helped here. Second, one should exercise control over the signs radiating from oneself. Each reinforced the other, producing the classic, and doubtless stereotypical, Victorian man and woman, conservative of dress perhaps, yet always aware of changes in fashion.[134] Displaying collective self-control and mastery of visual codes was a vital technique through which gender roles and social position could be maintained. As Simon Gunn has per-suasively argued, the middling classes always betray "the impulse to convert appearances in public into images of authority, to make social difference and the assertion of power over others visible by symbolic means."[135]

Specific guides to visual signification and urban detail flourished. The ancient art of physiognomy thrived anew following the publica-tion of Johann Caspar Lavater's *Physiognomische Fragmente* (1775–78), translated into English as *Essays on Physiognomy* (1789–93). Lavater em-phasized structural, permanent features of the face (size, angles) rather than more transient, volitional expressions. Anatomical configuration provided a key to individual soul or character. Physiognomic knowledge and training told one how to attentively use one's eyes to distinguish the true nature of others: "Precision in observation is the very soul of physiognomy. The physiognomist must possess a most delicate, swift, certain, most extensive spirit of observation. To observe is to be atten-tive, so as to fix the mind on a particular object, which it selects, or may select, for consideration, from a number of surrounding objects."[136] This knowledge was potentially open to all, creating the possibility of a "democratisation of observation" at the precise time that physiological optics was developing.[137] The "spirit of observation" would be enhanced and technologically embedded by eye testing, spectacles, and practices of ocular care, just as the spread of tests and eye care would have been incomprehensible without a corresponding proliferation of injunctions to examine and scrutinize.

During the nineteenth century, phrenophysiognomic inquiries developed in two directions. First, more formal sciences co-opted some of their assumptions while eschewing certain other elements, along with the nomenclature, increasingly associated with pseudoscience. Thus, the belief that inner states were revealed physiologically, involuntarily, and, eventually, evolutionarily underpinned, among other things, Mill's unsuccessful science of character (ethology), Darwin's *The Expression of the Emotions in Man and Animals*, Cesare Lombroso's criminal anthropology, and Francis Galton's composite photographs of criminal types.[138] The possibility of detecting the obscure traces of deviance had long stimulated physiognomic inquiry: in 1872's *Physiognomy Illustrated*, Joseph Simms claimed that, once the criminal facial type was fully recognized, it would simply function as "a sign-board denoting the rottenness within."[139] This, of course, could work only with an attentive public. Second, physiognomic assumptions can be seen in the popular press, art, the theater, novels, and music halls. Writing in 1856 on the trial of William Palmer, the "Rugeley Poisoner," Dickens observed: "Nature never writes in a bad hand. Her writing, as it may be read in the human countenance, is invariably legible, if we come all trained in reading it." "The physiognomy and conformation of the Poisoner," he continued, "were exactly in accordance with his deeds, and every guilty consciousness he had gone on storing up in his mind, had set its mark upon him."[140] Dickens's novels doubled as manuals for public character reading.

Thus, we might speak, cautiously, of the cultural salience of a certain empiricist-realist mode of attentive world reading, an inculcation of what Smiles termed *accuracy of observation*.[141] Attentive visual fixation on detail was characteristic of the scientific method, whereby the senses were practically and technologically disciplined in laboratories. Only thus could new entities have their ontological status fixed and defined.[142] The practical ramifications of this are worth emphasizing: scientists learned to use their eyes, and deport their bodies, along with various instruments, in ways that ultimately became habitual, repeatable, and normative.[143] The pathologist and cell theorist Rudolf Virchow, for example, told his students that they must "learn to see microscopically."[144] The epistemological aspects of this are equally significant. Consensus was reached in the laboratory among honest gentlemen, by using agreed-on and open methods, and truth was produced and able to circulate in the world beyond. Yet these truths, and the whole system producing them, could acquire social salience only if the public could trust them. As Shapin and Schaffer observe, the solution was a system of "virtual witnessing": a realistic mode of writing, including the language of humility and the description of failed experiments, was vital to establishing public trust in the absence of the possibility of large-scale experimental replication.[145] Later realist technologies include photography, cinema, and, of course, television.[146] As the *Lancet* noted in 1859, with reference to the use of photography in medicine: "Photography is so essentially the Art of Truth—and the representative of the Truth in Art—that it would seem to be the essential means of reproducing all forms and structures of which science seeks for the delineation."[147] Individual consumers of the report, photograph, or moving picture could feel as if their own senses were implicated in epistemological verification: scientific truths had a kind of transparency to them, in contrast with practices like alchemy and mesmerism, which relied on the inscrutable, opaque mental states of the practitioner.

This idea has been plausibly extended to provide a model for the trust inspired by liberal regimes themselves. Yaron Ezrahi notes: "The belief that the citizens gaze at the government and that the government makes its actions visible to the citizens is . . . fundamental to the democratic process of government."[148] Techniques of visibility echo those used by science: the publication of data, the availability of documents, the openness of process. Public bodies themselves operate according to the "principle of supervision," the notion that other bodies observe them, report on them, critique them, and make their findings visible to those subjects attentive enough to read the newspapers or even visit archives.[149] Of course, this is often overlaid and complicated by more enchanting visual spectacles, like the pomp of the monarchy, as Bagehot noted: the civil servant's dossier is wreathed in ermine.[150] In this sense, Britain's political system combines, in no easily analyzable way, the visual regimes of the church, the law court, and the laboratory.

My focus, however, is less on high politics and political institutions than on the more dispersed operation of power in urban space. Here, the built environment became a politicovisual problem precisely because it was perceived as blocked, gloomy, filthy, and demoralizing. Comprehending the problem was feasible only if space itself could be visually perceived, as a whole: hence the need for mapping.[151] Maps could display potential points of congestion and disorder. The democratization of cartography is also evident in the rise of pocket maps, which allowed one to find one's way about the city without asking others for directions. The pocket map is a small device facilitating autonomy. *Mogg's New Picture of London*, published in 1847, announced: "To avoid the inconvenience of taxing his friend to an attendance upon him in his peregrinations, it is indispensable that he provide himself with a good plan. . . . Like the clue of Ariadne, [plans] will conduct him through the labyrinth, and occasionally consulted, will enable him unattended to thread with ease the mazes."[152] The historian's eye should not be drawn to the clichéd evocation of labyrinths and mazes, but to the emergent language of "(un)attendance." Maps, along with street signs, house numbers, and streetlamps, allowed the individual, when alone in the city, to be secure, mobile, and autonomous. Again, very concrete and basic forms of freedom could be promoted by multiple, mundane visual technologies.

To conclude this chapter, I look at urban space in more detail and particularly at the relation between vision and the other senses and the social implications of these relations.

The Senses, Space, and Social Differentiation

It would be futile to analyse social tensions and conflicts without accounting for the different kinds of sensibilities that decisively influence them.
ALAIN CORBIN, *THE FOUL AND THE FRAGRANT*

Perceptual practices and discourses were not merely operative at the level of the liberal individual; they were common to groups of liberal individuals and defined against others. They were operative in particular physical spaces, and this operation was defined against other physical spaces. The liberal subject's visuality had, here as elsewhere, an exclusionary and oppositional dimension. In the literature of social investigation, novels, and public health, another perceptual regime was articulated: wherever these texts were consumed, this other sensory world was imagined. Here, the poor were presented as stinking, groping, nonvisual beings, indifferent to their own squalor. As Corbin has suggested, the formation of perceptual forms of social imaginary was essential to the production and reproduction of social difference.[153] Imagining the poor thus was not merely incidental to denying them rights, benefits, or votes. Corbin speaks here of the "visceral depths to which the nineteenth century social conflicts reached."[154] I want to explore this other perceptual regime, by looking at how liberal subjects sensorily described their experiences with others and the "imaginative sustenance" they drew from the experience.[155]

Slum visitors routinely recorded their experience as a profound shock to the senses. In *How the Poor Live*, the journalist George Sims described entering a room in London's East End: "The stranger, entering one of these rooms for the first time, *has every sense shocked*, and finds it almost impossible to breathe the pestilent atmosphere without being instantly sick."[156] Being a stranger was felt viscerally and perceptually, as the judicious distance and volition of the visual disintegrated, to be replaced by reliance on the hands and nose: "You have to grope your way along dark and filthy passages swarming with vermin. Then, if you are not driven back by the intolerable stench, you may gain admittance to the dens in which these thousands of beings, who belong, as much as you, to the race for whom Christ died, herd together."[157]

The monotonous drone of this discourse, with its motifs of dungeons and constriction, marks its ubiquity and social salience. For the respectable observer as well as the reader, the experience or thought of being plunged into fetid darkness was unnerving. An enduring image in slum and sanitary investigation was the investigator lighting a candle to ascend tenement stairs or to inspect the shivering poor in their abodes. The medical officer of health (MOH) for Glasgow, James Russell, described this in 1894: "In the brightest and longest summer day, the gas, or more usually a paraffin lamp, is kept burning and when often, in the extremity of poverty, there is no artificial light but a glimmer from a handful of red ashes, and, going in off the light outside, one hears voices, but can see no one."[158] Fifty years earlier, Engels famously observed how Manchester had evolved two separate spatial and sensory systems. The town's main thoroughfares "conceal[ed] from the eyes of the wealthy men and women of strong stomachs and weak nerves the misery and grime which form the complement of their wealth," misery located in the alleys and courts behind the facades.[159]

The respectable, Engels reminds us, routinely experienced the poor through the sense of smell. John Liddle, the MOH for Whitechapel, expressed this clearly: "when they [i.e., the poor] attend my surgery, I am always obliged to keep the door open. When I am coming down stairs from the parlour, I know at the distance of a flight of stairs whether there are any poor patients in the surgery."[160] One was poor because one smelled, and there was no need for physiognomy to discern anything subtler. Smell transgressed civilized distance between bodies: sweat and dirt evoked a doubtful morality.[161] Booth made this point when bluntly reporting the condition of a woman living in a back room at number 33 Parker Street in London: "No one would willingly go within five yards of her, so offensive was her breath."[162] Smell wafted not only from the bodies of the poor but from their animals, homes, and workplaces as well. Stinking dogs appear as frequently as midday candles in these narratives: canine effluvia, some claimed, could cause fainting.[163] Dilapidated sanitary arrangements became key foci of civilized disgust. Workers' bodies, meanwhile, absorbed the smells of the workplace and then transported them beyond its walls. Robert Roberts recalled how, in early-twentieth-century Salford, "some men working in the many offensive trades of the day smelled abominably, and people would avoid the public houses where they foregathered."[164] Noise was also important. Reports on the poor regularly described shouting, swearing, music, and barking, which destroyed the conditions necessary for attention:

The absolute requirements of a day's work in a business place are quite sufficient in themselves to tax the ordinary system; but when to those are added the efforts required to maintain the attention amidst perpetual distraction, then the strain to which these parts of the nervous system are subjected which are in connection with the auditory apparatus, no wonder that the number of cases of overwork and overstrain are on the increase.... If a man were exposed to a series of blows upon his skin in a walk through the city he would certainly call for the interference of the police; but he is helpless as to the shocks inflicted upon his brain through the auditory nerve.[165]

Noise has been defined by Jon Agar in Douglasesque terms as "sound out of place—sound which has transgressed the boundary around what is orderly."[166] Thomas Carlyle famously built a special study to cocoon his delicate brain from such auditory blows.[167] Inflicting needless noise on another was becoming a hallmark of incivility.

These manifold discourses created a set of perceptual oppositions, which I will simplify. On one side stood the liberal subject, an individual, distant but attentive, who surveyed and read the world, without ever ceasing, of course, to physically engage with it. In contrast to this person, we find the *desensitized* human being. Victorian writing on the poor was a continuous discourse on desensitization. Alcohol, Sims argued, was the cheapest, easiest way to annihilate sensibility: "Drink dulls their senses and reduces them to the level of the brutes they must be to live in such sties [sic]."[168] In turn, habitual drinkers "lose all relish for plain nutritious food, and their appetites can be stimulated only by something savoury and piquant."[169] The desensitized had a dysfunctional sensory apparatus. At the Great Exhibition, when the lower classes appeared in public on a mass scale, they were often represented as incapable of viewing objects from a distance: the *Illustrated London News* observed rather condescendingly that they were "more prone to touch, feel, and finger the goods than they ought to have been."[170] Chadwick argued that laborers' senses could become so unresponsive that they became physiologically incapable of appreciating environmental improvement: "The faculty of perceiving the advantage of a change is so obliterated as to render them incapable of using.... the means of improvement which may happen to come within their reach."[171] This produced a corresponding quality of sheer indifference to that which ought to inspire anger or disgust: "That which is most painful to the humane observer [is] an indifference to the evils which surround them."[172] In *The Bitter Cry of Outcast London*, this was most horribly manifest in incest: "The vilest practices are looked upon with the most matter-of-fact indifference."[173] If the respectable liberal subject was attentive, then the desensitized nonliberal subject suffered from abject indifference, the incapacity to respond to the world, which made such a being more object than subject. While the former wiped his spectacles with cambric and visited the optician, the latter appeared to disregard his senses altogether: one slum visitor noted that "eyes, mouth, and ears are often sore and inflamed through want of cleanliness and care."[174]

The causes of this obliteration of care and attention, contemporaries argued, were manifold. Environmentalist explanations were common. Workers in chemical factories, for example, became desensitized through prolonged contact with pungent, corrosive gases: "All delicacy of perception is lost to those who are constantly subjected to the influence of noxious smells and flavours."[175] Sailors, their skin tanned and leathern from exposure to sun, wind, and salt, likewise lost tactile subtlety. The slum dweller's indifference was routinely attributed to the cumulative effect of damaging environmental factors: "The healthy human body often becomes inured, after long exposure, to unpleasant odours, and at length hardly notices them, if always immersed in them."[176] The senses became damaged, but not irrevocably, and environmental remedies were available in the form of clean water, fresh air, and sunlight.

Some, however, saw desensitization as congenital, intrinsic, or immutable. Charles Shaw, the chief commissioner of police in Manchester, depicted cesspool emptiers as follows: "Often hardly human in appearance, they had neither human tastes nor sympathies, *nor even human sensations*, for they revelled in the filth which is grateful to dogs, and other lower animals, and which to our apprehension is redolent only of nausea and abomination."[177] Such descriptions became increasingly biological, racial, and evolutionary as the century progressed. Maudsley associated the inability of the insane to rationally perceive the world with arrested development: "[Their eyes] may be full and prominent, have a vacillating movement, and a vacantly-abstracted, or half-fearful, half-suspicious, and distrustful look. There may, indeed, be something in the eye wonderfully suggestive of the look of an animal."[178] This incapacity to be attentive also linked the desensitized with non-Western populations. Herbert Spencer revealingly observed that savage tribes had "acute senses and quick perceptions" but no ability to concentrate, se-lect, or synthesize: as their "mental energies go out in restless perception, they cannot go out in deliberate thought."[179] When located in evolu-tionary time, as they frequently were, such humans could be viewed as perceptually backward, or "previsual." Sensory use became a key criteria or marker of an individual or a culture's position on the evolutionary ladder, in turn informing colonial imaginaries.[180]

With sex and gender, the issue was slightly different, being more com-monly a question of oversensitization rather than desensitization, al-though working women, and prostitutes in particular, were sometimes thrown right across the spectrum from the former to the latter.[181] In 1674, the Cartesian Nicolas Malebranche argued that the extreme delicacy of women's neural fibers made them intellectually inferior to men.[182] The male liberal subject occupied an Archimedean point between the hysterical, acutely sensitive woman and the deadened, indifferent drunk. Explicit here was the idea that men and women used their eyes in differ-ent ways, not merely because of cultural prescription or environmental forces, but because of constitutional factors: "The eye of man is the most firm; woman's the most flexible. . . . Man's surveys and observes; woman's glances."[183] The difference between male and female eye move-ment, with suitable class inflection, was noted by Spencer. Women of the lower orders were, he claimed, "averse to precision."[184] Greater per-ceptual differentiation between the sexes fitted Spencer's version of evo-lution. In *Man and Woman*, Havelock Ellis argued that women generally had more responsive tactile senses, making them more nervous and sensitive to pain.[185] This easily merged into the highly pervasive and ideologically overdetermined image of the hypersensitive, or irritable, housewife, for whom, as one doctor intoned, "it is scarcely exaggerating to say that she can hear the cat walk across the kitchen floor."[186]

Eugenicists, among others, tried to prove that perceptual abilities were distributed normally within populations, by measuring things like the capacity to distinguish between different weights and sounds (the nor-mal could, thus, be superimposed over the Archimedean).[187] Here, Gus-tav Fechner's psychophysics was an obvious resource.[188] But the per-ceptual imaginary gained force less from scientific authority than from the sheer repetitive weight of discourse, studded with cliché and turgid metaphor, proclaiming that there were, indeed, two basic sensory orders, that the population was split into those who felt and saw one way and those who did not. Thus, unease or disgust at specific smells, noises, and sounds was magnified by the realization that, for some people, these things were not objectionable, which, in turn, viscerally reinforced a sense of deep, inescapable somatic difference.

"Disgust is always present to the senses, arguably more so than any other emotion," observes William Miller. There is, perhaps, nothing natural or inevitable about finding certain things repulsive or unsettling, although, as Miller indicates in a sympathetic critique of Mary Douglas, rules of disgust are not infinitely protean (tears, e.g., seem never to be disgusting).[189] But these rules change over time: "That which occupies the site of disgust at one moment in history is not necessarily disgusting at the preceding moment or the subsequent one."[190] For the liberal subject, the threshold at which certain things become repulsive has clearly and necessarily fallen: the prolix discourse on the disgusting testifies to this.[191] The historian cannot account for the emergence of liberal subjectivity without explaining these novel feelings and emotions, which grounded experience in the most familiar territory of all, the body.

Thresholds of tolerance are too vague and visceral to be truly scientific, although there may be attempts to scientifically measure them. They act as powerful forces of constraint, yet they are always spatially specific. What is perceptually apposite in a bathroom is intolerable in the street. Workshops can be noisy; libraries cannot be. The housewife's hypersensitivity makes her ill equipped to do much more than hunt for specks of dirt or muffle sounds with carpets and cushions. Perceptual norms informed the sociospatial structure of cities. One judge in 1879 noted: "What would be a nuisance in Belgrave Square would not necessarily be one in Bermondsey."[192] So thresholds of tolerance are multiple and elaborated concretely in practice: "There is no single threshold of tolerance identifiable since thresholds vary with the context."[193] These contexts became, over the nineteenth century, increasingly mediated by environment and technology: it is environmental manipulation that allows space to be so perceptually differentiated. Bazalgette's London sewers, for example, were the condition of possibility for the perception of environmental improvement felt in 1907 by Henry Jephson in *The Sanitary Evolution of London*: "Looking at the great river even now in its purified state, as it sweeps under Westminster Bridge, any one would shudder at the idea of being compelled to drink its water in its muddy and unfiltered state. How infinitely more repugnant it must have been when the river was 'the great sewer' of the metropolis."[194] The disgusting past, when people were not, we imagine, so repelled or inclined to shudder, is a counterpoint to the clean present: sanitary evolution and changing thresholds of tolerance, again, were inseparable from biological evolution itself.[195]

Distance was clearly fundamental here. Standing on Westminster Bridge, Jephson was several miles from the outfall sewers at Beckton and Crossness, while the Great Stink was a fast-fading memory. His senses never had to encounter untreated sewage being dumped at sea because of the giant sewage system that precluded contact and sustained the distance on which his repugnance was built. Such distance can be historically assessed by analyzing shifting building laws. In 1844, Parliament passed a law stating that no building for the purposes of an "offensive trade" was to be constructed within forty feet of the street and fifty feet from a dwelling, and vice versa.[196] Sound was policed by similar enactments. Here is an 1897 Birmingham bylaw: "No person shall sound or play upon any musical or noisy instrument or sing or shout in any street or public place (a) within 50 yards of any dwelling house after being requested to desist by any constable or by an inmate of such house . . . (b) within 50 yards of any church, chapel, or other place of public worship . . . (c) within 100 yards of any hospital, infirmary or convalescent home."[197] The sick got an extra fifty yards because silence was particularly important to the process of recovery. Hospitals, like cemeteries and abattoirs, were by this time sometimes being built on the outskirts of the city. This architectural, legal, and cultural history would allow what Lefebvre called *proxemics*, or the analysis of the production of sociospatial distance, and it cannot be disentangled from the slow formation of new sensibilities.[198] Suburbia, in particular, is essentially a mammoth apparatus of distantiation, a crafted space of sweet smells, quiet streets, upturned noses, gardenesque views, upholstered softness, and tinkling pianos.[199] Civilized disgust has its true home in suburbia.

The war on desensitization was fought on many fronts, with many weapons: biblical tracts, self-help literature, education, housing, sanitary systems, laws, and countless acts of personal example. To close, I offer an example whereby technological renovation produced a tiny shift in the threshold of tolerance, a little increase in sensitization. In the late 1860s, a series of old privies and ash pits in Tebbut Street, just off Rochdale Road in north Manchester, were removed. They were replaced with a new dry ash pit system. Residents of houses were not totally desensitized, being sick of the stench of the old privies, but the new system drove back their tolerance further, making them aware of another aromatic stratum buried beneath older ones:

Perhaps the most interesting and important information obtained was from the occupants of the houses, some of whom stated, that whereas before the alterations were made they never opened the windows of the back bedrooms at all, in consequence of the stench that came into the rooms from the privies and ashpits below, they now opened them daily and got the rooms ventilated; and that, though from habit they had become almost unconscious of the somewhat less noisome atmosphere in the rooms below, yet that since the alterations at the back of their own houses had been made, they had *become aware of* a disagreeable foecal smell in the houses of neighbours and friends.[200]

Becoming aware, becoming attentive, and becoming aware that others were not so aware and attentive: the capacity to take notice was softly instilled through numberless little acts of municipal engineering.

————

The history of Victorian visuality, then, is not a history of a monolithic, objectifying gaze. It is more multiple than this: Victorian visuality refuses to be captured by single paradigms. If the eye was measured, charted, tested, protected, and normalized in historically unprecedented ways, this was not because of any singular, dominant scheme of coercion or discipline. It was, rather, because vision, in various forms, had become integral to an enormous number of practices that were associated with liberal subjectivity: attention to detail, character reading, concentration, silent reading, the use of scientific instruments, social observation, efficient motility, and so on. These bodily practices, in turn, were always articulated in specific technological and architectural spaces. As Sims, Russell, and others made clear, it was far easier to exercise visual command and control, or to govern one's eyes in a productive way, in a wide street or a well-lit library than in a narrow alley, a noisy tenement, or a gloomy slum basement. The rest of this book will focus in much more detail on these technological and environmental dimensions of Victorian visuality.

Oligoptic Engineering: Light and the Victorian City

If cities could be transformed, the rest would follow.
BENJAMIN WARD RICHARDSON, *HYGEIA* (1876)

Benjamin Ward Richardson's model city of health, Hygeia, has been viewed, quite correctly, as a paradigmatic example of Victorian sanitarianism.[1] But Hygeia also demonstrated the centrality of a reconstructed perceptual environment to the sanitary, and well-governed, city. Richardson's city was devoid of anything distracting, unsettling, or disgusting. "Noiseless" wooden pavements were traversed by sober, taciturn, purposeful citizens: "The streets of our city, though sufficiently filled with busy people, are comparatively silent." Smoke was decarbonized before being "discharged colourless" into the atmosphere, while all kitchens were situated on upper floors to prevent the smells of cooking from percolating through domestic atmospheres. Animals were slaughtered in "narcotic chambers" in municipal abattoirs far from sensitive human eyes. Broad streets and low buildings ensured that Hygeia was "filled with sunlight," reflecting from "white or light grey stone" pavements, permeating houses and hospitals through large windows. Gas and distilled water flowed through generous subways.[2] In Hygeia, urban space, clear vision, and absence of noise or stench were inseparable from socioeconomic progress. Perception was rationally managed: nobody was overstimulated, and nobody was desensitized.

Richardson's plan showed how the city's problems with sanitation and government could almost be reduced to problems of perception, which were, in turn, soluble through careful attention to the material form of the city: its streets, pavements, subways, and walls. Hygeia was never built, and its perfectly managed space remained a pipe dream. But everything from which it was built—plate glass, wood paving, smoke-abatement technology—was contemporaneously being deployed across Britain (and in Europe and the United States) in an attempt to shape the sensory environment so as to promote decency, comfort, health, and productivity. In this chapter, I examine some of these material techniques as they were introduced into the British urban landscape and explore how they overlaid and intersected with older spaces.

The results, obviously, were far messier than the sanitary utopia of Hygeia. But this empirical complexity cannot occlude the discernment of certain recurring visual patterns or trends: the shaping of spaces to encourage public self-observation (the "oligoptic" arrangement), the em-bedding of points of supervision, the creation of networks of inspection, and the constitution of physical privacy. I first look in more detail at how visuality became a major urban problem. I then discuss the first two of the recurring visual patterns: the oligoptic and the supervisory. Finally, I discuss four material strategies designed, among other things, to make such modes of perception normative.

Invisibility, Darkness, Disconnection: The Problem of Urban Vision

Well before Richardson proposed a solution, the perceptual condition of urban space was an acknowledged problem. As industrial cities grew, houses accumulated in dense, irregular clusters around factories and workshops, often filling all available open space. This was the gloomy, constricted territory of desensitization, the spatial and sensory antithe-sis of Hygeia, where the cool distance and self-control of the visual was virtually impossible. This kind of visuality associated with liberal sub-jectivity, meanwhile, was being assembled in residential suburbs and commercialized city centers. The social consequences of this were clas-sically depicted by Engels in *The Condition of the Working Classes in England*. "Every great city has one or more slums, where the working class is crowded together," he noted. "In general, a separate territory has been assigned to it, where, *removed from the sight of the happier classes*, it may struggle along as it can."[3] The imagined perceptual differences between the classes were firmly anchored in urban geography itself.

This process, which Engels found most pronounced in Manchester, generated a double visual problematic. First, the "happier classes" were protected from "everything which might affront the eye and nerves," which had the effect of rendering poverty (and its physical and sensory manifestations) more unsettling whenever it was encountered.[4] Aside from Engels's concerns about lack of social interaction, there were quite explicit policing issues surrounding this visual separation. In the secluded zones of poverty, official urban knowledge shaded into tenebrous uncertainty, generating fears and fantasies of the unseen, unknown masses "secluded from superior inspection and from common observation."[5] These mysterious realms were, in both fact and fiction, spaces of crime and disorder. During political unease in the 1830s and the early 1840s, Britain's fledgling police forces were frequently frustrated by the maze of alleys into which rioters vanished. "Crime," noted Lord Aberdare in 1875, "is promoted by security from detection and punishment and the difficulties of the detection and the facilities for escape are multiplied in these Alsatias which are found in all our towns," something demonstrated by the disturbing ease with which Jack the Ripper furtively and murderously drifted through Whitechapel slums a decade later.[6]

The second aspect of this visual problematic was that inhabitants of slums, and particularly children, were denied the opportunity to see and imitate the example set by their social superiors. Example, stated Smiles, taught "without a tongue.... All persons are more or less apt to learn through the eye rather than the ear; and, whatever is seen in fact, makes a far deeper impression than anything that is merely read or heard."[7] At the 1884–85 Royal Commission on the Housing of the Working Classes, there were numerous pleas to open up culs-de-sac and blind alleys to allow circulation of wider "opinion." "I think there is nothing so improving to the lower classes as to see a good deal of the classes above them," noted the MP Edward Watkin.[8] Such disengagement from public opinion was compounded by the structure of much working-class living space, which was often itself communal and devoid of privacy. Two paradigmatic forms of such domestic space were the Scottish tenement (individual homes in larger houses with common stairs) and the English court (houses grouped around a shared space with water pumps and toilets). There were, for example, around two thousand such courts in Birmingham in the early 1840s.[9] This communal space meant, as Sims

Figure 2.1 Court with shared facilities, in St. Michael's Ward, Manchester. Eleven houses share a tap, closets, and ash boxes. From Marr, *Housing Conditions* (1904).

complained, that "poor artisans' children grow up with every form of crime and vice practiced openly before their eyes," which made governing through the delegated administration of shame almost impossible.[10] Moral improvement, he implied, was precluded by a dangerous mixture of promiscuity and disconnection, which meant that what was ideally private, offensive, and desensitizing spilled out into public space and that what was ideally public and improving was excluded (figure 2.1).

A corresponding set of physicomedical anxieties surrounded the physical absence of light itself (figure 2.2). Darkness, warned Florence Nightingale, generated sickness, depravity, and lassitude: "Where is the shady side of deep valleys, there is cretinism. Where are cellars and the unsunned sides of narrow streets, there is the degeneracy and weakliness of the human race, mind and body equally degenerating. Put the pale withering plant and human being into the sun, and, if not too far gone, each will recover health and spirit."[11] Lack of sunlight was blamed for many ailments, including scrofula, anaemia, phthisis, chlorosis, and rickets. In 1890, the doctor Theodore Palm demonstrated that a map showing the distribution of rickets would correlate almost exactly with one showing areas with deficient sunlight.[12] Darkness was a classic sanitary nuisance: it allowed microbes to flourish and dust to accumulate, it made cleaning difficult, and it was inimical to health. The medical officer of health (MOH) for Manchester, John Leigh, found children in the gloomiest parts of the city to be "blanched and flabby," and, by the 1920s, the city was providing sunlight treatment at several centers.[13] Darkness and narrow streets also damaged the eyes. One early-twentieth-century survey of fifty thousand London children found that, in poor neighborhoods, without "wide main thoroughfares," 29.01 percent of children had defective vision, a number falling to 18 percent in "more outlying suburbs."[14] The physiological development of the visual organs was

Figure 2.2 Narrow alley, without sunlight. From Gamgee, *Artificial Light Treatment* (1927).

again explicitly overdetermined by material surroundings. As Nightingale implied, mental defects also flourished in such environments. "A material, as well as a *moral* and *mental*, etiolation or blanching occurs," observed the physician Forbes Winslow, "when the stimulus of light is withdrawn."[15] Prolonged absence of sunlight, noted another doctor, had "a most depressing effect on the mind."[16]

Sunshine was a universal agent of vitality, a dehumidifier, deodorant, and disinfectant. Solar rays, Robert Koch demonstrated, could nullify the tuberculosis bacillus. Sunlight could, reputedly, neutralize snake venom, while "putrefactive liquids may actually be rendered sterile by simply submitting them to the action of sunlight."[17] Thermodynamic theory suggested that sunlight was the source of all terrestrial energy. "It is the foundation of all our vital energies and powers, as well as the material origin of all mechanical force of whatever description on the face of the earth," observed the *Plumber and Sanitary Engineer* in 1879, "excepting, perhaps, the tides, earthquakes and volcanoes, of which mankind makes very little use."[18] In schools, the calculated diffusion of benevolent rays might facilitate (however fancifully or metaphorically) the transformation of sunlight into brain waves: "Physical light actually becomes transmuted into mental and moral light and vigour-an example of the transmutation of energy that should command the attention of schoolmasters and school boards."[19] Darkness was physically incompatible with all forms of dynamism: material and intellectual, individual and social.

Organizing the City: Municipal Engineering

Visual disconnection and darkness were two dimensions of a larger-scale problem of early-nineteenth-century urban government. The doctor and social investigator James Phillips Kay depicted Manchester as physically disorganized, lacking systematic infrastructure, and altogether unregulated: "The greatest portion of these districts, especially of those situated beyond Great Ancoats-street, are of very recent origin; and from the want of proper police regulations are untraversed by common sewers. The houses are ill soughed, often ill ventilated, unprovided with privies, and, in consequence, the streets, which are narrow, unpaved, and worn into deep ruts, became the common receptacles of mud, refuse and disgusting ordure."[20] Such material disorder implied, for many contemporaries, defective administration. Tocqueville found "no trace of the slow continuous action of government" in Manchester.[21] This was equally apparent in London. Before the 1855 Metropolis Management

Act, which established the Metropolitan Board of Works (MBW), there was little orchestrated management of infrastructures and in some areas nothing that could formally be called *government* at all. The metropolitan reformer Joseph Firth recalled that, in 1855 in St. Pancras, "a large portion of the parish was without any sort of government whatever." Two sizable districts within it were "without lamps, or proper household conveniences, and the roads were in a wretched condition."[22]

Physical reorganization on a truly municipal or metropolitan scale could occur only as part of the general reform of local administration taking place from the 1830s. However flawed, the MBW did provide the first systemic management of metropolitan systems. Outside London, the 1835 Municipal Corporations Act led to substantial reorganization of local government, most immediately by significantly changing the social groups wielding authority.[23] Old self-appointing commissions were abolished and replaced by more liberal bodies, elected, representative (at least of the propertied classes), and, to some extent, accountable to the ratepaying public. With this liberalization of administration came new powers to ease the passage of legislation, especially that relating to the city's physical environment: there was a huge growth in permissive or adoptive legislation, while Parliament also began passing model clauses acts from 1845. Both forms of legislation were substantially cheaper and quicker to adopt than earlier forms, and they also had the appealingly liberal quality of being noncompulsory.[24] Municipal governments thus adopted them if they, and their constituencies, desired, thereby creating a genuine structure of local self-help: "The principle of self-government is found to be the only principle upon which a large and dense urban population, like that of Manchester, can be satisfactorily governed."[25]

Although municipal government remained undemocratic, politically fractious, parsimonious, and hardly immune to corruption, such reformed councils were the most important bodies overseeing the improvement of streets and housing.[26] The municipal management of roads, sewers, and lighting was one of the first recognizably social government techniques in Britain, something embraced by liberals like William Farr, who argued in 1843 that "over the supply of water—the sewerage—the burial places—the width of streets—the removal of public nuisances—the poor have no command . . . and it is precisely upon these points that the Government can interfere with most advantage."[27] This was most famously embodied in Chamberlain's "civic gospel" in Birmingham. Figures like Gomme, the clerk of the London County Council in the 1890s, implemented a kind of municipal utilitarianism, arguing that individuals perform best under conditions of minimal government, yet required the provision of certain basic services, like water or streetcars, in order to be respectable, healthy, and mobile: in other words, to be liberal subjects.[28] By 1900, for example, around 80 percent of municipalities owned their own water supplies.

These reformed municipal governments began, slowly and unevenly, to tackle some of the specifically urban problems of darkness and visual estrangement. Lighting streets became cheaper and legally simpler: it is worth remembering that, at one time, even fixing a lamp on a street required parliamentary permission.[29] Later in the century, building legislation would make demolition of obstructive or excessively dark buildings rather easier, on the grounds of public health.[30] The habitation of cellars was particularly attacked. Liverpool's first MOH, William Duncan, claimed that between thirty-five and forty thousand people lived in the city's cellars. Here is a description of one of his subterranean expeditions: "When the door . . . was closed both light and air were excluded. . . . On one occasion Dr. Duncan had to grope his way, at noon day, into a house in Thomas-street; on a candle being lighted the patient was discovered on a heap of straw in one corner of the room, whilst in an opposite corner a donkey was comfortably established."[31] This dark, constricted, and presumably stinking space was a paradigmatic milieu of desensitization. "So numerous were the inhabited cellars in Manchester," mused Leigh, "that it might well have been considered a city of cave dwellers."[32] Liberal subjectivity, of course, was resolutely nontroglodytic. Between 1868 and 1872, some 2,400 cellars were closed by Manchester's newly formed Health Committee: by 1874, only 108 remained, peopled mainly by the old. The habitation of cellars, however, remained legal, as long as the rooms fulfilled certain spatial criteria, including being seven feet tall, with at least three feet above street level, and possessing their own water closet. In the early twentieth century, inspectors bemoaned the lack of legal regulation of the physical aspects of subterranean rooms, from light levels to dampcoursing.[33] There were still over 1,500 occupied cellars in Dublin in 1913.[34]

One important consequence of local government reform was the appointment of municipal surveyors, engineers, and inspectors whose job it was to supervise and construct roads and buildings, a process evident from the later 1840s. James Newlands, for example, was appointed borough engineer for Liverpool on January 26, 1847.[35] Manchester appointed a city surveyor (1857), an improvement surveyor (1869), a city architect (1869), a building surveyor (1863), a surveyor to the paving and sewering department (1840), an assistant surveyor of highways (1863), an outdoor surveyor/draftsman (1865), a subinspector of piping (1867), a superintendent of street gas mains (1858), and a surveyor and drafts-man of gas mains (1860). Each of the city's outlying boroughs also had its own surveyor. In 1875, the city surveyor, John Lynde, received four additional assistants after complaining of his department's workload. Nationally, the process was sufficiently widespread for the Institute of Municipal Engineers to be founded in 1873. Municipal engineers were often concerned with questions of technological standardization. They were also, like illuminating engineers, concerned with questions of phys-iology as well as building. Fleeming Jenkin, professor of engineering at the University of Edinburgh, argued: "In respect to domestic sanitation, the business of the engineer and that of the medical man overlap; for while it is the duty of the engineer to learn from the doctor what con-ditions are necessary to secure health, the engineer may, nevertheless, claim in his turn the privilege of assisting in the warfare against dis-ease, by using his professional skill to determine what mechanical and constructive arrangements to secure these conditions."[36]

Finally, the density of the sprawling, growing city created specific legal problems relating to the distribution of sunlight. The common law of "ancient light," dating from 1189, was reworked under the 1832 Pre-scription Act, which stated that, if one had enjoyed a specific quantity of natural light for twenty years, "the right thereto shall be deemed abso-lute and indefeasible," a right procurable by continuous use or through various forms of grant.[37] In urban areas, however, it was impossible to rigidly uphold this law, which entailed legal acknowledgment that there was, quite palpably, less natural light in the town than in the country. Lord Cranworth noted, in the 1865 case of *Clark v. Clarke*, that "persons who live in towns, and more especially in large cities, cannot expect to enjoy continually the same unobstructed volumes of light and air as fall to the lot of those who live in the country."[38] This did not imply that developers had the right to build where and how they liked, but it did introduce an important, and very liberal, principle of compromise. One central problem in such cases was the measurement and proof of the "specific quantity" of light: it was notoriously difficult to quantify precise amounts of light entering rooms, so other techniques became common, such as measuring the quantity of sky lost as a consequence of the erection of new buildings (figures 2.3 and 2.4). Most commonly, the more subjective formula of "substantial privation of light, sufficient to render the occupation of the house uncomfortable, or to prevent the plaintiff from carrying on his accustomed business in the premises," was used.[39] In 1877, the *Architect* concluded that the result was surprisingly frequent legal protection of the individual's right to light: "In many

AS IT WAS.

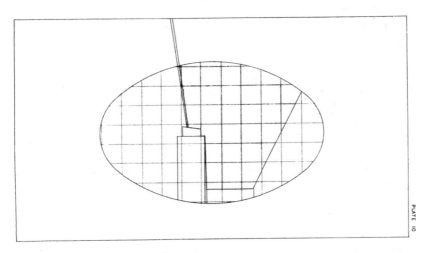

Figure 2.3 View through a skylight from behind the dispensing counter in a shop. From Fletcher, *Light and Air* (1879).

AS IT IS.

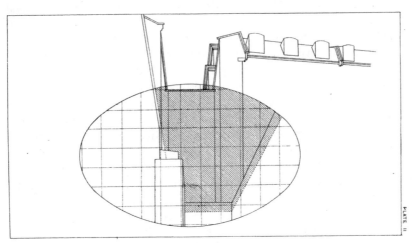

LOSS OF SKY 97⅓ PER CENT.

Figure 2.4 The same view as in figure 2.3 above, following the construction of an adjacent building, showing the percentage of sky lost to view. From Fletcher, *Light and Air* (1879).

important and complicated cases we shall . . . have cause to congratulate ourselves, in the interest of the public health, that the law, instead of looking with natural jealousy upon the enforcement of a servitude, now leans to the right of light in preference to the right of building."[40]

Model Cities, the Oligoptic, and the Supervisory

These various concerns about engineering, vision, and light were often brought together in a systematic and clear way by designers of model communities. For example, James Silk Buckingham's Victoria (1849), one mile square with a population of ten thousand, was proposed as an explicit spatial and visual solution to the problem of governing industrial society. "Would it not be possible," Buckingham asked, "to remodel society, by systematic association, on a better plan than the present?" Buckingham explicitly connected architecture, visibility, and conduct. Better humans, he urged, would be produced in conditions where their actions were permanently open to public display, which, in turn, involved a complete repudiation of the unplanned Manchester model. Such visual functionalism was expressed with quixotic clarity: "From the entire absence of all wynds, courts and blind alleys, or culs-de-sac, there would be no secret and obscure haunts for the retirement of the filthy and the immoral from the public eye—and for the indulgence of that morose defiance of public decency which such secret haunts generate in their inhabitants."[41]

Here was a city of total ocular connection, with no dark corners, cellars, or secrecy, and the enforced visual normalization of decency— alcohol and tobacco, e.g., were banned). The omnipresent "public eye" was symbolized by a three-hundred-foot-tall octagonal tower positioned at Victoria's geometric center, capped with a spire, a clock, and an electric light, the latter a rare technology in 1849. This plan appears consistent with the demands of socialists like Engels, liberals like Smiles, and philanthropists like Octavia Hill: the classes would be visually united, and the free circulation of opinion, as well as maximum exposure to imitable conduct, would be materially secured. Yet, with its omnipresent, inescapable, moralizing gaze, there was something almost panoptic about Victoria, although it lacks the detail of Bentham's plan. It was a city where privacy aroused acute suspicion, a city with an explicitly coercive public eye. Similar concerns were maintained in Hygeia, which was, to repeat, the technological antithesis of Kay's Manchester, with each house supplied with municipal energy and bathed in sunlight. Unlike in

Buckingham's paternalistic project, alcohol was not explicitly banned. Rather, people rationally and voluntarily eschewed drink: "A man *seen* intoxicated would be so avoided by the whole community that he would have no place to remain."[42] The public eye persists, but there is more emphasis on the individual visual agency from which it is composed, while the Smilesian concern for imitation and shame remains integral to urban visuality. Finally, as befitted the sanitary vision, privacy was ceasing to be purely associated with suspicion and secrecy and becoming integral to securing cleanliness, health, and decency.

In *French Modern*, Paul Rabinow analyzed French plans from the later nineteenth century that particularly emphasize these systemic and sanitary aspects of urban design. Tony Garnier's Cité Industrielle is described as "a grid of intelligibility, this time for modern welfare society," and "an urban parallel to Bentham's panopticon."[43] Rabinow does not mean that Garnier's city was itself panoptic: rather, he implies that organized, governable space is being rethought à la Bentham, but on an urban scale. Yet the term *panopticon* does connote totally controlled and planned space as well as carrying pronounced visual implications. We must remember, again, that none of these cities was ever built, not only for reasons of cost and practicality, but also because such total oversight was regarded as illiberal. The *Sanitary Record*, for example, swiftly dismissed Richardson's project as "communistic."[44]

Cities might be thought of and known as totalities, but they were not rebuilt thus. The vast, if piecemeal, schemes to rebuild and improve the streets, houses, and public spaces of British cities in the nineteenth century aspired neither to completely reorganize urban space nor to saturate that space with a despotic and undetectable "public eye." Widening a thoroughfare, for example, was designed not to fix bodies or to control their motions but to stimulate organized, but free, movement and circulation. Constructing bylaw housing with backyards and private toilets was designed precisely to provide "obscure haunts" into which one might withdraw for the sake of decency. As urban space was lived and experienced, there was no perfect, total view of it, even for policemen or doctors: rather, there was a desire for maximum spatial interconnection of demonstrably public space (the street, the park) and a simultaneous attempt to divide clearly these public spaces from private, domestic ones. This multiplicity of interconnected visual spaces has been termed *oligoptic* by Latour, and, although the term is certainly not unproblematic, it captures enough of the visual logic I see in nineteenth-century British urban reconstruction for it to be useful here.[45] The term is entirely analytic, but (unlike Latour) I am using it to capture some salient elements of a historical process.

Oligoptic space is not simply a splintered panopticon but a very particular kind of viewing arrangement. The panopticon provides the total, permanent, and architecturally enclosed overview of a single space and its occupants. By contrast, an oligoptic space lacks a central, dominant vantage point. This makes an oligoptic space an arena within which a small group of people observe each other: it is a place in which mutual oversight takes place. Moreover, looks, gazes, and glimpses are distributed relatively symmetrically. Unlike in the panopticon, which is "a machine for dissociating the see/being seen dyad," in an oligoptic space the viewed can always return the glance because all are viewers and viewed, and, hence, one can always verify whether one is being watched.[46] This is, therefore, a space of many, trivially chaotic, and unpredictable lines of sight, as opposed to Bentham's formal, radial configuration. While the panopticon traps inmates in a "cruel, ingenious cage," oligoptic space allows one the freedom to move and look as one wishes.[47] Finally, the panopticon deprives the fixed subject of privacy, while oligoptica, local points (streets, parks, offices) where the few see the few, function as heterogeneous, interlinked outsides of multiple private spaces. Freedom of public perception is paralleled by a freedom to withdraw from view when one chooses, behind a locked door if necessary.

In his description of the "exhibitionary complex," or the visual arrangement of the museum, Tony Bennett captures the essence of oligoptic space. The crowd regulates itself by monitoring itself: "As microworlds rendered constantly visible to themselves, expositions realised some of the ideals of the panopticon in transforming the crowd into a constantly surveyed, self-watching, self-regulating, and, as the historical record suggests, consistently orderly public—a society watching over itself."[48] Likewise, in theaters, it was important that "the audience should see each other, so as to allow all who wish it an opportunity for personal display, and for scrutinising the appearance of others."[49] The "ideals of the panopticon" are, in some sense, spatially secured here, in that conduct is visually regulated. But both means and form are fundamentally different. One could, of course, always voluntarily leave the exposition and escape the gaze of fellow visitors. The possibility—indeed, necessity—of withdrawal from the "public eye" is especially true when we consider the gestating form of domestic architecture during the period. There was a pronounced tendency for what Martin Daunton calls the replacement of "cellular and promiscuous" domestic arrangements typical of the court or Scottish tenement by "open and encapsulated" ones.[50] This was the space of communal disconnection described by Engels. By contrast, bylaw housing erected after 1850 had enclosed, private facilities. Houses were constructed along parallel sides of streets connecting openly to other urban thoroughfares. This was an architecture in which interconnection and privacy were mutually constituted, generating a firm threshold integral to the visual constitution of the liberal subject.

The freedom to look and be looked at, the mobility of the gaze, and the capacity to withdraw into utter privacy: all marked urban space as categorically nonpanoptic. Yet to claim that society simply visually organized itself, as if autochthonously assembled from a myriad of self-monitoring liberal monads, is probably as misleading as a myopic obsession with the faceless, disciplinary gaze. Absolutely egalitarian visual arrangements, without provision for any form of oversight, are deeply risky and fragile. Here is a tiny, petty, and, hence, eminently representative example. In 1884, the yard of Embden Street School, in Greenhays, Manchester, was opened as a public playground. Publicity, however, led to the place becoming "the resort of idle fellows, mere 'toughs' who *assume the right of the public* and exercise it by coarse violence, insult and even personal attack."[51] The space was closed to the public shortly afterward because a "consistently orderly public" failed to spontaneously self-organize. Spaces where public assembly took place on a reasonably significant scale almost always required the appointment of overseers, supervisors, porters, watchmen, wardens, or doormen. These supervisors or superintendents had a privileged, but humble, verifiable, and certainly not omniscient, point of perception, often at a point of entrance and exit: a booth, lobby, gatehouse, office, or desk, from which they watched, assisted, answered questions, and occasionally disciplined.[52] Supervision acts as a generic visual technique overlaying and organizing oligoptic space. This principle of supervision is very different from the Habermasian version outlined in the previous chapter, being concrete, engineered, embodied, and elaborated in specific spaces.[53]

Supervisory space became an integral feature of many buildings. In 1872, the *Builder* described Bethnal Green Museum thus: "The main approach, through the western entrance, leads into the central hall, which commands almost a complete view of the whole of the building."[54] Any official could, therefore, assume a nearly total view of the free comportment of voluntary entrants. In his analysis of later-nineteenth-century mills, Richard Biernacki identifies a "circumferential" visual arrangement, whereby factory owners and superintendents occupied central, elevated rooms from which laborers' movements were observable. This design promised a commanding view of an oligoptic space, but one

Figure 2.5 General reading room, Anderston Library, Glasgow (built 1904), an oligoptic space with a point of verifiable supervision. From *Municipal Glasgow: Its Evolution and Enterprises* (1914).

that was limited in time (to working hours) and in space (to the shop floor only).[55] Leisure hours and visits to the bathroom or the canteen were generally unmonitored. In libraries, the "alcove" system, where readers occupied desks between bookcases positioned at a right angle to walls, provided excessive opportunity for nefarious practices and was, consequently, often abandoned: "Supervision of the tables in the alcoves from the catalogue counter is impossible, and opportunities for theft and mutilation of books are provided, without much chance of discovery."[56] Yet this supervisor never achieved omniscience: his or her physical presence or absence was absolutely verifiable, spatially and temporally circumscribed, and never total (figure 2.5). One early-twentieth-century guide to reference libraries made the point colorfully: "It is not necessary to be able to see into every nook and cranny of the department from one point in the room; even if this could be attained it would be of little practical value, as the reference librarian is neither Argus nor stalk-eyed crustacean!"[57]

The decline of the corridor in school and hospital design was part of the same trend, allowing a better view of self-controlled, rather than shackled, bodies: "In the design submitted by me, the hall is the centre of civilisation to the school [*sic*]; into it every door to every room in the building opens, and from the chair is commanded every part of the building and all the people in it. The hall is lit like the nave of a church, with a range of clerestory windows above the roofs of the classrooms surrounding it. It takes away the necessity for close, stuffy, dark and dreary passages, and gives a cheerful hall and promoter of light and ventilation."[58] One paradigm of this arrangement was the school playground, where children freely played and fought, coalesced into cliques, or remained lonesome or aloof, in a little world always potentially visible from surrounding schoolrooms, something consciously being worked into building designs in the early nineteenth century.[59] In hospitals, the corridor was replaced by the open ward system. Not only was this a means of enabling the clinical gaze, but, along with various visual techniques (scouts, inspection windows, supervision rooms), it also allowed nurses to see most of the ward most of the time.[60] Different perceptual patterns (supervisory, clinical) thus intersected in the same institutional space.

The physical reconstruction of the Victorian city, then, had palpable, multiple perceptual dimensions that cannot be reduced to the disciplinary or the spectacular: the visual organization of space was more multilayered than this. I want now to return to Hygeia and look at how very particular material systems and techniques were used, across Britain, to attempt to secure, at least partially, conditions of self-observation and supervision, clear vision, and silence. I will look at four of Richardson's favored techniques: the widening of streets, smoke abatement, the use of glass, and soundproof paving.

Street Widening

It has been estimated that, between 1865 and 1915, some £25 million was spent on widening British streets.[61] Street widening facilitated traffic circulation and policing, but it also functioned to promote the circulation of opinion and communication, as the 1838 select committee established to discuss plans for metropolitan street improvement explained:

There were districts of London through which no great thoroughfares passed, and which were wholly occupied by a dense population composed of the lowest class of persons who being entirely secluded from the observation and influence of better educated neighbours, exhibited a state of moral degradation deeply to be deplored. It was suggested that this lamentable state of affairs would be remedied whenever the great streams of public intercourse could be made to pass through the districts in question. It was also justly contended that the moral condition of the poorer occupants would necessarily be improved by communication with more respectable inhabitants, and that the introduction at the same time of improved habits and a freer circulation of air would tend materially to extirpate those prevalent diseases which not only ravaged the poorer districts in question, but were also dangerous to the adjacent localities.

Lines for new metropolitan streets were frequently chosen thereafter to demolish slums and introduce air, light, and interclass communication.[62] This was historically novel: John Nash, for example, admitted in 1812 that Regent Street was planned as "a boundary and complete separation between the Streets and Squares occupied by the Nobility and Gentry, and the narrow Streets and meaner houses occupied by mechanics and the trading part of the community."[63]

The coordinated widening of old streets and building of new ones was initially hampered by the same problems of urban government that Kay and Tocqueville lamented. In 1849, the exasperated Board of Health grumbled: "Even single streets are divided, often longitudinally, and paved and cleansed at different times under different jurisdictions. In the parish of St. Pancras . . . there are no less than 16 separate paving boards, acting under 29 Acts of Parliament."[64] Reformed municipal administrations attempted to coordinate and organize streets. In Manchester, the borough council took over the building and control of the city's highways in 1851, founding the Paving, Sewering and Highways Committee to regulate and repair them. Liverpool did this in the later eighteenth century.[65] Local and national legislation gave authorities the power to remove obstructions from streets and to organize their naming and numbering. In 1828, the unreformed local government of Manchester decreed that no new street should be under twenty-four feet wide: this figure had increased to thirty feet by 1841. During its first eighteen years of existence, the MBW renumbered 120,000 houses, renamed 1,500 streets, and named over 2,400 new ones. This process made the city more legible and negotiable, for pedestrians, cabdrivers, and postmen.

Such street improvement was sometimes evoked to symbolize the advent of a more planned, progressive, modern society. In his 1883 *Laissez-Faire and Government Interference*, the Liberal MP George Joachim Goschen argued that organized, supervised traffic regulation had triumphed over naive individualism. Traffic laws and superintending policemen organized urban circulation:

Till some years ago the street traffic in the metropolis regulated itself. The rules of the road were held to suffice. The stream of vehicles passed to and fro under a very wide application of "laissez-faire" and "laissez-passer." But when blocks became more frequent, collisions more numerous, street accidents more and more a common occurrence, the cry arose for the police. Society in the shape of two policemen stationed itself in all the important thorough-fares. Coachmen were stopped, drivers directed, foot-passengers assisted, refuges constructed. Freedom of passage ceased. The principle of individual liberty yielded to organised control.[66]

Such "organised control" might also involve more careful calculations of the play of light and shade along streets. The Boston architect William Aktinson's researches into street axes and window aspect made the distribution of natural light within cities significantly more calculable than previously. The east-west street was, he concluded, "a street of extremes, cold in winter, hot in summer. The diagonal street is very much better off" (figure 2.6).[67]

Standardization, calculation, and "organised control" often remained ideals rather than realities, however. More critical contemporaries were well aware of this. The advent of the MBW, for example, had failed to produce satisfactory harmonization of streets between and within vestries and parishes, as Firth noted in 1876: "Probably, London roads generally are now better than they were twenty years ago, and they will bear comparison with some towns abroad. But like everything else they illustrate by their excellence and by their deformity the lack of unity in government, and in many cases the absence of adequate scientific knowledge on the part of the surveyors."[68] Some parts of London had adopted the steamroller, he complained, while others had not. Similar discrepancies in upkeep, street cleaning, and street naming and numbering lingered. In 1874, one writer in the *Builder* complained that "the proportion of street-ends without any name up at all in them is very considerable, and in many other cases, the name has become nearly illegible, and seems never to be renewed," while numbering occasionally went round corners from one street to another. One house allegedly had the number $2\,^{3}/_{4}$.[69] Homogeneity remained elusive and relative, often because of eminently liberal principles of local self-government.

To illustrate these points, I look at the Deansgate improvement in Manchester in the 1870s. Deansgate was Manchester's central thoroughfare, running from Knott Mill in the south to the city's cathedral and Victoria Station in the north, and, by the second half of the century, it had become too narrow for the city swelling around it. "After making allowance for width, there is no street in Manchester that has so much traffic as Deansgate," observed the local magazine, *Sphinx*, in 1869. "It is a street in which not only thousands walk, but also hundreds

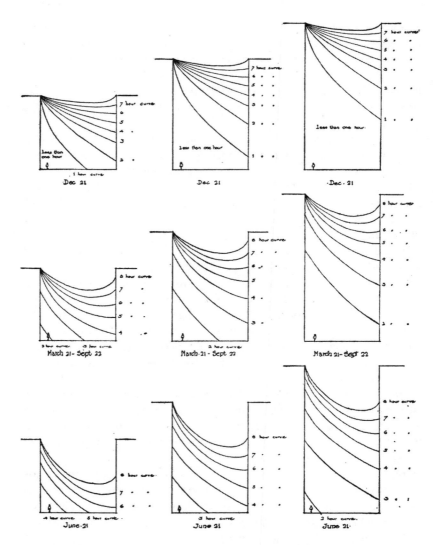

Figure 2.6 Light curve for a diagonal street, showing the amount of illumination received at "the four typical periods of the year," corresponding to equinoxes and solstices. This cross section is of a street running southeast and northwest, looking northwest, or of a street running northeast and southwest, looking southwest. From Triggs, *Town Planning* (1909).

stand; and the footpath is so narrow that one-half of its passengers are driven into the road." It had become a space of congestion and loitering. The corners around the Wood Street slums were "marked by lounging ruffians."[70] Jerome Caminada, recalling his detective career, described the area as "the rendezvous of thieves . . . a very hot-bed of social iniquity and vice."[71] Blood from local slaughterhouses trickled into the streets. During the 1849 cholera epidemic, Leigh identified these tightly packed nests of houses as among Manchester's unhealthiest.[72]

Deansgate had become a problem: disorganized, depraved, and diseased, an impediment to urban circulation, a place where distant, objective observation was impossible. Widening the street would simultaneously open the area to public observation, remove slums, and free circulation of air, light, and traffic. In 1867, the council announced that the entire thoroughfare would be widened to "a uniform length of twenty yards."[73] The First Deansgate Improvement Bill, giving the council powers of compulsory purchase, became law on July 12, 1869, covering the area from Victoria Bridge to John Dalton Street.[74] A second act, passed in 1875, extended the improvement south to Knott Mill. The first improvement was, according to the *Manchester City News*, approaching completion in March 1873.[75] The deputy town clerk declared the improvement completely finished on January 26, 1880: "The widening of this important thoroughfare having been completed, the Improvement Committee are prepared to sell their surplus lands in Deansgate, which comprise excellent sites for warehouses, shops and other premises, in plots to suit purchasers."[76] Within ten years, the project's success had become a standard feature of histories of Manchester's improvement. "The Act," claimed the *Manchester Guardian* in 1890, "cleared away the old rookeries . . . and broke up the black spots which lay within an arrow's flight of the gilded chambers of the Town Hall."[77]

Manchester's largest single act of civic reconstruction in the period was, however, substantially more convoluted than such bare announcements suggest. The project was plagued by accidents, strikes, material shortages, bad weather, and innumerable squabbles over property. Deansgate rapidly became a rather bleak, bare space, with half-built properties, empty shops, and demolished frontages giving it a "very prejudicial" appearance, according to one local businessman in 1870.[78] In 1881, a year after the project's supposed completion, the *Manchester City News* described Deansgate as a "wilderness."[79] The street was patently failing to become a glittering social and economic focal point, and the council found itself pressed with requests to erect property itself in order to attract business. The *Manchester Critic* derided the improvement as a "disgrace" and fundamentally "hollow and deceptive," with new frontages merely further disguising the moral and sanitary mire behind.[80] The improvement threatened to magnify the visual disconnection and desensitization that Engels condemned. Moreover, only the south side of the street was significantly affected by the improvement, leaving slums to the north largely untouched. Reporting on the sanitary condition of the area in 1881, Leigh observed that, despite the demolition of 1,260 houses, the mortality rate of the Deansgate sanitary district remained the city's worst. Some passages were still only two feet, two inches wide, he complained, concluding that the region's topography remained insalubrious: "There are still a large number of courts and passages in this district. A large number of the streets are narrower than is now required by the byelaw of the Council of the city of Manchester, 1868. The passages that do exist are very narrow and the spaces between the backs of the houses is [*sic*] small and insufficient."[81] Reformers were still documenting the city's narrowness and darkness in the early twentieth century (figure 2.7).

Urban Opacity: Smoke

If the air circulating through streets was unclear, gray, and gritty, then no amount of widening could prevent perception becoming clouded and movement uncertain. The "smoke nuisance" persistently compromised vision throughout the period, from Michael Angelo Taylor's 1819 parliamentary antismoke campaign to the formation of the Coal Smoke Abatement League of Great Britain in 1905.[82] "The evil of smoke," lamented *The Times* in 1845, "has reached a most intolerable height."[83] The term *smoke* referred here to any cloud of particles remaining after industrial or domestic combustion, and smoke assumed particular characteristics according to locality, containing, in varying proportions, "carbonic oxide, carbonic acid, carburetted hydrogen, carbons in sewer gas, sulphurous and sulphuric acid, sulphuretted hydrogen, sulphide of ammonia, bisulphide of carbon, nitrous and nitric acids, phosphoretted hydrogen, and sulphur compounds—and also in the neighbourhood of alkali works, hydrochloric acid."[84] Added to this was urban dust (tar, ashes, flakes of paint, cement), animal waste ("tardigrads, anguillulae, polygastrica, bacteria, vibriones, monads, diatoms, infusoria, and microzymes," not to mention "wings of insects, legs of spiders, bits of spiders' webs," dung, urine, saliva, and sweat) and vegetable matter (spores, pollen, cells, seeds,

Figure 2.7 The
persistence of
darkness. Houses
near a railway
bridge in Ardwick,
Manchester, that
are only six feet
from railway
arches. From Marr,
Housing Conditions
(1904).

fibers).[85] When fogs formed, this protean cloud bound with water vapor, thickening into a brown, opaque paste.

Wherever smoke thrived, life wilted. "Vegetation is stunted and unhealthy, and threatens to disappear altogether from the face of the land," grumbled the *Lancet* in 1874.[86] During London's 1879 fog, Rollo Russell, the meteorologist and antismoke-campaigning son of the Whig prime minister Lord John Russell, reported: "Many of the fat cattle exhibited at the great show at Islington died of suffocation."[87] Industrial pollutants infiltrated the body, silently coating its viscid surfaces: "The fog, laden with dirt, fills the lungs with cold and repugnant vapour; it fouls the membranous lining of the bronchial tubes, tends to block up the air-cells, and, while at once irritating the air-passages, and impeding respiration, lessens the oxygenation of the blood, and thus the heat of the whole body."[88]

If the first half of the nineteenth century had been the great age of fevers, then the second was characterized by pulmonary and respiratory disorders.[89] Smog had moral consequences: alcohol consumption peaked at times of foggy disconnection. Domestic cleanliness became almost impossible. Soot and dust infiltrated the home through windows, doors, and flues, "cover[ing] every article of furniture, darken[ing] and spoil[ing] all drapery, curtains, carpets, table-covers &c." Gloves, it was claimed, were the only way to prevent Mancunian hands from blackening.[90] Smoke, Russell concluded, "defeats attempts at cleanliness and neatness even among the most scrupulous of the poor."[91] At the first meeting of the Manchester Association for the Prevention of Smoke in 1842, its chairman, the Reverend John Molesworth, claimed that Manchester's residents "saw, tasted and felt" their city's smoke.[92] The Smilesian canon of character traits—thrift, perseverance, sobriety—was often altogether unrealizable in such dusty, obscure environments.

Most germane here was the devastating impact of fog and smoke on vision—and particularly oligoptic and supervisory vision. During London's "Great Fog" of 1813–14, which lasted seven days following an intense frost, the physician Thomas Bateman observed: "All objects at a few feet distant from the eye [were] invisible: houses, railings and trees, and even the cobwebs hanging over them, became thickly spangled with . . . freezing humidity."[93] During another London fog, Russell found that "it was not possible during a greater part of the day to see across a narrow street." He "measured the distance at which objects became visible, and found it to be four and a half yards."[94] On New Year's Eve 1888, William Marcet, the president of the Royal Meteorological Society, described fogbound people "groping their way through the streets as if they had lost their sense of sight."[95] The following year, the *Spectator* argued that fogs annihilated social perception and interconnection, causing people "to dread more and more the possibility of becoming a separate and painful centre of darkness, instead of an almost unconscious sharer in the prodigal overflow of light."[96] Opacity overwhelmed the entire city, magnifying disconnection and social dislocation, and destroying motility and the circulation of opinion.

The descent of smoke and fog obscured familiar landmarks, generated illusory masses, and forced involuntary, collective reliance on the senses of proximity. While walking with his son in Hyde Park one murky December afternoon, Marcet recalled: "Every now and then we were conscious of being near a tree and overhanging branches . . . yet there were no trees and no branches when we struck at the simulated objects, or felt for them with our hands. This phenomenon, confirmed by two independent observers, is, I think, worth recording."[97] When objects were visible, their indistinct outlines might make them appear larger than they really were.[98] Such spatial and formal distortion was compounded by the effect of atmospheric density on sound waves. Policemen swore that Westminster Abbey's bells were audible across greater distances during fogs, citing Tyndall's 1874 experiments with whistles and organs as scientific proof of the impact of atmospheric thickness on sonic transmission.[99]

There were two basic ways to assail the smoke menace: through technology and through the law. The old apparatus of courts leet and common law proved invariably useless in thwarting industrialists, so private bodies and pressure groups were forced to find new strategies. The first of these was the development of "smoke clauses" in local improvement acts in the 1840s, stipulating that industries would utilize smoke-consuming furnaces: these clauses were formalized in the Towns Improvement Clauses Act (1847) and the Smoke Nuisance Abatement (Metropolis) Act (1853). Such acts were, however, hamstrung by problems of definition and quantification: precisely what constituted a smoke "nuisance" remained unclear. More specific were the Alkali Acts of 1863, 1874, and 1881, which tackled the problem of hydrochloric acid, defining a maximum legal level of impurity (0.2 grains of alkali per cubic foot), thus aiming to produce measurable truths about atmospheric tolerability.[100]

Aside from new developments in calibration and laboratory analysis, a great number of smoke-abatement devices were patented. New forms of domestic fire, modeled on the "closed-grate" system popular in continental Europe or utilizing gas power, were also promoted, although the English love of open fires persisted. The physician Alfred Carpenter fantasized about communal gas heaters being connected to sewers to regulate domestic atmosphere: "It is not unlikely that in the future, rows of houses may be heated by one longitudinal furnace . . . which might be connected with the sewers, and assist in the dissipation of fog."[101] From 1876, the Noxious Vapours Abatement Association encouraged the "intelligent firing" of stoves. Some inventors, like Josiah Parkes (1820) and Charles Wye Williams (1839), devised furnaces that were designed to drag air into the device for more effective combustion. Other inventors, notably John Juckes (1841–42), devised hoppers and grates to deliver an endless supply of fresh fuel to stoves. Finally, a range of flues, giant chimneys, sprays, and pumps were constructed to purify smoke after combustion.[102] There were even proposals for huge fans to blow smoke away from cities, aerial sprinkler systems, and electric charging of the atmosphere to create rain. Many of these designs tempted manufacturers with promises of improved efficiency and profit, and exhibitions of them were held at South Kensington (1881–82) and Manchester (1882), organized by Ernest Hart, the editor of the *British Medical Journal* and founder of the *Sanitary Record*, and Octavia Hill.

Mechanical stokers and automatic firers were, it seems, relatively successful, but many such contraptions failed, often because, while reducing smoke, they compromised the smooth running of engines. As the *Engineer* admitted in 1896, the steelmen of Sheffield, a notoriously smoky city, would have reduced smoke levels long before if was profitable to do so.[103] Charting the shift in legal thresholds of impurity provides one way of assessing changing perceptions of tolerance, but there is little doubt that these shifting perceptions must be balanced against, and, indeed, related to, the persistence of the smoke problem across the century. The various acts proved easy to evade, and industries routinely relocated to areas beyond the reach of such legislation. Their basically liberal premises—lack of compulsion, small penalties, emphasis on "best practicable means," hesitancy about enforced inspection—and limited extent plainly failed to produce tangible improvements to the atmosphere of cities.[104] Additionally, certain technical solutions to excessive smoke suffered from practical problems: furnaces clogging with vitrified ash, chains of endless grates snapping. In 1881, George Shaw-Lefebvre, the future president of the local government board, complained that London fogs "were of denser and of longer duration than formerly, even invading the summer months."[105] The liberal commitment to pure vision was largely overridden by equally liberal commitments to freedom of trade and permissive legislation.

Technologies of Lucidity: Glass

Glass is, of course, an ancient invention: archaeological evidence suggests that it was being used at least four millennia ago, although hardly on a large scale.[106] In medieval and early modern Europe, windowpanes were sufficiently precious to be removed and hidden when residents left their homes. In the eighteenth century, the secure glazing of larger areas was facilitated by the development of sheet glass and mechanical grinding. The critical nineteenth-century innovation was plate glass, which, wrote the *Builder* in 1870, "far surpasses in transparency and elegance the small panes formerly used."[107] Plate glass was considerably more uniform than sheet glass and more easily purged of bubbles and turbidity. There were two basic types of plate glass: rough plate, used for skylights and bottles, and polished plate, which could be integrated into shop and cabinet design or silvered and made into mirrors. Although it was acknowledged that "no type of glass" was "perfectly transparent," the greenish-blue tint of the best plate glass was held to have "no deleterious effect whatsoever, the majority of persons being entirely unaware of its presence."[108] Analytic chemistry enabled glassmakers to control translucence and color.[109] Plate glass was also physically stronger, and its brittleness, strength, and rates of decay could be measured and predicted, meaning greater areas could be spanned with enhanced clarity.[110] This efflorescence of glazing was also stimulated by Peel's abolition of excise duties on glass (1845) and the repeal of the window tax (1851).

The glasshouse was an important site of vitreous experimentation. James Loudon, in his *Encyclopaedia of Gardening* (1822), contended that glazing could be used to refract maximum quantities of sunlight onto plants. His optimism seemed boundless: "There is hardly any limit to the extent to which this type of light might be carried." He fantasized about producing an "artificial climate" englobed within vast pellucid orbs above cities, banishing the British weather to an artificial exosphere.[111] Something of this hubris was evident in the schemes of Joseph Paxton, who, apart from designing the Crystal Palace, proposed an eleven-mile-long Great Victorian Way around London.[112] Toughened plate glass generated a great upsurge in glass architecture (winter gardens, arcades, markets) after 1850. The Leeds Corn Exchange, with its giant glass roof, is a particularly fine example of this. "There is a general disposition to build with more height between the floors, and to employ plate glass more freely," noted *Engineering* in 1866.[113] The glass was strong enough to support human weight; deck or basement lighting flourished, deflecting solar radiation into basements, and helping prolong the legal existence of subterranean work and habitation (figure 2.8). Finally, the development of wired glass by Pilkington gave glazed portions of structures greater protection from fire.

Observers were forcibly struck by the lucidity of plate glass. "We can see objects through it without distortion or obstruction of any kind," commented the architect Aston Webb in 1878, favorably comparing it to older forms of sheet glass.[114] The limpid glass display cabinet enabled shoppers and museumgoers to inspect clothes and exhibits without being able to touch: it could teach people to survey and inspect. During the Great Exhibition, *The Times* noted: "We want to place everything we can lay our hands on under glass cases, and to stare our fill."[115] The vogue for public aquariums in the 1870s would have been impossible without plate glass for the tanks.[116] This viewing experience was familiar to those

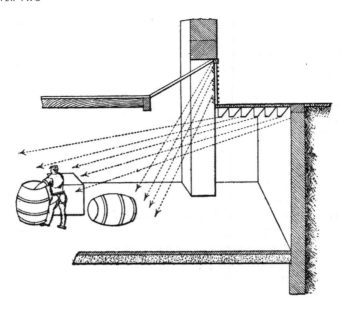

Figure 2.8 Prismatic basement lighting. From Thwaite, *Our Factories, Workshops, and Warehouses* (1882).

who frequented other urban displays: "The animals which these tanks will contain are seen not only through glass frontage, but in precisely the same way as the contents of the glass cases in our museums are."[117] The glass case organized a particular form of viewing experience, one common to multiple urban commercial spaces.

Flowers flourished under glass, which repelled damaging radiation while admitting the light necessary for photosynthesis. Oligoptic engineering was never fully separable from sanitary engineering. Human beings, it followed, would also thrive in a lighter climate, which led to investigations into the orientation and aspect of buildings (figure 2.9). "The sun's rays impart a healthy and invigorating quality to the air, and stimulate the vitality of human beings as they do that of plants," observed John Haywood in 1873.[118] Some builders recommended orienting houses to the southeast, to maximize the amount of sunlight without "bleaching and warping" furniture and decorations; aspect diagrams facilitated these calculations.[119] Others recommended glass roofing to admit sunlight into garrets and attics.[120] Some circular hospitals were constructed with sunrooms, warmed by steam pipes and girdled by annular promenades, above the wards.[121]

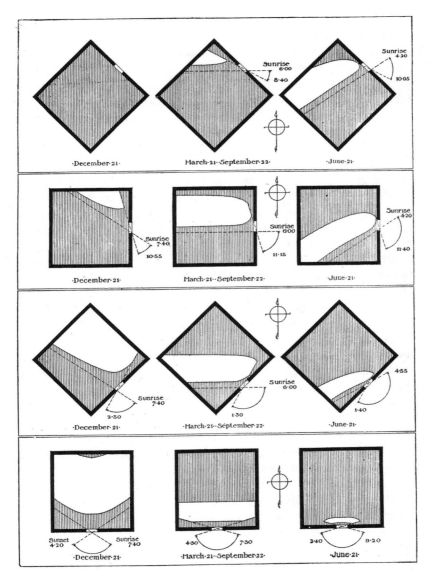

Figure 2.9 Calculating the illumination provided by windows of different aspects. The room is twenty-four feet square, illuminated by a window three feet, six inches wide, eight feet high, and two feet above the floor. From Atkinson, "Orientation of Buildings and of Streets" (1905).

Fenestration became quantifiable. "In this climate," stated the architect Douglas Galton, "adequate light will not be secured with less than 1 square foot of window surface to about 100 to 125 cubic feet of the contents of the room," depending on local conditions of site and aspect, although in hospitals the figure should be lower (one square foot per fifty or seventy cubic feet).[122] The 1894 London Building Act stated that, in all new rooms intended for human habitation, the window space must occupy not less than one-tenth of the floor space.[123] Specific visual practices could also be stimulated by the judicious deployment of glass. Skylights, for example, were recommended for operating theaters and art galleries. There was "no doubt," declared the (entirely unbiased) architects John and Wyatt Papworth, that the small skylit gallery built by their father was "the best lighted gallery in England, and perhaps in the world." This success was achieved by avoiding excessive light: "The reason many galleries fail of success [sic] is, that they are over lighted."[124] Overlighting threatened to destroy perception by dazzling or straining the eye. "The aim of the architect," wrote Carl Pfeiffer, "must be . . . to arrange his buildings that they will supply a sufficiency of light and heat, and yet keep both light and heat under perfect control."[125]

Overlighting was a particular problem in schoolrooms, where light was often admitted "frontally," directly hitting children's faces: "This method of lighting the room is very injurious to the eye, because, firstly, the retina becomes fatigued by the full glare upon it, and the diffused light renders the comparatively dark images of the printing and writing more difficult to be perceived."[126] Such retinal exhaustion was compounded by the geographic situation of many schools: German research indicated that childhood myopia was more common in schoolrooms located on bottom floors, along narrow streets, and opposite tall edifices.[127] School windows should be large and devoid of opacity, frippery, or tracery. Architects generally recommended lighting from the left, to ensure that (for the nonsinistral) the shadow of the writing hand and arm would not fall across the page. At the International Education Conference held in Brussels in August 1880, it was resolved: "Class-rooms should be lighted during the day by windows, *on one side only*, and to the left of the pupils."[128] The glass should be protected: "All windows facing towards a street, or otherwise exposed to stone-throwing, should be covered externally with strong wire."[129]

Glazing promised control over the transmission of sensory data: buildings allowed the entry of light but not of sound or smell, while glass cases enabled fish to be perceived by the eyes only. The judicious, meditative distance of the visual was encouraged, as Sennett notes: "Fully apprehending the outside from within, yet feeling neither cold nor wind nor moisture, is a modern sensation, a modern sensation of protected openness by very big buildings."[130] Glazing combined visual connection with olfactory, tactile, and sonic disconnection: a particularly visual modality of privacy. In hospitals, for example, the cellular lucidity of vitreous architecture avoided "the unpleasantness often felt in associating with strangers" yet enabled oligoptic monitoring to function as "the patients would also, to a certain extent, watch each other when awake."[131] Hospital wards thus combined the oligoptic with the supervisory and the clinical gaze. Finally, glazing might reify asymmetries of vision. School doors often had upper portions glazed to allow taller teachers to observe pupils' activities through them, and careful deployment of interior glass became indispensable to workhouse and asylum design.[132] Dwarf blinds could allow asymmetrical visual connection between inside and outside while maintaining the influx of light: "[They] are intended to keep off the observers of the neighbours or passengers, and are chiefly used in towns; they are generally made with light mahogany frames, wither filled up with moveable perpendicular lath, on the Venetian principle, or else with wire-gauze or perforated zinc. In all cases they are intended to allow the inmates of the room to see outwards, whilst at the same time they prevent the reverse from taking place."[133] This was not, again, panoptic since any humans caught in the field of vision would have no idea that they were being observed and this field was itself heavily circumscribed.

Such calculated perceptual economies, however, had obvious limits. Some schools and hospitals may have taken advantage of advances in glazing, but most did not. Natural lighting of schoolrooms, complained the *Sanitary Record* in 1879, "almost universally depends on accidental circumstances."[134] This situation was hindered by Education Department rules that still insisted on light hitting the faces of both children and teachers, an arrangement universally condemned among architects as fatiguing to the eyes. Similarly, the lucidity about which Webb enthused was restricted to cabinets, cases, and wealthy homes. In 1908, one expert concluded: "For the glazing of ordinary windows, sheets are often employed which produce the most disturbing, and sometimes the most ludicrous, distortions of objects through them." The public's toleration of this he found most puzzling.[135] The uneven spread of technologies of lucidity, then, perhaps succeeded only in sharpening contrasts between the perceptual environments of the wealthy and the poor and, hence, reinforcing the sensuous dichotomies that so troubled Engels.

Engineering Silence: Soundproof Paving

Clear vision is invariably experienced by virtue of not only particular viewing conditions but also the absence of distractions or impingements on the other senses, perhaps most notably noise. As these various techniques for improving visual perception were being mobilized, noise was becoming a serious urban problem, for two conjoined reasons. First, the city's growing population, bustle, and technological complexity generated historically novel noises, from railways and industry in particular. Second, shifting perceptual norms made previously tolerable sounds (street music, animals) potentially less endurable for certain social groups. In 1876, the *Sanitary Record* observed that bells, buzzers, cart wheels, whistles, steam trumpets, dogs, and vocal costermongers all precluded the possibility of silent, reflective work and contributed to the "amount of nervous disturbance now so rapidly on the increase."[136] Street surfaces were among the primary causes of such aural irritation. London's Strand gained the reputation of being "the most intolerably noisy thoroughfare in the metropolis."[137] Another writer concurred: "For continued noise, concentrated noise, we must contend that that of the Strand is unique, and without rival." The insufferable clatter was caused by the street's paving stones: "The way is paved from Exeter Hall to Temple Bar with granite boulders."[138] Because of their impact on the nerves, street surfaces were a medical as well as an economic question: "The noise of the London streets cannot fail to injure those individuals who possess a peculiarly nervous organisation."[139] This could trigger, in turn, "indulgence in alcohol" to combat the disturbance to sleep or concentration.[140] As with smoke, intolerable perceptual conditions could lead to alcohol abuse. If one was not already desensitized, in other words, one had to desensitize oneself to bear such noise.

By the 1870s, stone blocks and macadam (a compacted surface of smaller stones) were both falling out of favor as street surfaces. Although granite paving was very durable, the gaps between blocks accumulated filth and often caused horses to trip.[141] This unevenness accentuated vehicular clatter, leading to experiments with noiseless wheels and tires as well as less plangent surfaces: "Probably the desirability of decreasing the noise caused by the vibration of vehicles has had more influence than anything else in attempts to supersede the hard granite."[142] Many experiments—with, for example, rubber, metal, brick, glass, cork, tar, and crushed seashells—were undertaken to attempt to create the perfect street surface: durable, silent, and comfortable for horses.[143] In Manchester and Liverpool, clinkers from refuse-destruction units were used to pave some streets.[144] The two most important substances, however, were asphalt and wood. By 1873, the *Lancet* was referring to the "battle of the pavements," and, over the next twenty years, extended trials took place across London and in provincial cities.[145]

Asphalt is a bituminous substance occurring naturally or produced by the distillation of petroleum. Its use for paving in Britain dates back to around 1838, but the first large-scale experiment occurred in Threadnee-dle Street, London, in 1869. Observers were struck by its soundlessness. Here is a report from the *Engineer* in 1872: "Consequent to the laying of the Val de Travers asphalte the roar of Cheapside has given place to the mere clatter of horses' hooves, as if a regiment of cavalry had taken the place of the usual wheel traffic. . . . Let any one ride in an omnibus going westward through Cheapside, and notice the kind of shock occa-sioned by the transition from asphalte to granite."[146] Institutions desir-ing silence were immediately interested. The architects Mills and Mur-gatroyd of the Royal Exchange in Manchester wrote to their city council in 1871 expressing a desire for repaving outside the building as "it is important that it should be as far from noise as possible."[147] However, asphalt paving developed a reputation for slipperiness and hardness. Horses' pelvises fractured when their hind legs splayed as they slipped: in 1889, the Horse Accident Prevention Society (Slippery Roads) was formed to campaign against the new surface. Horseshoes wore out in three fewer days on asphalt than wood.[148] The engineer Henry Allnutt stated that asphalt was "at times dangerous to walk on": "It is like walk-ing on ice. . . . A few days ago a little girl was killed at the Plumstead Board School by falling from a see-saw on asphalte pavement."[149] The particular frictional (and sonic) qualities of asphalt did, however, make it ideal for skating, and the "rinkomania" of the 1870s provided tempo-rary commercial solace for companies that had overinvested earlier in the decade.[150]

Wood paving was likewise lauded for its sanitary qualities (figure 2.10). Hygeia was paved with firm wood encased in asphalt, while Chad-wick listed wood pavements among his arsenal of sanitary technologies, claiming that it was less dusty and absorbent than other pavements.[151] Numerous kinds of hardwood, including hemlock, mahogany, and eu-calyptus, were used in later-nineteenth-century trials, leading to the City of London being "traversed, from east to west, by an unbroken line of wood paving" by 1892.[152] In addition to being gentle on horses, wood paving was pleasantly quiet. Hospitals were often girdled with such pave-ments, owing to the nervous decline witnessed in convalescents ex-posed to street noise: "The boon which will be thus secured to sufferers

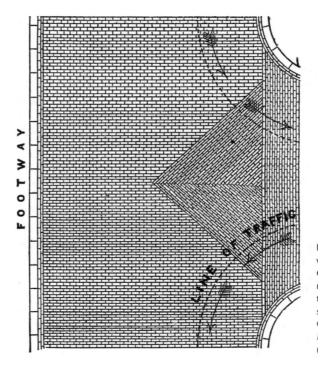

Figure 2.10 Stayton's wood paving in Chelsea, showing the method of blocking at the intersection of streets. From Law and Clark, *Construction of Roads and Streets* (1877).

in . . . hospital in the shape of increased quiet will necessarily be of in-calculable value."[153] The streets around Manchester's St. Mary's Hospital were relaid in 1877, after Thomas Radford complained to the Paving, Sewering and Highways Committee that "the patients have suffered very much from the noise" of the old road. The committee consented to pro-vide Manchester Grammar School with wooden paths in 1882.[154]

While promising to protect delicate convalescents and anxious clerks from clatter, wood paving proved insufficiently durable and sanitary for busy urban streets. Grouting slowly powdered, creating an undulat-ing surface that accumulated water and dung between the blocks.[155] In hot weather, clouds of stinking particles escaped from subsurface spaces. Wood also swelled according to temperature and humidity, and changes in its form could disturb curbstones. Wood pavements could not durably produce silence without simultaneously creating sanitary problems. As the *Engineer* indicated in 1894, once creosote protection eroded, there was "little or nothing to prevent the fibres and pores of the wood from be-coming saturated with rain water and other far more noxious liquids."[156] The surface became pitted and puckered, smeared with "slimy slop."[157]

In Paris, Pierre Miquel analyzed the microbial contents of strata of wood paving at the Chaussée d'Antin. Wood exposed to wheels contained over 45 million bacteria per gram of material. At three centimeters, the level dropped to forty-two thousand. By the mortar bed it was fifty-two thousand. And it rose again, to 12 million, in the sandy foundations.[158] In a post-Pasteurian age, these figures did not make wood paving attractive. Despite wood pavements being described as "long established in favour" in one 1909 text and their continued use in America, use of the surface fell into terminal decline thereafter, although some wooden street surfaces still existed in the 1950s.[159]

In the later nineteenth century, Victorian streets often presented something of a patchwork of surfaces: "Horses are rendered timid by the numerous transitions which they now undergo from one kind of paving to another."[160] Asphalt, after its uncertain start, gained in popularity. In Manchester, for example, only 968 square yards had been laid for early trials, and it was soon suggested that "no trial can give a fair result unless a considerable area is paved."[161] Analytic chemistry could calculate the optimal ratio of asphalt to pitch.[162] Larger trials literally cemented asphalt's reputation as a reasonably cheap, generally durable, moderately quiet surface. It would be reductive to regard its success as purely due to its generation of silent, clean streets since the availability of cheaper materials and the development of the automobile are of at least equal and, probably, greater significance. But asphalt pavements, as well as wooden ones, were technologies designed to generate silent traffic. They created potential archipelagos of silence, slender islands of civic life materially designed to minimize noise.

Attempts to municipally engineer physical conditions conducive to visual connection, brightness, clarity, and privacy were almost unfathomably messy and compromised. The Deansgate widening took over a decade and left many of the area's problems untouched, while urban smoke appeared as omnipresent in 1900 as it was in 1840. Extensive use of plate glass remained limited to shops, warehouses, and respectable residences, while noise persisted in urban streets as wood pavements crumbled beneath the feet of horses and pedestrians. The historian might conclude that oligoptic engineering was a monstrous failure and that the visually improved city never materialized. It is more appropriate, however, to consider the catalytic role played by the perception of failure, combined with shifting thresholds of tolerance, in fueling a limitless desire for more improvement.[163] The sight of rusting scaffolding and torn-up pavements, narrow alleys, and thick smoke acted as constant empirical, visual proof of the need for further improvement. While the perfectly improved city was an utter impossibility, it was always possible—indeed, easy—to use technological or practical failure as the justification for further improvement and better planning. How did one create a city of cleaner air? More experiments, more technologies, more exhibitions. What were the solutions to the ills of Deansgate? More inspection, more plans, more figures, more reports. Simon and Inman's 1935 *The Rebuilding of Manchester* closes with a chapter entitled "The Urgent Need for a Plan," reminding us that the pattern of improvement was cyclic, or dialectical, rather than linear.[164] One need only look at the sheer volume of works devoted to the smoke nuisance in the early twentieth century to see this.[165]

What, then, became of "urban visuality" during this period? I have argued that one important, if sometimes tacit, dimension of urban improvement in the Victorian period was the attempt to create public spaces where society was free to observe itself. Escapable visual self-regulation, albeit with relatively permanent but unobtrusive and non-panoptic supervision, was integral to the organization of liberal space. This project was always in process rather than being completed and perfected, to be sure. But something more than sheer material failure and accident accounts for the sense of incompleteness and delay characterizing the oligoptic city. In the case of street widening, for example, property owners regularly upheld their own freedom to do as they wished with their land, rather than sell it for the theoretical good of others. Building London's Embankment, for example, required the purchase of over 550 properties, at the cost of over £2 million.[166] With smoke, industrial productivity and domestic cheer—which were, of course, expressions of the liberty of trade and manufacturing and the liberty of the individual, respectively—were generally regarded as more important than lucid perception or even functional pulmonary systems. As the *Engineer* observed in 1896, the "doctrine of black smoke and good business being inseparable" was firmly entrenched in British society (figure 2.11).[167]

Alain Corbin has argued that, at the precise historical juncture when the experience of noise and stench was, for the respectable classes at least, fast sinking below the horizon of tolerability, industry managed to represent itself as, if not exactly salubrious, then at least more tolerable than other practices raising sensory problems in civil society.[168] He is surely correct. Moreover, domestic fires were not legally regulated until

Figure 2.11 Photograph of smoke from domestic chimneys taken in Manchester on June 23, 1922. From Simon and Fitzgerald, *Smokeless City* (1922).

the 1956 Clean Air Act: the private realm was left entirely free to pollute the public. Thus, smoke was, and continues to be, substantially harder to remove from the public stage than blood or dung heaps. Smoke might irritate, choke, and occlude, but it seldom disgusts. Its pervasive presence suggests that not every threat to public perception was equally vilified and that liberal commitments to economic dynamism and noninterference with trade might override equally liberal commitments to health, cleanliness, and lucidity.

When we consider the geographic pattern of visual improvements in nineteenth-century cities, two further conclusions suggest themselves. First, improvements often reinforced the socioenvironmental distinctions that they were supposed to help obliterate. As critics of the Deansgate improvement maintained, street widening frequently left old moral miasmas putrefying behind a newer, sleeker facade. Silence and clean air, meanwhile, remained the preserve of respectable suburbanites. Second, perceptual control was vastly simpler when undertaken within the walls of institutions than outside in the more unruly streets. The library and museum were far more successfully oligoptic, in my terms, than the street or even the park. The ultimate failure of these latter spaces to function durably in a truly public fashion surely has as one of its causes the absolute impossibility of the total management of perceptual conditions.[169]

THREE

The Age of Inspectability: Vision, Space, and the Victorian City

Eternal vigilance is the price of artificial complexity.
LANGDON WINNER, *AUTONOMOUS TECHNOLOGY*

"Inspection has become one of the characteristic features of modern government," observed Benjamin Kirkman Gray in *Philanthropy and the State; or, Social Politics* (1908). "The street lamp and the inspectors' reports are different symbols of one political faith, *viz.*: that lighted streets will do away with crime, and inspectors' visits will ensure obedience to the law." The omnipresence of inspectors, he continued, "represent[s] the growing determination of society to guide and control the intercourse of its members."[1] In his two-volume work *Local Government in England*, published in 1903, the Austrian statesman Josef Redlich also discussed inspection at some length. Inspection was integral, he observed, to Bentham's rationalized model of government, in which "the light at the center radiates to the very circumference of the State." Without inspectors, center and locality would be altogether disconnected: "The inspectors are the eyes and ears of the central government, but they are also the organ through which the central government acts directly upon the local authorities."[2]

Historians of nineteenth-century British government and administration have generally concurred with the views of Gray and Redlich. Although inspection was by no means a nineteenth-century invention, its enormous growth and institutionalization is regularly seen as integral to the development of the modern British state.[3] In *The Age of Equipoise*, William Burn suggested that the mid-Victorian period might be characterized as the "Age of the Inspector."[4] For Albert Venn Dicey and later scholars like Oliver MacDonagh, William Lubenow, and David Roberts, the emergence of formal centralized inspectorates was not only critical to, but in many ways constitutive of, the "revolution in government" through which the British state slowly found appropriate ways of seeing.[5] This "occurrence of profound significance," which "spread like a contagion," was an essential technique through which institutions and practices subject to some form of state regulation—factories, schools, workhouses—were exposed to routine observation.[6] Herman Finer, in his *Theory and Practice of Modern Government* (1932), called this "the social microscope." Beginning with royal commissions in the 1830s, he noted, "an apparatus of exploration was invented for the social field, mightily influential in its sphere as the invention of the microscope had been in physics and medicine."[7]

The focus in these many works has almost invariably been the central state inspectorates formed from the 1830s to regulate, for example, factories, education, mines, and prisons.[8] A substantial body of literature has also addressed more private evangelical and philanthropic groups that inspected the homes of the Victorian poor.[9] Rather less space has been devoted, however, to the various forms of inspection undertaken by nineteenth-century municipal and local governments.[10] Moreover, this literature has never really asked what this "apparatus of exploration" tells us about the specifically *visual* operation of power in nineteenth-century Britain, beyond metaphors of radiating light or social microscopy. In this chapter, I address both these absences by examining the activities of nuisance and sanitary inspectors, whose routine work carried them into the homes and workplaces of the Victorian city, and by discussing the very particular visual form that their inspection took. The form of this inspection was not panoptic, but neither was it oligoptic or even supervisory in the fixed sense outlined in the previous chapter. The activities of inspectors formed a distinct pattern of their own, extensive, mobile, and delimited. Their perception did not radiate through spaces or fields; instead, it circulated through networks themselves designed increasingly to make inspection easy and unintrusive. These networks, in turn, left large expanses of Victorian urban life largely uninspected, and this strategic invisibility structured the experience of privacy and self-inspection so integral to liberal subjectivity.

Nuisance and Sanitary Inspection

The legal term *nuisance* has a long history, stretching back to medieval times, when it referred to anything broadly annoying or offensive, from eavesdropping to slaughtering.[11] The consolidatory 1875 Public Health Act defined and grouped nuisances into eight basic categories (premises, privies, animals, deposits, overcrowding, workplaces, fireplaces, and chimneys), and it made the appointment of "inspectors of nuisances" a legal requirement for local authorities.[12] In the 1891 Public Health (London) Act and the 1897 Public Health (Scotland) Act, the term *sanitary inspector* was used instead of *inspector of nuisances*, which Christopher Hamlin sees as evidence of the gradual medicalization of the concept of nuisance.[13] By 1900, many activities that might formerly have been regulated by inspectors of nuisances, like the loud barking of dogs, fell to the police to handle.[14] That said, the older conception of public annoyance remained potent, and the distinction between nuisance and sanitary inspection was never entirely clear, so I use the terms broadly synonymously here.

Permissive legislation facilitating the appointment of nuisance inspectors was passed from the 1840s. Thomas Fresh was appointed as Liverpool's inspector of nuisances in 1847, while Glasgow formed its Committee on Nuisances in 1859.[15] Manchester had nineteen inspectors of nuisances by 1870, each with his own district. On the establishment of the Metropolitan Board of Works (MBW), London's vestries and district boards were bound to appoint fully qualified medical practitioners "to inspect and report upon the sanitary condition of the parish or district."[16] Some cities appointed special inspectors for specific tasks, like the inspection of weights and measures or explosives. Early inspectorates often worked closely with the police: in Liverpool, William Duncan made use of four police officers to make a survey of cellar dwellings.[17] Police officers were forbidden from also acting as nuisance inspectors from 1873, but some areas of nuisance regulation, like those of lodging houses and slaughterhouses, were still undertaken by the police late in the century.[18] By this date, women were being employed more frequently as inspectors: Dublin, for example, appointed four female inspectors in 1899 to advise on the cleaning of houses and the correct feeding, cleansing, and clothing of children.[19]

What, exactly, did these inspectors inspect? While older forms of nuisance inspection were remarkably eclectic, including everything from kite flying to the state of public urinals, the sanitary inspector's orientation, as Hamlin suggests, was becoming more medical. In *The Sanitary Inspector's Handbook*, Albert Taylor argued that the sanitary inspector should be "the eyes, ears and nose" of the local sanitary authority, responding to complaints about specific nuisances as well as undertaking general, routine inspection.[20] This still gave the inspector a very broad remit, relating mainly to housing, lodging houses, noxious trades, slaughterhouses, infectious diseases and factories or workshops. To give some sense of the sanitary inspector's work, I examine the first four of these foci in turn.

House-to-House Inspection

In *Disease and Civilization*, François Delaporte argued that the 1832 cholera epidemic generated historically unprecedented "inspection of [Paris's] smallest nooks and crannies," especially the housing of the city's poorer regions.[21] His insight is entirely applicable to Britain, where, following the epidemic, Kay urged the "importance of minutely investigating the state of the working classes," to reveal "the secret miseries which are suffered in the abodes of poverty, unobserved by those to whom he may come to advocate the cause of the abandoned."[22] Finer's "social microscope," in other words, had a distinctly medical or epidemiological origin. Inspection of all types, Kay suggested, should be mobilized alongside the regeneration of streets and houses. Thus, he supported the work of charitable and religious groups while also advocating something considerably more systematic and medicalized, for example, the constitution of boards of health to combat physicomoral ills through detailed investigation of the domestic conditions of the poor.

What is of particular historical significance is the slow shift from house-to-house inspection occurring intensely but only during times of epidemic to its becoming a permanent, routine, and pervasive dimension of urban government.[23] John Simon, the first medical officer of health (MOH) for the City of London, is particularly important here.[24] In 1848, Simon urged Daniel Whittle Harvey, the City police commissioner, to inspect and suppress insanitary practices in the City. The following year, Harvey dutifully reported that he had inspected 15,010 houses, of which 2,524 had "offensive smells" and 1,120 had privies in an offensive state. Twenty-one cellars, he had discovered, were being used as cesspools. By the 1870s, this sort of habitual house-to-house inspection had become the mainstay of local sanitary administration and could be remarkably intense and extensive. "Where they possibly can do so," urged the public health chemist Alexander Winter Blyth, MOHs "should have a minute inspection of all the houses in their district made once, at least, in every five years." In the autumn of 1866, he noted approvingly, nearly 10,000 houses were examined in Merthyr Tydfil in South Wales. The data generated "standpoints of reference whence to mark the improvements made," allowing the MOH "to note the dark spots that call for amendment by referring to this 'Dictionary of Habitations.'" Blyth reported that this inspection was gradually permeating rural regions: in Gloucestershire, in 1874, "the condition of the whole or part of 72 parishes" and "8546 separate premises" had been inspected.[25]

Despite the putative neutrality and universality of house-to-house inspection, those houses in slum areas, the zones of desensitization, remained far more subject to inspection than others. Houses in crowded regions or "with appearance of dilapidation," noted Taylor, "will demand more frequent inspection," implicitly reflecting the fact that the inspector's eye was generally turned toward the poorer quarters of town.[26] In Hackney, according to its MOH, John Tripe, the "better class" of houses was inspected only on complaint.[27] MOHs regularly composed lists of particular houses to be inspected, especially in cases of infectious disease, which again tended to shift the focus toward slums, courts, and less salubrious streets. These inspections involved entering the home and scrutinizing sanitary arrangements. In his 1847–50 report on Liverpool, Duncan was candid about this potential transgression of privacy. The inspector, he noted, was "inspecting not merely the exterior but penetrating into the interior, of the dwellings of the poorer classes—many causes of disease and mortality must be discovered which would otherwise have passed unnoticed."[28] In order to observe any structural cause of disease, concurred Taylor, "all inspections and enquiries made by the sanitary inspector should be thorough—a call made at a house for a few minutes and returned as an inspection, is misleading."[29]

This made inspection a potentially slow, painstaking process. The sanitary engineer Gerard Jensen recommended that inspectors start in the cellar and slowly work up to the roof and chimney, inspecting floors, cisterns, sinks, baths, closets, and drains en route. Drain inspection was helped by the deployment of various simple tests, using smoke, water, air, or aromatic liquids. Such liquids, for example, were injected into drains: if the aroma manifested itself anywhere along the pipe, there was a leak.[30] Similarly, the level of ventilation should be assessed. Air direction and flow were measurable by observing and recording the amount of smoke "disengaged from smouldering cotton-velvet, and less perfectly by small balloons, light pieces of paper, feathers, &c.," which enabled the direction of airflow to be ascertained.[31] Anemometers could then be used to calculate the rate of airflow through rooms. Obstructions should be removed: the inspector might have to remove birds' nests, for example, from ventilation pipes.[32] Furnishings might also be scrutinized since the relation between certain forms of decoration and disease had become established: lead, antimony, and arsenic were all found in paint. Cornelius Fox, the MOH for East, Central, and South Essex, recounted a case where twenty-six ounces of white arsenic was found in the "chintz and lining" of the bedroom of one sickened, depressed man.[33] The chemical composition of wallpaper was, however, perhaps of less concern than its tendency to harbor moisture in air pockets or conceal patches of damp or dirt.[34] Inspectors were, finally, expected to estimate the amount of light that rooms received: "As a 'rough and ready' average, the window space should be one-tenth of the floor area of the room."[35]

Inspecting Lodging Houses

The lodging house was long acknowledged as a problematic urban space, with its reputation for overcrowding, crime, and poor hygiene, its "atmosphere of gin, brimstone, onions, and disease."[36] No account of the city's seamier side was complete without a lurid description of nocturnal conditions in lodging houses.[37] Although the term *lodging house* was ambiguous and sometimes used synonymously with the term *tenement*, the basic defining feature of a lodging house was the sharing of bedrooms by strangers, often for short periods of time.[38] Lodging houses were widespread in cities with large migrant or transitory populations: Jephson, in *The Sanitary Evolution of London*, estimated that there were around five thousand lodging houses, with eighty thousand residents, in London alone.[39] The 1866 Sanitary Act permitted their registration and inspection and included clauses enforcing privy accommodation and limewashing. Under the 1875 Public Health Act, no house was to be registered as a common lodging house until it had been inspected.[40]

In Liverpool, a port with many lodging houses, formal registration began in 1846 and inspection itself in February 1848. By the mid-1850s, there were five specialized lodging house inspectors appointed by the city's health committee.[41] In Glasgow, the city's tenements, which were often not strictly lodging houses, were famously regulated by the ticketed system, dating from 1863. Houses of three rooms or less, of capacity smaller than two thousand cubic feet, were measured and their capacity etched on a metal ticket on the door or lintel. "This is done," noted James Russell, "by affixing tinplate tickets on the outer door, stating the cubic contents, and the proportionate inmates allowed, at the very low rate of 300 cubic feet per adult or two children under eight years." In 1888, there were 23,288 ticketed houses in the city, of which 16,413 were one roomed and 6,875 two roomed.[42] As Russell suggested, the main focus of inquiry was the amount of available air space per lodger, and three hundred cubic feet was a standard figure in many British cities by the later nineteenth century.[43] Inspectors were also to ensure that lodgers opened windows to ventilate sleeping space.[44]

The inspection of lodging houses was undertaken by nuisance inspectors, specialized inspectors, or the police. If common rooms were established, the police were legally allowed to enter them: if they were not, the house was inspectable only by a nuisance inspector.[45] Manchester's lodging houses, for example, were regulated by the city's watch committee rather than its health or nuisance committee. During the 1884–85 Royal Commission on the Housing of the Working Classes, two London police inspectors, John Bates and James Powell, testified that the police scrutinized lodging houses far more regularly than the sanitary authorities did: "We visit once a week, and, in a great many cases, twice a week, where we consider it necessary. We compel them to scrub the bedroom floors once a week, and we compel them to sweep them daily."[46] Police officers were given warrants for entry, and local government board by-laws stated bluntly that "lodgers shall afford free access to officers to any or all of their rooms for the purpose of inspection."[47] Lodging houses were, thus, the most easily and regularly inspected domestic space in the Victorian city: the privacy of a lodger palpably mattered less than that of a gentleman.

Industrial Nuisances

William Blackstone defined a *nuisance* as an "annoyance" that "worketh hurt, inconvenience, or damage" on a community.[48] This more public sense of the term became particularly relevant in an age of rapid industrialization, when historically novel levels of noise and stench were generated by workshops and factories. Here, the history of nuisance regulation connects directly to the history of perception and of shifting thresholds of tolerance. "The subject is entitled to protection against things which are offensive to the senses," intoned Chadwick, "from which no injury to the health or other injury can be proved than the often overlooked but serious injury of discomfort, of daily annoyance, as by matters offensive to the sight, as by allowing blood to flow in the streets; by filth, by offensive smells, and by noises."[49] With the emergence of industrial chemistry, the number of nuisances grew exponentially, to include the manufacture of sulphate of ammonia from gas liquor, the distillation of coal tar, the melting of pitch and asphalt for road surfaces, varnishmaking, fat melting, gluemaking, manuremaking, and coffee roasting.[50] Most of these nuisances were subjectively registered through smell: one public health official composed a taxonomy of "effluvian nuisances."[51]

The regulation of industrial nuisances, however, proceeded very differently than that of lodging houses. Some specific chemical industries were regulated by the Alkali Acts (discussed in the previous chapter). Although some justices might side with local property owners against industry, the ability of nuisance inspectors radically to alter industrial practice was hugely limited.[52] Slaughterhouse owners, for example, might reasonably argue that their property was originally built at a "decent" distance from housing but that urban growth meant that houses were built closer to them, creating the conditions under which nuisances were experienced. Court cases were successfully fought using this argument.[53] Elsewhere, the premise of "best practicable means" was well established. Henry Letheby, who succeeded Simon as the MOH for the City of London in 1856, argued that, since better technology had become available to prevent nuisances, sanitary authorities should act as technological advisers to industry. Public health and business interests, he implied, were entirely reconcilable through technology.[54]

The role of the nuisance inspector here, then, was to record and advise rather than to regulate. I illustrate this with the example of smoke. The 1847 Towns Improvement Clauses Act included smoke-abatement clauses and permitted towns to appoint smoke inspectors. In that year, for example, Manchester's nuisance committee appointed a special "inspector for the suppression of the smoke nuisance," and, in 1850, it was reported that, during the previous year, he had observed 510 chimneys.[55] In 1867–68, Inspector Hurst of the smoke subcommittee occupied himself watching and recording the emissions from 1,709 chimneys.[56] In some towns—for example, Nottingham—the police were entrusted with the job of inspecting chimneys. By the early twentieth century, special inspectors were appointed for smoke inspection "in most large manufacturing towns."[57] Local authorities composed particular guidelines for inspectors, guidelines that usually involved the monitoring and documentation of the duration for which smoke of particular colors was visibly emerging from chimneys. These regulations were not nationally standardized. In Birmingham fifteen minutes of black smoke was allowed but in Brighton only two.[58] Inspections and the data they produced could be utilized to produce startling facts: in one London week in 1879, Carpenter noted, there were only twelve actual minutes of sunshine.[59] But, as we saw in the previous chapter, such disquieting data never translated into anything like effective pollution regulation, not least because factory owners were often content to pay small fines and keep polluting. This might be compounded by the attitudes of local magistrates. In Sheffield, for example, the MOH complained that judges regularly refused to punish polluters, owing to a systematic bias toward industry.[60] To conclude: industrial pollution was quite heavily inspected but very lightly regulated.

Inspecting Food

Like nuisances in general, unwholesome food and drink had been subjected to various forms of regulation for centuries. Medieval kings deployed ale tasters to protect them against poisoning, while acts were passed sporadically thereafter to curb the mendacious and adulterous practices of vintners, butchers, and bakers.[61] Something of this sovereign politics of food lingered into the nineteenth century: Engels recorded a case of twenty-six "tainted hams" being publicly burned in Bolton.[62] Like the bodies of ancien régime bandits and regicides, tainted meat was marked and spectacularly condemned. Meanwhile, the Customs and Excise historically took charge of thwarting imports of adulterated food.[63]

Rapid early-nineteenth-century urban growth created unprecedented opportunities for adulteration.[64] In the early 1850s, the chemist Arthur Hill Hassall undertook an exhaustive series of inspections of, and tests on, London's food, and his reports, published regularly in the *Lancet*, were instrumental in making adulteration a major public health concern. Hassall concluded that it was almost impossible to locate pure food in London. His major weapon against adulterators was the microscope. "This is certainly the most practical and important use which has ever been made of that instrument," he stated, "for by its means hundreds of adulterations have been discovered, the detection of which is beyond the power of chemistry." Hassall depicted himself as a master observer, able to detect poisonous chemicals in confectionary and mites in cheese and flour. He also advocated simple tests that ordinary consumers could perform on their own food. Adulterated coffee, for example, could be detected simply by mixing it with water: coffee floated, while chicory did not.[65]

Few consumers, however, had the time, money, or wherewithal to perform such tests. What was needed was a more systematic network of inspection and analysis that would prevent adulterated food appearing on the market. As Thomas Walley, the principal of the Edinburgh Royal Veterinary College, stated: "A purchaser with a few pence in his pocket, and the cravings of hunger gnawing at his stomach, is not likely to exercise great discrimination in the purchase of the necessaries of life; nor can he afford to pay an expert to teach him what to choose, or what to avoid. He, perforce, buys that which to his uneducated senses is most likely to satisfy the predominant feeling of which he is cognizant—hunger."[66] The development of the institutionalized inspection and analysis of British food was a protracted and complicated process, developing alongside networks of sanitary and nuisance inspection without ever being entirely distinct from them. Under the 1875 Sale of Food and Drugs Act, all local authorities were obliged to appoint public analysts to perform tests, of greater sophistication than those undertaken by Hassall, on samples of food sent by inspectors. Numerous local figures, including police constables, market inspectors, and inspectors of weights and measures, were legally able to act as inspectors of food. Francis Vacher, the MOH for Cheshire, recommended that "in every district there should be at hand an officer appointed solely for this duty," formally appointed as a nuisance inspector, but actually inspecting only food.[67]

Of all foodstuffs, meat perhaps attracted the most focus, especially since the slaughterhouse was notoriously difficult to inspect and regulate, often being little more than a converted shed or cellar. "There is, perhaps, no trade which requires more constant supervision than that of the butcher," grumbled Blyth.[68] Under the 1847 Towns Improvement Clauses Act, nuisance inspectors received powers to enter slaughterhouses and butchers' shops at reasonable times, examine living animals and carcases, and seize any food they considered unfit for human consumption. Some urban areas appointed specialist meat inspectors: Leeds did so in 1858 to curb the widespread local practice of carting diseased meat into town at night. In Liverpool, there were four inspectors of slaughterhouses by the early 1860s.[69] By the late nineteenth century, municipal veterinary officers were employed in large urban areas: one of their tasks might be to provide expertise for meat inspectors.[70]

Meat inspectors were expected to familiarize themselves with the physical form of both carcase and meat. They should be able to detect bad meat, which, according to Letheby, was "wet, flabby, and sodden, with the fat looking like jelly or wet parchment."[71] If seized, it should ideally be transported to a "bad meat depôt."[72] Walley gave detailed directions for the inspection of the pancreas, liver, kidneys, and lymph glands for signs of disease. Dropsical or tuberculous lesions were most common in areas dense with connective tissue (back, breast, or diaphragm). The pleura, meanwhile, were often stripped from animals that had suffered or died from tuberculosis.[73] Udders, which were sold (boiled and sliced) from cookshops to the poor, "should be carefully examined in every instance."[74] The various royal commissions on tuberculosis in the late nineteenth century and the early twentieth recommended that meat inspectors be tested on the basic signs of animal health and disease as well as "the names and situations of the organs of the body."[75]

This system was undoubtedly geographically uneven, but, by 1900, the scope of food inspection was increasing while rates of adulteration were demonstrably falling.[76] Glasgow, for example, once had a reputation for being flooded with tubercular or bad meat. However, following the discovery that sixty anthracic cows had been slaughtered and sent to the dead-meat market in 1882, the corporation began a clampdown on the trade. The seizure of tuberculous cows was empowered by the 1890 Police (Amendment) Act, which led inspectors well beyond the city's limits, to remote barns and byres, even unearthing tuberculous organs buried in farmers' gardens.[77] In 1900, the corporation veterinary officer, A. M. Trotter, along with four assistant veterinary surgeons and sixteen meat inspectors, began coordinating the inspection of milk and meat.[78] In his first annual report, Trotter noted that 2,449 visits had been made to stations, 632 to wharves, 7,153 to shops, and 124 to sausage factories.[79] The circuits of urban meat supply were being slowly subjected to regular inspection. All farms supplying milk to the city's fever hospitals were compelled to give their cows tuberculin tests.[80]

———

Over the course of the nineteenth century, then, sanitary and nuisance inspection proliferated and became more specialized. "The task of Inspecting London," noted William Moyle in 1901, "long since a many-sided and most important operation, becomes vaster and more complicated and as a consequence, an army of inspectors daily pass London in review."[81] Inspection was not, however, performed by an "army of inspectors" alone. It was also performed by physical elements and devices engineered to make certain objects, spaces, and systems (drains, houses, meat) strategically visible, or inspectable. Building the inspectable city was as much a question of engineering, inscription, and equipment as it was of training a corps of mobile, conscientious, attentive inspectors.

Making Cities Inspectable: Agglomeration, Accessibility, Legibility, Portability

The first way in which inspectability was engineered was through what might be termed the *principle of agglomeration*. This principle underpinned the architecture and practice of clinical medicine, for example. By agglomerating many patients in a single space, as at the Moorfields eye hospital, it became possible to discern medical identities and differences that remained below the threshold of visibility when the sick remained isolated in the home or small hospitals.[82] Agglomeration was also evident in a very different space, the municipal abattoir. Before large abattoirs were built in the late nineteenth century and the early twentieth, animals were killed in tiny slaughterhouses scattered erratically across urban space, and the meat was then dressed and prepared by butchers. There were over fourteen hundred such slaughterhouses in London alone in 1874.[83] It was impossible to subject these spaces to permanent inspection: "The whole philosophy of meat inspection is this, that the further removed from the entirety of the animal to which it originally belonged, the more difficult it is to determine the wholesomeness of the part. . . . The only time when effective inspection can be made is at the place and at the time of slaughter."[84]

The abattoir was, among other things, a space designed to simplify the inspection of live animals, slaughter itself, carcases, and meat (figure 3.1).[85] Inspection and slaughter would, theoretically, occur alongside one another in the same, sanitized space. There would be no slaughter without inspection, and refractory butchers would be disciplined: "Nothing less than the most open public inspection at any hour of the day can ever keep the will in condition for permanent sanitary action."[86] By 1908, there were around 135 public abattoirs in Britain. In addition to allowing inspectors to peruse the surfaces of organs, this kind of agglomeration could also facilitate a kind of delegated mutual inspection, or oligoptic cross-monitoring, of butcher by butcher, overlaid by fixed supervision and circulating inspection: "It is found that in England some butchers object to the open hall system, as each sees the class of cattle his trade rivals purchase and kill. From the customers' point of view this is certainly an advantage, as it naturally tends to prevent the butchers buying ill-nourished and diseased animals, hoping to place the meat on the market without proper inspection."[87] Urban infrastructures might also be physically agglomerated. Algernon Henry Grosvenor, the chairman of the South London Dwellings Company, noted that the sanitary arrangements of model housing were "as far as possible . . . grouped together so as to be easily open to inspection, and the drains and pipes have been kept as far possible [sic] outside the building, so that any defect may be easily detected and quickly rectified."[88]

Subterranean infrastructures were, however, uninspectable unless they were engineered with a second principle, that of *accessibility*, in

Figure 3.1 Meat inspection at Smithfield, London, with agglomerated carcases. From Moyle, "Inspecting London" (1902).

mind. Much routine inspection involved the tiresome examination and monitoring of the subterranean apparatus on which urban existence had come to rest: water mains, sewers, gas pipes, electric cables, and telegraph wires. "Originally sound design and construction will not permanently secure the object aimed at," observed the *Sanitary Record* in 1878. "Pipes corrode and break, joints give way with time; moreover, when alterations are made, defective work is not infrequently introduced. Hence, periodical inspection is absolutely necessary to safety."[89] Yet these systems were simultaneously designed to be unintrusive, buried beneath roads or in walls. Hence the development of the "access pipe" system, whereby pipes with movable saddles or liftable lids allowed drains to be monitored without smashing them apart (figure 3.2). In 1872, William Eassie, an engineer and the future secretary of the Cremation Society, observed approvingly: "It is only eighteen years since pipes giving easy access for the inspection of their contents have been introduced. Previous to this the joints were all socketed dark and fast underground, and the drains sealed to all interference except that which necessitated destruction."[90]

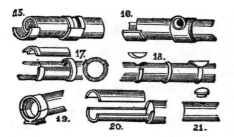

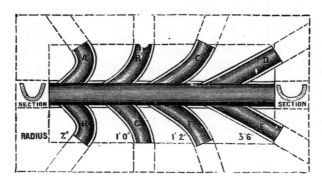

Figure 3.2 Engineering visual accessibility: the access pipe. From Eassie, *Healthy Houses* (1872).

Figure 3.3 Engineering visual accessibility: the inspection chamber, showing the arrangement of branch pipes. From Taylor, *Sanitary Inspector's Handbook* (1914).

By this date, the principle of strategic accessibility was becoming evident at critical points of sanitary networks, with inspection plates being fitted, for example, in road traps, water closets, and urinals.[91] Later in the century, the fully formed inspection chamber was appearing on "important bends and junctions" of sewer networks.[92] Such chambers, or "manholes," as they were often known, were places where multiple branches of the drainage system intersected for the purpose of inspection: they were made compulsory by the London County Council by the very early twentieth century (figure 3.3). They "afford[ed] a ready means of access for examination, and testing of the main drain and its branches, and of removing any stoppage which may occur in the same, without the necessity of opening up the ground, or disturbing the drains in the least."[93]

A third principle of inspectability was that of *legibility*. The provision of vital visual information was clearly essential to the autonomous but organized mobility of the oligoptic city. Street signs, house numbers, and traffic lights all provide basic data enabling easy negotiation of streets for the attentive, literate, noncolor-blind subject. The use of small signs like the tickets on the doors of Glaswegian tenements made inspection of tenements and lodging houses quicker and easier. Following the 1877 Canal

Boats Act, any inhabited boat was to have the word *registered* painted "white on a black ground in a conspicuous position on the outside of one of the cabins of the boat."[94] Many technologies creating legibility were designed for both inspectors and the wider public: the gas meter, discussed in the next chapter, is an example. We should observe the way in which, through strategies of inspectability and inculcation of norms of attention and scrutiny, formal inspection merged seamlessly with the more general activities of liberal subjects.

Legibility was particularly pronounced in the case of food labels, which were traditionally either nonexistent or very difficult to read. Hassall grumbled that they were "usually printed in inconspicuous characters" or located on "some obscure part of the package," meaning that only the most scrupulous or fussy of individuals would bother to inspect them.[95] Legislation slowly began to insist on customers being given clearer information about what they were actually buying. The 1887 Pure Beer Act, for example, stated: "Every person who sells or exposes for sale by wholesale or retail any beer brewed from or containing any ingredients other than hops and malt from barley, shall keep conspicuously posted at the bar or other place where such beer is sold or exposed for sale a legible notice stating that other ingredients are contained in such beer."[96] Under the 1889 Sale of Horseflesh Act, butchers selling horsemeat had to advertise the fact with "legible characters of not less than four inches in length, and in a conspicuous position," displayed "throughout the whole time" that such meat was being exposed for sale.[97] Similar legislation was passed for blocks of margarine, milk carts, and ice-cream barrows.

The fourth principle of inspectability was that of *portability*, or the use of mobile apparatus. Inspectors seldom used simply "eyes, ears and nose" alone, employing instead mobilized devices designed to ease inspection and recording. Basic measuring devices (which included not only consecrated instruments, like thermometers, but also everyday objects that were not obviously instruments, like balloons and feathers) were vital for calculating the atmospheric conditions of houses, as were pocket notebooks and pencils. Each kind of inspector carried appropriate equipment. Meat inspectors could not adequately scrutinize carcases without scissors, scalpels, and knives, while sanitary inspectors used inspection lanterns and mirrors to bring the insides of drains into focus.[98] Electrical inspectors were incapable of inspecting without their pocketbooks full of formulae and tables, while smoke inspectors might find cameras useful for producing portable records of smoke levels.[99] This equipment was simple, fragile, losable, and mutable.[100] It made inspection surgical in

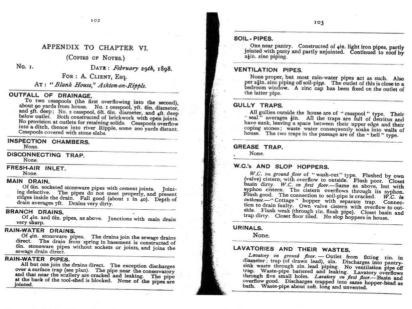

Figure 3.4 Model sanitary inspector's notes. From Jensen, *Modern Drainage Inspection* (1899).

its calculated, if sometimes clumsy, focus on discrete but vital elements of urban systems: pipes, fuses, chimneys, carcases. Often, techniques of portability, legibility, and inspection were combined, as in the loosely conjoined rise of the bus ticket, the portable torch, and the omnibus inspector.[101]

Having made measurements and calculations, inspectors then recorded their findings: "[An inspector] should keep a book, called the 'Sanitary Inspector's Journal,' in which he should enter a continuous record of his inspections, and of the sanitary condition of each of the premises inspected, and the result of any action or actions taken" (figure 3.4). A sanitary inspector should also "be able to draw" so as "to better explain some matters in his reports."[102] These reports should include formal charts into which discrete elements of information could be entered: date, time, number of house, name of inspector, and so on. All records and reports should be duplicated. Some inspections used the postal service: for example, samples of food might be sent from a rural place of provenance to a town laboratory for analysis. The MOH for Shrewsbury, a Dr. Thursfield, developed a writing frame allowing inspectors with poor eyesight to record their findings without the need of an amanuensis (his device was also used by the blind).[103]

114

The tactics of inspectability—the strategic agglomeration of inspectable entities, the engineering of visual access to vital systems, the deployment of written signs to make things legible, and the use of portable tools to extract information from inspected entities—remind us that networks of inspection were, indeed, networks in the sense implied by actor-network theorists: the combined product of human agents (inspectors) and material ones (institutions, visual portals, signs, apparatus).[104] There were no fields of inspection, but there were many circuits or networks through which inspectors and their inspection circulated. In order to expand this point, I now move from the technical to the political or moral structure of inspection.

Delimiting Perception: The Politics and Tactics of Inspection

The home and the workplace were both routinely entered by inspectors and were sometimes physically designed with such inspection in mind. Such a structure of inspection, of course, might appear totally illiberal. Had not Pitt the Elder famously declared that even "the King of England . . . dares not cross the threshhold of the ruined tenement"?[105] Numerous latter-day Pitts saw fit to equate inspection of all kinds with statism and Continental despotism. The 1858–61 Newcastle Commission, for example, contrasted British liberty with the experience of nations "habituated to the control of a searching police, and subjected to the direct action of the government."[106] The archreactionary Colonel Charles Waldo Deleat Sibthorp, who opposed railways and water closets with Victorian gusto, reacted thus to proposed public health legislation in 1847: "He condemned the inquisitorial power of the inspectors, which would almost authorise them to go to the house of the Lord Mayor of York and see what he had for dinner, and whether he went sober to bed, which he was sure the right hon. Gentleman always did."[107]

Sibthorp evoked the image of inescapable inspection destroying the last frontiers between state and subject, leaving even the aristocracy unfree to eat, drink, and sleep as they pleased. There was, to be sure, some truth to such accusations. The British home was, Peter Baldwin argues, probably less "inviolable" by government officials than its French equivalent.[108] Certain kinds of individuals, like prostitutes and lodgers, meanwhile, found it substantially harder to evade inspection than the likes of Sibthorp.[109] Particular forms of inspection could, theoretically, be regularly derided as oppressive. The visits of vaccination inspectors to houses were particularly resented. One early-twentieth-century writer to the *Vaccination Inquirer*, notes Nadja Durbach, commented that it was better "to find a burglar in the house than an inspector, as he departs at once, is rather cheaper, and does not pretend to be acting for the 'common good.'"[110] In imperial contexts, the British instituted inspectorates that were substantially more disciplinary than domestic ones. Lord Cromer, for example, referred to the introduction of "systematic English inspection" after the British military occupation of Egypt in 1882.[111] Wherever the British went, it seems, they took their principle of inspection with them.

It is highly tempting to generalize from such examples and reduce the entire apparatus of inspection to a form of panopticism: preventive medicine, for example, has been depicted as "the panoptic overseer of communal life."[112] Bentham described the panopticon as a *"field of inspection"* that could feasibly be "dilated to any extent," thus increasing what he called the *"inspective force"* of the particular institution subjecting humans to observation: panoptic inspection could, theoretically, be limitless.[113] Inspection might be the paradigmatic visual practice of a society of surveillance, whereby multiple human practices are subjected to "a faceless gaze that transformed the whole social body into a field of perception [with] thousands of eyes posted everywhere."[114] Inspectors, it would seem, formed a mobile and dispersed network of total oversight. This conclusion would be erroneous, however. In order to fully comprehend the visual structure of inspection, we must look, not simply at what inspectors saw, but what they did not see, why and how they did not see it, and the connections between these many forms of limitation and liberalism.

There were many problems administering and organizing inspection, at both the national and the local levels. First, small numbers of inspectors were often given the task of inspecting enormous regions. William Rendle, the MOH for the vestry of St. George-the-Martyr, Southwark, put it bluntly in 1865: "It is absurd to expect one officer, in a part only of his time, to inspect regularly and efficiently at least between two or three thousand houses inhabited by the poorest; to see that shifty agents do the work ordered, and that dirty, impudent, ignorant people do not destroy the work when done."[115] This problem was, perhaps, acutest in rural areas. In 1885, R. A. Selby of the Agricultural Labourers' Union described the inspection of Wiltshire villages: "The sanitary inspectors are very lax in the villages. We do not find that they take the interest that they would take in the towns."[116] In many areas, inspectors became burdened with additional duties, like overseeing vaccination, collecting rates, and testing weights and measures.[117] Such multiple appointments might, according to the optimistic Blyth, be "conducive to economy and efficiency, provided the officer appointed is a man of suitable character and energy."[118]

Inspectors were not, however, automatically endowed with either quality—and frequently lacked both. At least two MBW inspectors, for example, were dismissed for drunkenness.[119] According to the *Mining Journal* in 1867, some inspectors' visits consisted of "merely of half an hour's saunter on the pit-bank, and a conversation with the overman and his deputy."[120] Lord William Compton noted of sanitary inspection in Clerkenwell: "By the accounts of the poor people it seems to me that no sanitary inspectors ever went into any of the rooms. The sanitary inspector used to walk through the passage into the back court, and then walk out again, if he went at all."[121] This state of affairs was regularly compounded by low pay: George Reid, the MOH for Staffordshire, complained in 1896 that inspectors were poorly, and unevenly, paid and had insecure tenure.[122] This might leave inspectors open to bribery. In 1871, both sanitary inspectors in St. Leonard, Shoreditch, were accused of taking bribes.[123] Elsewhere, more blatant forms of corruption might be found. In 1851, the local board of health for Cowes on the Isle of Wight appointed an inspector of nuisances who was also the secretary of the local water company.[124]

Overwork, apathy, low pay, and systematic confusion: given such palpable flaws in the system, it is unsurprising that private groups sometimes inspected inspectors and reported piously on their manifold failings. In 1887, for example, the Manchester and Salford Sanitary Association inspected the Ancoats area of the city and the activities of the sanitary inspectors theoretically inspecting it. "We have frequently met the inspectors in the leading thoroughfares," reported John Thresh, the investigation's leader, but "we have never on any occasion had the pleasure of meeting one in any of the filthy passages or courts." Thresh concluded that inspection of Ancoats at least "is intended to be a farce; and so it would be were not the results so tragic."[125] In 1873, the same body, suspicious of official data, had hired an independent observer to calculate the level of Mancunian smoke emissions. This observer counted sixty-nine offenses in a week, in contrast to the city's official inspectors, who noted only four.[126] However, we should not automatically assume that "official" and "voluntary" modes of inspection were necessarily at odds.[127]

Inattentive inspection was not necessarily, however, a consequence of inapplication, laziness, or, as the Victorians would have it, lack of character. Inspection was often a physically demanding, laborious task. The head of the Alkali Inspectorate, Robert Angus Smith, referring to

inspecting emissions from factories along the river Tyne in 1864, commented on "the difficulty of mounting to the summit, and when there
of working calmly at a height of 125 feet on a platform slenderly railed
under a strong wind and even rain": "One may occasionally stand for an
hour under these conditions . . . and as a rule it may be said that inspectors who are not equal to sailors in climbing cannot make examinations
at the summit of the towers."[128] This was obviously not a solid, circumscribed "field of inspection" but a precarious, dangerous point. Mines,
similarly, were anything but transparent to inspectors, which meant
that owners found hiding children or illegal activities incredibly easy.
Inspectors might have peered into the labyrinthine depths of the earth
and penetrated the bottoms of sulphurous clouds, but this did not mean
that they necessarily saw clearly, well, or frequently.

Such spatial problems were exacerbated by lack of public cooperation.
In 1878, Hussey Vivian, a Swansea copper smelter, complained of the
infringements to liberty that inspection entailed: "I need not say that
there is and must be a great dislike upon the part of any manufacturer
to have inspectors running over his works with power to go where they
please, and spy into everything they like."[129] Owners of factories were
often reluctant to provide statistics for inspectors.[130] The 1876 Royal
Commission on the Working of the Factory and Workshop Acts heard
how lookouts were deployed to warn workshop owners of imminent
inspection: "Some of us, in visiting one such village, in company with the
sub-inspector, saw a little girl flying down the village street, and calling
into the windows of the workshops, the tenor of which was not hard
to divine."[131] Inspectors might, finally, suffer the indignity of physical
assault: in 1878, for example, an inspector was attacked while writing a
report on a nuisance on a Cambridgeshire publican's premises.[132] Three
years earlier, the 1875 Public Health Act made anyone preventing or
obstructing inspection liable to a penalty not exceeding £5.[133]

Rather than being peerlessly omniscient, inspection might appear simultaneously intrusive, ineffective, and detested. This would, again, be
misleading. It is easy to overestimate the resistance to inspection. According to Hamlin, the system was successful, and this very success can
be gauged by the relatively low number of cases that were brought before
the courts.[134] Certainly, inspectors were not encouraged to be particularly litigious. As Taylor observed: "It is by no means necessary or desir-
able that in every case of a nuisance brought under the cognizance of the
inspector, the law should be threatened or invoked."[135] Rather, the inspector ideally acted as the gentle upholder of the norms of decent, polite
society. His or her visit might have been a minor inconvenience, but it

seldom appears to have been experienced as a visceral violation of the basic liberties and rights of Englishmen. The inspector, generally, was tolerated. A. T. Rook, the superintendent of the Manchester City Council Nuisance Department, informed his town clerk in 1884: "During the past 11 or 12 years . . . not a dozen complaints have been made to me of the conduct of the inspectors."[136] Given the volume of houses inspected, this is a significant figure.

A central reason for this general tolerance was the tactics used by inspectors to cultivate trust and cooperation.[137] Particularly vital were identifiability and politeness: "The Sanitary Inspector should carry about with him a written document setting out his position as an officer of the local sanitary authority and fully authenticated by the authority. . . . He should bear in mind that he has *no right to force an entry*, and that a polite request almost invariably meets with the desired permission, for an appeal to the law will be found to be of rare occurrence—comparatively speaking—and the Sanitary Inspector is considerably helped in his duties by a polite bearing."[138] One should always know, in other words, precisely who the inspector was and what he or she was inspecting. Some areas, like Marylebone, issued pamphlets listing names and addresses of public officers, including analysts, inspectors of weights and measures, and vaccinators.[139] The slow introduction of written examinations for inspectors was another element of this process.[140] The gaze of the inspector was never faceless or nameless, and it often did not arrive unexpectedly. The use of written notification (postcards, e.g.) to inform tenants of the time and date of future inspections became common: "It is found in practice that this card is appreciated by householders and that the work proceeds more smoothly."[141] This practice extended to the delivery of written notices following the identification of nuisances, in which the procedure, timetable, and potential penalties were clearly delineated.

No sanitary inspector's manual was complete without paragraphs on tact and courtesy. "Civility and kindness," declared Taylor, "must characterise all [the sanitary inspector's] actions, and rude behaviour or supercilious officialism should find no place in his conduct."[142] The interactions between inspector and owner or tenant were delicate negotiations, in which the inspector should be cool, calm, and, above all, persuasive: "The inspector will be able to do a good deal by persuasion, which in many cases is preferable to the application of the law."[143] The cultivation of trust involved the basic assumption that the public would be helpful and cooperative if treated with cordiality and respect. This might extend to employing inspectors with particular language skills.

In Leeds in 1899, for example, an inspector was appointed who spoke Yiddish and could interact with Polish and Russian immigrants in the city's Leyland slum area.[144]

David Vogel has argued convincingly that cooperation and negotiation have characterized British environmental regulation since the Victorian period.[145] This tendency appears more representative than friction or animosity, although grudging acceptance or indifference was, perhaps, more widespread than anything. In Glasgow, Trotter observed how spontaneously helpful the meat industry had been in establishing conditions of inspectability. Large consignees of pork had volunteered to send their products to abattoirs for inspection: "The importance of this departure cannot be over-estimated, and, more especially, when it is remembered that these firms do this voluntarily, and at considerable expense and trouble."[146] Similarly, inspectors were reminded that overinspection would, ultimately, reduce the supply of meat and inflate its price, with damaging economic implications.[147]

The basic ethical dimension was the deliberate avoidance of any kind of intrusion into the personal, private life of the inhabitants of property. This kind of injunction appears regularly: "He should never pass any remarks, or appear to notice *anything he sees which does not concern him*, for people resent any inference in their private affairs, and do not like to think the inspector is prying on them."[148] Very simply put, the inspector should examine a specific set of *things* and not the *people* using them. This made sanitary and nuisance inspection fundamentally different from the private or philanthropic work of people like Octavia Hill, which explicitly entailed personal connection with, and moral judgment of, the poor.[149] Russell put it very nicely in observing how, when visiting houses inhabited by criminals, the sanitary official "takes no notice of the evidence of gruesome business he sees about": "His eyes are only for the leaking roof or the damp wall."[150] The inspector's eyes should focus on the apparatus securing healthy, comfortable, private existence—waterproofing, gully traps, ducts, fuse boxes, and cisterns—rather than private existence itself. This apparatus, situated in corners, behind walls, or in basements, was seldom examined by inhabitants, as Stockman noted of gullies: "Most householders, particularly where servants are kept, never look in these gullies; the consequence is they become very foul by the accumulation of soap and grease, and are, when in this state, often the cause of complaint."[151] Inhabiting private space involved its own forms of habitual blindness that the inspector's gaze periodically rectified.

Another set of tactics surrounded the times of legitimate entry. Inspectors could enter property between nine in the morning and six in the evening, by politely asking to be allowed into the building. Only in exceptional circumstances would premises be forcibly entered.[152] Certain businesses might legitimately be entered at nights or on weekends, but nocturnal entry of housing raised important questions relating to decency and privacy. J. Eisdell Salway, vestry clerk for the parish of Chelsea, stated that the inspection of housing outside regular hours was "an intrusion upon a man's freedom": "Night visitation would undoubtedly create friction."[153] Bedrooms in particular would be difficult to enter, some argued, for moral reasons. This prescription did not, however, apply to lodging houses or ticketed houses, where nocturnal inspection was absolutely necessary to verify that overcrowding was not taking place. In Glasgow, there were around forty thousand nocturnal visits yearly in the early 1880s. One inspector, Alexander McCallum, in the Eastern Police District, discovered people hidden in cupboards, under beds, and on roofs.[154] Asymmetries of inspection, then, had a fundamentally temporal nature.

Sanitary inspection could be simultaneously profoundly intrusive and completely tolerated. In 1886, for example, three chemists undertook investigations into the level of carbonic acid, organic matter, and microorganisms in the air of Dundee's homes. To procure their samples, they undertook nocturnal investigations, between 12:30 and 4:30 A.M., since domestic repose meant that the atmospheres were least disturbed during these hours. Over several nights, they visited around sixty houses. Each visit took about thirty minutes and involved the completely unforeseen arrival of three chemists and their measuring contraptions: "The houses were visited without warning of any kind to the inhabitants, so as to avoid the risk of having the rooms specially ventilated in preparation for our visit. In every case but one we were most civilly received, and willingly allowed to collect the necessary samples of air, measure the room, and obtain such information as was required. We were, in fact, agreeably surprised to find that so little objection was made to our visit."[155] This is, admittedly, a single example, but it clearly demonstrates the degree to which inspection of numerous kinds had become tolerated as a legitimate aspect of urban government.

Although these houses were visited "without warning of any kind to the inhabitants," it was completely apparent to their inhabitants who these men were and what they wanted to inspect, even if the precise point of the inspections might have remained enigmatic. This cultivation of identifiability marked the inspector as fundamentally dissimilar to that traditional, shadowy foe of liberty—the spy. Before the foundation of a functional police force, informants, spies, and agents provocateurs were, as E. P. Thompson observed, used routinely by magistrates during periods of unrest, like the Luddite risings.[156] Such figures collated information precisely by concealing their true identity, unlike inspectors, who procured their data after revealing who they were. Bentham's 1818 scheme for a ministry of police was rejected by a parliamentary commission because it threatened to "make every servant of every house a spy on the actions of his master, and all classes of society spies on each other."[157]

Creating an enduring distinction between inspection and spying was difficult, especially for state inspectorates. In its early years, the factory inspectorate was accused of spying on the political activities of workers, colluding with industrialists, and inflaming existing troubles. The factory reformer Richard Oastler noted in 1844: "Inspectors, superintendents, or any other government officers . . . are always the spies of the government."[158] These accusations seem not to have been entirely the result of libertarian paranoia, as extant letters between inspectors suggest.[159] Other forms of inspection did necessitate elements of disguise rather than surprise. According to the *Analyst* in 1876, the food inspector had to be a "man of resource, with a good deal of the spirit of a detective in him," to avoid becoming known to retailers.[160]

But even food inspection was, as Trotter suggested, more often characterized by cordiality and cooperation than by fractiousness and furtive spying. The network of inspection and analysis that developed after 1875 was consciously distinct in personnel, objective, and modality from the older Excise Department, which maintained a bloated staff of nearly five thousand officers around the country as well as between sixty and seventy analytic chemists in London.[161] Hassall criticized them as being "driven to adopt a system of espionage, and to the rude and inquisitorial proceeding of entering forcibly upon suspected premises, and of seizing any adulterated articles or substances employed in adulteration, and which, perchance, they might find in the course of their search." This mode of inspection was characterized by spying, impoliteness, violence, and confiscation. It was, Hassall carped, fundamentally illiberal: "Here is interference with the freedom of trade and the liberty of the subject with a vengeance."[162] Joshua Toulmin Smith, the inveterate anticentralizer, certainly agreed. When Excise agents raided his Birmingham home in 1846, looking to seize an illegal still, he was sufficiently outraged to raise his case in Parliament.[163]

The inspector was, finally, supposed to be a paradigmatically attentive subject, even if such attentiveness was facilitated by engineered inspectability: "He should have an observant eye, a quick ear and a sensitive nose, and be able at once to detect any defective or faulty sanitary arrangements of dwellings and other buildings."[164] Let me again use the example of meat inspectors. Ocular discipline was required for the operation of microscopes, while the ability to discern chromatic nuance was essential to distinguish healthy from unhealthy meat. Diffused darkness, for example, was usually caused by refrigeration and was seldom harmful, while magenta might indicate rinderpest or tuberculosis.[165] But inspection was never a purely visual practice. Inspectors needed to listen well since "on visiting slaughter-house lairs a very slight change in the breathing sounds may be sufficient to draw attention to an infected animal."[166] Specific animal diseases and the presence of particular drugs were often revealed by smell, while tactility was equally important: one should be able to detect whether meat "pits on pressure."[167] Indeed, according to Vacher: "A really skilful inspector might almost be trusted to examine a roomful of carcases blindfold, and pick out the diseased ones."[168]

These manifold qualities of the ideal inspector were, to be sure, often but partially realized. But the critical point remains: the system of inspection was founded on principles of consent, cooperation, trust, and attentiveness. It was also severely circumscribed in its scope. The nineteenth-century individual was not, in general, subjected to seemingly limitless, inescapable inspection. Inspection might have been detailed and, occasionally, seemingly obsessive, but it was always spatially localized. To repeat: it never formed a "field" of perfectly known space. Rather, it assumed the knotty and reticulated topography of the network, with points of acute focus and lines of visual connection leaving vast swathes of space (remote farms, obscure mines, secret slaughterhouses, occupied bathrooms, criminal minds) ignored. In addition to this was a very systemic modality of uninspectability that was increasingly being engineered into the city's very form. Networks of inspection left large realms of human existence in very calculated obscurity. These were the zones of uninspectable privacy integral to the oligoptic city.

The Necessity of Obscurity

Liberalism is often characterized as a particularly self-critical form of government, forever conscious that it was possible to overgovern and interfere with the privacy and liberties of the individual citizen.[169] Inspection—which was thorough, regular, and expansive, on the one hand, and delimited, tactful, and negotiated, on the other—absolutely exemplifies this government ambivalence and, moreover, demonstrates how this ambivalence took visual form. Just like it might be possible to govern too much, it might also be possible to see too much and to undermine the autonomy of the individual. Freedom worked only if there were clearly defined limits to what, and who, was seen, which generated restless and necessary insecurity about how people used mental and physical privacy.[170]

In order to fully understand this, we must return to the emerging built form of the city, particularly domestic architecture, the "insides" around which oligoptic spaces clustered. The open, undifferentiated arrangements of courts and tenements created a kind of communal visibility that was regarded as inimical to the formation of self-consciously decent subjects. Over the course of the century, there was a pronounced trend, across the social spectrum, toward producing functionally differentiated homes, with bathrooms for washing, bedrooms for sleeping, and kitchens for cooking (figure 3.5).[171] As the architect J. J. Stevenson put it: "Keeping pace with our more complicated ways of living, we have not only increased the number of rooms in ordinary houses, but have assigned to each a special use."[172]

Internal differentiation involved abolishing shared facilities and reconstructing them as entirely private spaces deep within domestic space itself. Sydney Waterlow, the chairman of both the Improved Industrial Dwellings Company and the Highgate Dwellings Company, noted: "In our tenements, every one being self-contained, there is a separate copper, a separate sink, a separate coal cellar, and a separate closet, and when the woman shuts the door, everything she has is within, excepting her drying ground for her clothes, and that is on the roof. There is a strong feeling and desire to have everything to themselves."[173] Self-enclosure, ideally, enclosed domestic sensory experience firmly within the walls of the home. In the Peabody dwellings at Bermondsey, it was claimed that "the noise in one room could not be heard in the other, and when water was spilt in one room it could not run through to the one below."[174] Despite its regional diversity, terraced housing followed this pattern of self-enclosure as well as containing thoroughly private rooms.[175]

The toilet is the most exemplary space here. The process through which the toilet became a private installation, a space of intimate, voluntary, and complete isolation, has been charted in detail elsewhere.[176] Most bylaw housing built in the Victorian period was equipped with some form of private ash pit, pail closet, or water closet.[177] Under the 1844 Metropolitan Building Act, all privies were to have doors to secure a basic degree of privacy and dignity for the occupant. The lavatory, as well as the bathroom, as Corbin suggests, was becoming a "real room."[178]

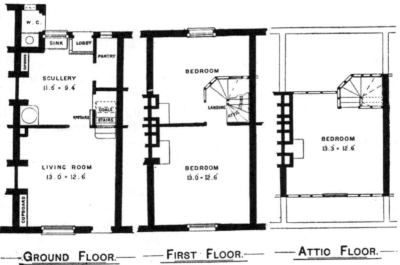

GROUND FLOOR.————FIRST FLOOR.————ATTIC FLOOR.

Figure 3.5 Functional differentiation of domestic space. Workmen's cottages, Birmingham, with separate bedrooms and private water closet. From Bowmaker, *Housing of the Working Classes* (1895).

This process was equally evident in the workplace: the orders on sanitary accommodation following the 1901 Factory and Workshops Act stated that "every sanitary convenience shall have a proper door and fastenings, and be so enclosed as to secure privacy."[179] These rights slowly expanded thereafter: all adult Britons are today legally entitled to clean, private toilets, with hot and cold water, when at work.[180]

When toilets were built within the house, they required ventilation, which was usually achieved via an opening window, which also admitted light and, potentially, vision into the room. The law of light and air, however, said nothing at all about vision or privacy: it could not prevent, for example, railway passengers or residents of opposite houses from peering into one's house.[181] These windows, then, had to be obscured, with blinds, by making them frosted or opalescent, or even by resorting to the use of tissue paper or paint on a solution of Epsom salts.[182] Light and vision were physically disaggregated. The toilet, as Tom Crook has observed in a fascinating study, "made individuals obscure rather than transparent," particularly when equipped with a lock or "vacant/engaged" bolt.[183] Doors, locks, and partitioned and obscured glass created security of self-inspection. These rooms slowly became equipped with the tools of self-scrutiny and grooming: mirrors, brushes, soap, combs, an arsenal of narcissism unquestionably as important to the development of the "modern individual" as diaries, self-help manuals, and confessional literature.[184] This risky private space would, in turn, become the epicenter for concerns about masturbation, sexual encounters, and drug use.[185]

Public conveniences, meanwhile, developed from the 1830s in railway stations, restaurants, and other public spaces.[186] There were 827,820 visitors to the Great Exhibition's public lavatories, erected by George Jennings, who in 1858 began campaigning for a series of underground "halting stations" in London. The *Engineering News* described a state-of-the-art public convenience at Charing Cross, in London, in 1894. It consisted of twenty-eight urinals, twelve water closets, and two washstands and was built underground, making it "non-objectionable as a feature of the landscape" while maintaining "privacy." Light, not vision, was admitted through prismatic pavement lenses, and a ventilating chimney was artfully dissembled inside an ornamental lamppost situated immediately above the toilets. The urinals were free, with water closets and washstands operating "by dropping the requisite coin into a slot outside of each door." The whole apparatus was, thus, largely, and mechanically, self-regulating. "A single attendant in charge," with minimal supervisory responsibilities (maintaining cleanliness, answering questions, preventing disorder), was "all that is necessary."[187]

The private bedroom was, for many, a primary spatial index of decency. In 1877, Ellice Hopkins wrote condescendingly of her own, middle, class: "Our large houses, our separate bedrooms, our greater education, make us . . . more particular in our ways than you."[188] In these spaces, private and intimate habits were entirely invisible to outsiders. The contrasting image of filthy beds or rooms crowded with entire families in a state of promiscuous disarray was an abiding trope of slum literature. The building of separate bedrooms in bylaw housing was an obvious solution. A more ad hoc, but quite common, arrangement was the introduction of curtains, screens, or other forms of partition to divide this space into intimate zones, something often undertaken in lodging houses. Wooden screens were used in certain model lodging house projects to divide large rooms into multiple, discrete sleeping areas.[189] This was legally recognized in the 1875 Public Health Act, which stated: "Any room in a common lodging-house set apart for the use of two or more married couples for sleeping purposes must be fitted with a wooden screen between each bed to hide it from the view of any other occupants of the room, allowing sufficient access to the bed it is wished to screen."[190] Such partitions usually came down close to the floor, leaving a gap for ventilation, and they might be equipped with bolts. The local government board also recommended such partitions for canal boats and hop pickers' seasonal housing. These partitions in turn generated anxieties about the misuse of privacy as well as concerns that the partitions themselves would be stolen.

This concern for privacy extended to other spaces. In hospital space, there was no straightforward shift from an open and intimate eighteenth-century style to a privatized, individuated nineteenth-century one. As Lisa Cody has shown, eighteenth-century London's lying-in hospitals often provided privacy for women, while shared beds could still be found in later-nineteenth-century workhouse infirmaries.[191] But various nineteenth-century designs, notably the "pavilion style," pioneered by (among others) Florence Nightingale and the architect Henry Currey, did aim at more pervasive individuation of clinical space. The first such structures in Britain were Blackburn Infirmary and the Royal Marine Barracks Hospital, Woolwich (both 1858).[192] Each patient was a sanitary island, with an individual bed, bathing in a sea of light and fresh air. Nurses could cast their eyes along wards, but there was a corresponding drive to preserve patients' privacy when necessary. The clinical, supervisory, and oligoptic gazes, as well as the principle of physical agglomeration, therefore, existed alongside a countermovement to preclude gratuitous, demeaning scrutiny. Already in the eighteenth century, patients had begun to object to disrobing in front of doctors—and particularly in front of groups of medical students. This might be carried to startling lengths. James Reid, Queen Victoria's physician, famously never administered a full physical examination: indeed, he never actually saw her in bed until she was dying.[193] Most representative was the development of the curtained hospital bed. Although some sanitarians, like Nightingale, objected to curtains because they interfered with air currents, they became more common as the century progressed.[194] By the late nineteenth century, St. Bartholomew's in London, with lines of neat beds "each with its red quilt and light check curtains," was becoming typical of this arrangement.[195] Local government board regulations for workhouse hospitals in the early 1880s recommended using "two or three screens (on wheels) large enough to completely surround a bed when a patient is being bathed in the ward, or is very ill or dying."[196]

Death, according to Philippe Ariès, has over the last couple of centuries become "dirty."[197] Where it was once a relatively public and everyday experience, it is now largely hidden and unspoken, a secret event occurring typically in hospitals. The proliferation of hospitals was clearly essential to this deep and enduring obfuscation, as was the development of curtains and individualized hospital rooms (operating theaters, mortuaries). Public viewing of the corpse, for the purposes of either funerary ritual or inquest, was declining rapidly.[198] The 1875 Public Health Act included clauses regulating public mortuaries, which must shield the dead body from public view and curb the stench of death.[199] Meanwhile, the practice of retaining the corpse in the home, often to avoid the indignity of a pauper funeral, became for insensitive and uncomprehending visitors an irrefutable sign of desensitization: such individuals had become "so used to discomfort that they do not look upon a corpse in the room they live, and eat, and sleep in as anything very objectionable!"[200]

This slow disappearance of the corpse was paralleled by the concerted drive to eradicate spectacles of pain and death. The termination of public executions is, perhaps, the most-well-documented event here. "Punishment," as Foucault stated, became "the most hidden part of the penal process."[201] Public reveling in animals' pain also became intolerable for the civilized: it was a display of supreme indifference. The 1835 Cruelty to Animals Act banned outright such public practices as bearbaiting, while the 1871 Fairs Act allowed corporations to curb the bawdy and often bloody activities of fairs. Similarly, the spectacle of slaughter (not meat eating itself, of course) was gradually seen as being incompatible with urbane existence. Even in the later decades of the century, however, the act of killing animals was not always hidden. In 1874, sanitarians

Figure 3.6 Private slaughterhouse, open to the curious gaze of children. On the left, a boy can be seen looking through the doorway. From Cash, *Our Slaughter-House System* (1907).

in London grumbled that slaughter was still sometimes "screened from public view by canvas only, or not at all."[202] Slaughterhouses held youthful audiences transfixed (figure 3.6). One 1875 report noted: "In some localities it became almost a pastime for young children of both sexes to frequent the slaughter-houses, and witness the death-struggles of the butchers' victims. This familiarity with scenes of blood was justly considered as having an immoral influence."[203] The visual form of slaughterhouses was slowly subject to legal regulation. The 1844 Metropolitan Buildings Act stated that slaughter must take place no less than forty feet from streets and fifty feet from houses.[204] This act was frequently evaded, as the 1874 London Slaughterhouse Act made clear. Regulation 11 stated unconditionally: "No slaughtering may be done within public view."[205]

Through an uncoordinated combination of legislation, engineering, inspection, and social self-observation, death and pain were slowly consigned to shadowy realms: hospital, mortuary, execution chamber, and abattoir (figure 3.7). This private organic landscape, where life was experienced and undergone in its visceral depths, was traversed by the most tenuous lines of inspection. The access of outsiders was strictly controlled. As Mill noted in his early essay "Civilization," visual access to pain was increasingly limited: "All those necessary portions of the business of society which oblige any person to be the immediate agent or

Figure 3.7 Cocooning the organic. Public abattoir at South Shields, almost entirely closed to the world. The abattoir opened October 24, 1906. The entrance is shown at the top, the slaughtering hall at the bottom. From Cash, *Our Slaughter-House System* (1907).

ocular witness of the affliction of pain, are delegated by common consent to peculiar and narrow classes: to the judge, the soldier, the surgeon, the butcher, and the executioner."[206] The cocooned nature of the organic made a proliferating number of modern forms of voyeurism and scopophilia materially possible. In his study of prostitution in France, Corbin notes that the hegemony of bourgeois discretion and privacy "provided the basis for the erotic power of violating someone else's privacy." Brothels combined the private form of the house with the physical paraphernalia of voyeurism (holes drilled into walls and cupboards, semitranslucent draperies).[207] Voyeurism has been banalized by the twentieth-century cultural obsession with sex and death, carried to predictably ever-expanding extremes in pornography and horror movies. With the Internet, the whole ensemble has been made privately inspectable: it has been agglomerated, miniaturized, and made technically accessible and portable.

As a counterpoint to this, urban governments and private groups struggled, with frequently limited success, to introduce edifying spectacle into the city, for the purposes of making uplifting inspection potentially accessible to all. Perhaps the zoo, more than anything, epitomizes this rather poignant and slightly pathetic process. In 1831, the bedraggled remnants of the old royal menagerie were donated, in a desperate, magnanimous act, to the new Royal Zoological Gardens, which had been founded in Regent's Park in 1828. Originally intended for an elite audience, the Zoological Gardens were opened to the public in 1846, the year before the first recorded use of the abbreviated moniker *zoo* by Macaulay. Zoos were apparatuses of inspectability: animals were agglomerated, subdivided by species, made visually accessible through cages, huts, and runs, and rendered legible through signs and labels. They had an instructive, didactic purpose: one might stroke or feed animals, but the primary relationship was visual. At Rosherville Zoological Gardens in Kent, which opened in 1837, visitors were encouraged to learn the names and forms of the various beasts and birds "That those visitors who are not conversant with Zoology and Botany may not be left to wonder, as at other gardens, what is the nature of this animal or that plant, some of the most distinguishing facts in the natural history of each will be written on labels near it."[208] This involved complicated, at time brutal, organic tuning. According to the Reverend Charles Girdlestone, when the new Regent's Park monkey house was built in the 1840s, the builders made it airtight, leading to the deaths of at least fifty monkeys.[209] At the other end of our period, William Brend, a lecturer in forensic medicine, observed in 1917 that smog had produced a rachitic, tuberculous assemblage of animals to complement the human inhabitants of the surrounding city.[210]

————

The various networks of inspection, from state factory inspection to local nuisance inspection, along with the constitution of realms of privacy and instructive and wholesome spectacle, formed a distinct economy or pattern of perception. This system was geographically uneven, and compromised by numerous factors: lazy inspectors, obdurate factory workers, and aggressive landlords. It was also, of course, not the only such perceptual pattern operative during the period, and it overlaid, complemented, and clashed with others. Nonetheless, given the sheer volume of routine inspection undertaken in later-nineteenth-century Britain, does the period deserve the appellation *the age of the inspector*? This expression is altogether too humanistic: it ignores the innumerable strategies deployed to make things inspectable: agglomeration, accessibility, legibility, portability. The phrase *the age of inspectability* is more symmetrical since it also captures the networked and material elements of the process.

The fundamental features of this system and its political dimensions are worth reiterating. State inspectorates were important, to be sure, but the kinds of inspection that I have discussed were essentially local in nature: inspectors routinely answered to their local officials (MOHs, councils) rather than central ones. This respect for the locality and its idiosyncrasies was an integral aspect of liberal rule, found in the writings of Mill and the policies of Gladstone. Local forms of inspection allowed communities and towns to inspect and know themselves: metropolitan inspectors were simply not needed to monitor the quality of their food supply and the standard of their sanitary arrangements. Mediating institutions like the local government board, and the state inspectorates themselves, prevented the system from being absolutely locally autarkic and self-regulating: rather, it was a form of supervised local freedom.

Local inspectorates, in turn, could not fully function without the permanent assistance of an alert, sensitive public. Any aggrieved person could, for example, pass information about a nuisance to an inspector, who would then visit the offending premises and serve notice if necessary. This information might be passed anonymously, which introduced elements of facelessness and unidentifiability into the system. Inspection could not fully function, then, without an active citizenry attentive and attuned to the appropriate environmental conditions for contemporary existence. Nuisance inspection in particular relied tacitly on normative notions of the liberal subject's sensorium, which should be delicate and discerning. The very existence of the system illustrates how shifting thresholds of tolerance and perception were becoming institutionally embedded.

In short, to view inspection as centralizing and generally invasive and disciplinary, as Gray, Redlich, Finer, and many following them have done, is to miss a set of countertrends that complicate and undermine such a conclusion. The consensual, negotiated, and legally governed nature of the system reflected liberal beliefs in personal privacy and the rule of law. There was, ideally, nothing inquisitorial or arbitrary about checking the limewashing of dairies or the drainage of a lodging house. If one disagreed with an inspector's conclusions, there were formal rights of appeal. Penalties were not harsh, and, invariably, a warning preceded a fine. Inspectable entities were clearly delineated, and usually material, aspects of existence that, most agreed, required periodic monitoring for the good of collective existence. Most important of all, regimes of inspection left large areas of private, individual existence altogether uninspected, or, in a quite formal sense, free: central or local government had a very clear sense that it had a right to see only so much. Inspection was designed to be thorough but spatially circumscribed and unintrusive: the inspector examined one's water mains or cows, not one's conscience, political beliefs, religious proclivities, or morals. Inspection was not only not panoptic but also very consciously not panoptic. The modality of inspection, I think, demonstrates very clearly the perennial anxiety that Victorians felt about governing too much.

In his memoirs, *The Work and Play of a Government Inspector* (1909), Herbert Preston-Thomas argued, slightly playfully, that the life cycle of the average British subject was entirely subject to inspection:

Throughout the journey of life they find an Inspector perched on every milestone. As soon as the child is ushered into the world by the legally qualified doctor, the Inspector of Registration arrives in order to see that its advent has been duly chronicled. Then follows the Inspector of Vaccination to ascertain that Dr Jenner's rite has been properly performed. Then comes the question of education, and the Inspector of School Attendance appears on the scene. One Inspector examines eyes and talks of spectacles; another examines little Mary and talks of free breakfasts. When a modest cottage is built the Inspectors swarm. The Inspector under the local bye-laws objects to devices for economy in structure; the Inspector of Highways makes all sorts of stipulation as to roads; the Inspector of Nuisances is for ever girding at the dustbin. The victim invests his money in business, and is faced by the Inspector of Factories who insists on the machinery being duly fenced and the hands not being overworked.

If savings are placed in a colliery, up steps a new Inspector, so that any explosion may be followed by a Report. If, oppressed by too much inspection, the wretched man takes refuge under the shadow of Bacchus, he may have to encounter an Inspector of Police, an Inspector of Inebriates, or an Inspector of Lunatic Asylums. *Even after death he may not be free from inspection,* for is there not the Inspector of Burial Grounds?[211]

Preston-Thomas clearly ignores—probably in the interests of rhetorical effect—the domains of uninspectability that had been secured over the previous century. But let us focus on his list of inspected entities. It can be split into two: inspections of human subjects and inspections of technologies. The humans are babies, schoolchildren, inebriates, lunatics, and corpses: the technologies are house structures, roads, dustbins, machinery, and mines. Most of the former are inspections occurring only once in a lifetime, at critical organic passage points or "milestones" (birth, vaccination, death), or else are carried out on sections of the population requiring particularly intense institutional inspection. The latter, by contrast, are routine and regular inspections of physical objects: leaking roofs, damp walls, pipes, and mains. It was simply impossible to leave these systems altogether to their own devices. Since the functioning of society, and particularly the private practices of liberal subjects (cooking, washing, reading, mobility), now relied so heavily on such systems, their management had become utterly integral to government. The technological securing of collective and private capacities constituted the vital dimension of the aforementioned "growth of government." Inspectability, in other words, was a characteristic of technological or vital systems, not human organic existence, much of which took place in conditions of calculated and regulated opacity. Here was a palpable ontology of indirectness, the epitome of liberal rule.

One might conclude that Victorian government, at multiple levels, was turning its rather blinking, squinting, myopic eyes away from its subjects and onto the mammoth technological systems that sustained them. I want to focus now on two of these systems: gas and electricity networks. These were networks that sustained nocturnal perception, yet they were also designed to be hidden, self-regulating, and inspectable. There were specific visual problematics generated by complex, extensive, ramifying, underground networks through which energy circulated. The following chapters aim to address both these perceptual dimensions of gas and electricity: the inspectability of the systems and the perception that the systems secured.

FOUR

The Government of Light: Gasworks, Gaslight, and Photometry

The whole difference between the gigantic process of the gas light operation, and the miniature operation of a candle or lamp, consists in having the distillatory apparatus at the gas-light manufactory, instead of being in the wick of a candle or lamp. FREDERICK ACCUM, *DESCRIPTION OF THE PROCESS OF MANUFACTURING COAL GAS* (1820)

By a photometer, a gas-engineer means an assemblage of apparatus of different kinds, including meters, pressure-gauges, thermometers, balance, and other accessories, a remarkable feature of the collection being a great elaboration of cabinet-work and velvet curtains. ALEXANDER PELHAM TROTTER, "THE DISTRIBUTION AND MEASUREMENT OF ILLUMINATION" (1892)

The chemist Frederick Christian Accum waxed orotund when discussing gaslight. "Like the light of the Sun itself," he effused, "it only makes itself known by the benefit and pleasure it attends."[1] In 1809, he appeared before a parliamentary select committee considering the application of the newly formed London and Westminster Gas Light and Coke Company for incorporation. He was asked on what grounds he based his claim that gaslight was brighter than that provided by traditional lamps. This most loquacious of characters struggled to explain himself. He had tried measuring its brightness with Count Rumford's photometer but found it impracticable. Without a proven instrument to appeal to, he was forced to argue, in convoluted terms, that gaslight was brighter because its flame was larger than that produced by a candle or oil. His words were recounted contemptuously by William Matthews in his *Historical Sketch of the Origins and Progress of Gas-Lighting*, first published in 1827: "Having found that each of the cock's-spur lamps [lit by gas] has a flame equal to three inches in length, which makes of course nine inches of flame, and having found also that the parish lamp, which burns with a bright flame, is about half an inch, it follows that nine inches of gas flame are equal to eighteen and a half inches of parish light."[2] Such calculations were preposterous, implied Matthews, recalling with derision Accum's claim to be able to measure a flame in three dimensions: "I measured the whole dimensions of it, not only the length but the breadth and the thickness; and, in plain English, I measured the whole figure of it."[3]

On this evidence, Accum appears as something of a charlatan, but, in struggling to express himself, he revealed several early-nineteenth-century epistemological conundrums surrounding illumination. When one measured light, what exactly was one measuring? What instruments should be used? What should be the unit of measurement? How bright should streetlights be? In 1809, there was no defined standard against which to measure gaslight, no fixed unit of measurement, and no consensus about the kind of apparatus one should use to make the comparison. In 1892, when Alexander Pelham Trotter, the future president of the Illuminating Engineering Society as well as the first man on record to cycle a mile in under three minutes, described the "assemblage of apparatus" composing a contemporary photometer, he was depicting part of the gas network as important, stabilized, and established as burners and mains.[4]

Light had not only been, in Schivelbusch's felicitous expression, "industrialised."[5] It had also been *quantified*. Accum's "gigantic process" now included measuring instruments, notebooks, conversion charts, observers, and numbers as well as gasworks and fittings.[6] The gas network, then, was a double system. There was the physical infrastructure itself, and there were numerous regulating, monitoring, inspection, and recording processes necessary to maintain its smooth government. These regulatory agents (photometrists, inspectors, meters, governors) were sometimes rather inefficient, even ramshackle, but they aimed to subject both mains and light to an acceptable degree of control and predictability.[7] This chapter examines both aspects of the system. The first part provides an overview of the emergence and growth of gas infrastructure. The second part explores the history of photometry. Despite the enormous effort that went into making light a standardized, predictable commodity, photometry remained a problematic practice, and perfect regulation remained an elusive, impossible goal. No mention is yet made of incandescent light (gas or electric) or the visual capacities and experiences produced by artificial illumination, which are covered in the next chapter.

Work, Mains, Meters, and Burners: Assembling Gas Networks

The attention of savants and natural philosophers had long been drawn to the flammable qualities of certain gases. Marco Polo witnessed the burning of natural gas in Baku, on the Caspian Sea, in 1272.[8] In 1618, the French doctor Jean Tardin described a fountain of natural gas and explained how to distill coal in a retort.[9] In 1667, Thomas Shirley recounted how gas erupting from the ground near Wigan "did burn like oyle," while, in 1739, vapor gushing from a Lancashire ditch was reportedly "so fierce that strangers have boiled eggs over it."[10] These incidents were isolated, however, and no serious interest was shown in systematically harnessing the illuminatory qualities of gas until the work of the engineer William Murdoch in the late eighteenth century. Recalling the 1792 lighting of his Redruth home with coal gas, Murdoch stated: "My apparatus consisted of an iron retort, with tinned copper and iron tubes, through which the gas was conducted to a considerable distance; and there, as well as at intermediate points, was burned through apertures of varied forms and dimensions."[11] Gas was generated, distributed, and burned via a miniature network. Murdoch demonstrated his gaslight at Birmingham's Soho foundry in 1798 and 1802 and then at another foundry, in Manchester, owned by Phillips and Lee. In 1808, he estimated that each light was equivalent to that supplied by between two and a half and four candles.[12] Frederick Winsor demonstrated gaslight at the Lyceum Theater in London in 1804 and in Pall Mall in 1807.[13]

Rather than gushing sporadically from terrestrial fissures, gas was now deliberately manufactured. There were three main pieces of manufacturing equipment: retorts (receptacles for burning coal and producing gas), cleansing apparatus, and reservoirs (sometimes called *gas holders* or *gasometers*), which together formed the "gasworks" (figure 4.1). By 1820, there were 960 retorts in London. They were made from iron, clay, or brick and required the attention of a reliable stoker.[14] Coal was tested and graded in terms of purity: later in the century, its illuminating potential would be calculated in candles per ton.[15] The first reservoirs were often old brewer's vats, casks, or treacle barrels and were rather flimsy: "The reservoir was encumbered with a heavy appendage of chains, wheel-work and balance-weights, and from the construction of the machine, it was

Figure 4.1 Gasworks, 1812. At the right of the image is shown a furnace, with retorts inside. From Newbigging and Fewtrell, eds., *King's Treatise* (1878–82).

necessary to guard it from the impulse of the wind, the action of which on the gas holder, would have rendered the lights which the machine supplied with gas, unsteady."[16]

This system had yet to achieve autonomy from the elements: gasworks were, with good reason, soon associated with explosions and poisoning, a point developed in the final chapter. Between retort and reservoir were interposed purification chambers, where noxious elements (tar, ammonia, carbonic acid, sulphureted hydrogen) were removed from the coal gas, the aim being to produce pure carbureted hydrogen and reduce the nuisance caused by gas manufacture. By midcentury, condensers, purifiers, and scrubbers drew impurities from the vapor. Gasworks accumulated vast mounds of waste materials, which, from the 1840s, were absorbed into other industrial processes, including the alkali industries, dyeing, refrigeration, and road surfacing.

Mains, mostly built of iron, were the basic technology linking reservoirs to the point of use, although there were early experiments involving the delivery of gas in portable copper bottles.[17] Early gas mains were often made from recycled material: the Gas Light and Coke Company bought its first iron mains secondhand from two water companies, while some pipes were made from old gun barrels or hollowed tree trunks.[18] Winsor's Pall Mall mains were made of lead, while the Cambridge University and Town Gas Light Company experimented in the 1840s with tile mains, which, predictably, leaked.[19] Accum recommended that mains be laid at least eighteen inches underground to prevent them being disturbed by traffic and suggested that small reservoirs, often called *syphons*, be built along with them, into which moisture could trickle and collect. Particularly leaky networks might require emptying every couple of weeks.[20] The softening and deterioration of soil from leaking water and gas could necessitate the laying of extra clay in cities to prevent pipes from sinking through the earth. In particular, earth mixed "with ashes, slag, vitrified cinders, clinkers or chemical refuse in the presence of moisture . . . play[ed] havoc with iron" and might lead to the fitting of protective wooden troughs. The laying of larger mains, meanwhile, entailed the use of special equipment (figure 4.2).[21] Gas mains were, finally, not used for light alone: in the early nineteenth century, a special pipe running from the gasworks at Fulham to Hurlingham delivered gas for hot-air balloonists.[22]

When running pipes from the mains into homes, builders rarely encountered physical spaces designed with gas in mind, as the eminent gas engineer William Sugg observed: "Few buildings have ever been constructed in such a scientific manner as to admit of the use of gas in the best way now known."[23] Tin, he argued, was the best material for domestic fittings: if highly polished, the pipe could be thinner and cheaper than its iron equivalent (figure 4.3). Most domestic tubing, like that in the surrounding streets, was, however, composed of iron (Scottish fittings were usually made of tin, French ones of lead). Sugg advised all tenants, when moving into a new home, to have their domestic gas arrangements tested before signing the lease and laying floorboards. When moving out, tenants sometimes took tin pipes and fittings with them since the cost of tin did not depreciate: this infrastructure was not yet materially coextensive with the physical shell of buildings. By the early twentieth century, gas fitters were required to complete forms detailing the particular sizes of tubing used, before the pipes were officially inspected.[24]

Fittings and mains wound back to the gasworks itself, which by the 1870s might be several miles distant from homes and provide energy for thousands of lamps. To make this distribution network reliable, gas had to become a predictable, measurable commodity. The basic problem was that every rise and fall of the mains produced a corresponding change

Figure 4.2 Equipment for laying gas mains. From Newbigging and Fewtrell, eds., *King's Treatise* (1878–82).

in the speed and pressure of the gas. These two variables played a critical role in determining the quality of flame issuing through a burner. Simply put, the behavior of gas in mains was not independent of gravity and altitude, and, hence, the quality of light that customers received might be heavily influenced by geographic location. Accum acknowledged this problem with rather more clarity than he did the question of intrinsic brightness: "The velocity of the gas in the mains and pipes of supply, is in the first instance as various as there are differences in the altitude and extent of the mains and pipes of supply. A main at one place will furnish with a certain pressure of gas, a flame one inch high, while at a different altitude it will furnish a flame double that height."[25] Ideally, he argued, gasworks should be built at the geographic nadir of the district of supply, from which gas could steadily rise upward. The behavior of gas was generally calculable: for every ten feet of elevation, it gained a tenth of an inch of pressure.[26]

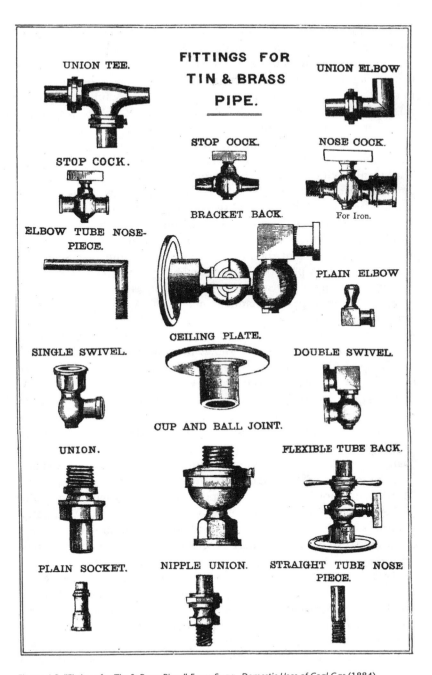

Figure 4.3 "Fittings for Tin & Brass Pipe." From Sugg, *Domestic Uses of Coal Gas* (1884).

The actual industrial and domestic use of gas varied, albeit reasonably predictably, across the course of a day, making supply a complicated process, requiring carefully designed networks and attentive managers. Accum recommended the construction of wider pipes at higher elevations in order to reduce pressure. But a potentially more useful set of devices was available: governors, or self-regulating feedback valves, which had been initially developed (in the eighteenth century) to automatically control the motion, speed, and pressure of certain complex physical instruments or systems (windmills, chicken incubators, steam engines).[27] The steam engine governor, for example, mechanically regulated the amount of steam entering the engine. It consisted of two balls attached to a flywheel, which sped up and slowed down, and, consequently, rose and fell, along with the engine itself. The centrifugal force of the motion moved a lever, which admitted more or less steam, thus maintaining a constant rate without the need for any human guidance. Such self-regulating systems prevented errors caused by human inattention or tiredness and might be regarded as exquisite signs of engineered intelligence. "The most perfect manufacture," noted Andrew Ure in his triumphant paean to automation, *The Philosophy of Manufactures*, "is that which dispenses entirely with manual labour."[28] Ure himself developed, patented, and named the thermostat, which automatically regulated factory temperature and, doubtless, added to his belief that the worker of tomorrow would be little more than an overseer.

These principles of technological self-regulation were soon applied to works, mains, and burners. The Gas Light and Coke Company fitted its first governors in 1816.[29] In gasworks, feedback devices were used "for the purpose of adjusting the supply to the demand," the velocity of flow changing as the number of burning lights across the system waxed and waned (figure 4.4).[30] Gas engineering manuals often recommended that "district governors" be positioned steplike along rising lengths of main, to prevent gas being wasted at higher parts of the network.[31] Similarly, governors could be positioned on each floor of a building, to maintain equal pressure throughout. Sugg observed that the pressure might be eight-tenths of an inch in the basement of a house but thirteen-tenths in servants' quarters in the attic: a governor would prevent waste in the latter. Governors could be adjusted to maintain minimal illumination levels in passageways, kitchens, or warehouses, "where glasses are not usually required" and perception of detail was unnecessary.[32] Positioning governors within the lamp itself prevented flickering when gas surged: hence the development of the combined governor-burner, or rheometer. The Peebles lamp governor, composed of a needle and cone,

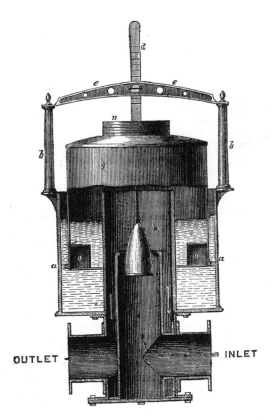

OUTLET →   ← INLET

Figure 4.4 Governor for
gasworks. From Hughes, *Gas
Works* (1885).

operated by "*silently regulating* the quantity [of gas] passing at all times."[33]
This silent self-regulation meant that attention could be turned toward
the illuminated surface and away from the flame itself, which could retreat from focus. A silent, self-regulating machine complemented the
attentive, self-regulating human reading beneath its lambent glow.

Feedback and measurement machines were also used to record quantity
of consumption and, hence, to charge customers. This was not a problem
with candles and oil lamps as the fuel and light were purchased simultaneously and the rate of consumption was tangible and visible. The disaggregation of fuel from flame raised the issue of how consumers should
be charged for gas. In the early years of gaslighting, customers paid a flat
rate via the "contract system," a "fixed sum per annum per burner of given
dimensions burning for a specified number of hours each evening after
sunset." This system often led to accusations of prodigal or illicit usage:

Figure 4.5 Wet meter. The revolving wheel is broken into four compartments. Gas enters at "B" and forces the wheel round. Chamber 1 is full of gas. From Hughes, *Gas Works* (1885).

"Often the gas tap was fully turned on, allowing the flame to rise sometimes to nine inches in length."[34] This, in turn, necessitated the appointment of inspectors to patrol streets identifying lamps burning beyond the prescribed hour. They rapped at the doors of recalcitrant users and had powers to extinguish the gas at the service pipe outside the house.[35] Lists of houses where contracts were paid allowed them to detect customers not paying for gas at all. The wastefulness of the system was countered with very personal, intrusive, and inefficient forms of inspection.

The gas meter was the technological solution to this state of mutual suspicion. The first functional model was developed by Samuel Clegg in 1816. It consisted of a drum divided into sealed, water-filled compartments that the gas turned as it passed. Such meters often leaked and froze, so dry meters were developed as a potentially more durable solution. A patent for a dry meter was granted in 1820, but the first successful model was Nathan Defries's model of 1838, which used the principles of common kitchen bellows. Early forms of dry meter gave rather unsteady light, while the "bagginess of the leather" made their action irregular, a problem solved by the use of multiple diaphragms.[36] By the 1840s, metered gas, using either wet (figure 4.5) or dry models, had almost entirely replaced the old contract system, and the two forms coexisted thereafter, although, by the early twentieth century, the dry meter was predominant. The meter was promoted as a quintessentially self-regulating apparatus: "The measuring of gas is not an interrupted process, and no attendant is required. The machine performs the whole of the work, and keeps a record of its own doings."[37] It became, for gas companies, the ideal method for making consumers regulate their own energy use: "By the consumer burning by meter, attending to the height of the flame of his burners, and not keeping them lighted longer than he may find necessary, he will soon find that a considerable saving of expense will be the consequence, on comparing what he may have to pay by meter, with what he might have been in the habit of paying for his gas when he burnt it by contract."[38] The consumer was encouraged to read the meter regularly and calculate consumption of light: wanton draining of urban energy supplies could be checked by translating municipal or corporate interests into private concerns for thrift. The gas consumer should find ways to curb excess, plan for the future, and be responsible. After all, by informing attentive householders of the precise levels, and rates, of energy being consumed, the meter made wasting fuel irrational and preventable: "It is very useful to compare a month's consumption with the corresponding months of previous years; if the number of lights and burners be the same, it will be ascertained whether the gas has been burnt carefully or carelessly; and the future consumption may be regulated accordingly."[39]

Nonetheless, this image of the self-regulating customer, like that of a self-regulating network, was more ideal than reality in midcentury. Private consumers had to be first prevented from abusing meters. Extracting water and tilting meters were particularly common ways of receiving more gas than was registered. The 1847 Gasworks Clauses Acts included penalties for tampering with or damaging meters. More routinely, customers had to be taught to read, and persuaded to trust, meters. The issue of reading was particularly salient: "The great percentage of dissatisfaction with gas consumers is with parties who look at the meter as a kind of *automaton*, and the officer who takes its index as a *conjurer*, who can and does make it say whatever he chooses."[40] This image recalled Wolfgang von Kempelen's notorious mechanical chess player of the late eighteenth century.[41] Gas bills, many feared, did not reflect the actual quantity of gas burned. Even the *Journal of Gas Lighting* admitted in 1880 that "the majority of gas consumers are unable to tell by an examination of the meter, how much gas has been consumed."[42]

One way to dispel distrust was to create a universal unit of gas. The Sale of Gas Act of 1859 defined the cubic foot as the legal standard for the measurement and purchase of gas. Yet, as one metering expert admitted, this did not solve some fundamental problems surrounding gas, which was "an aeriform body, invisible, highly elastic, varying in volume with every barometric change, very complex in its chemical constitution, affected by every change in temperature, liable to condensation, and to be absorbed by water, of which it is also an absorbent."[43] One particular cubic foot of gas, in other words, was not necessarily equivalent to another, in terms of purity, composition, or density. Meters, by contrast, seemed much more amenable to standardization, so the act focused more on them. Following the 1859 act, all meters had to be officially stamped, and there were penalties for forgery. The range of acceptable deviance of the mechanism was defined and set in favor of the consumer. Board of Trade officials were required to determine the accuracy of meters and to certify and stamp only those that did not stray from exact registration by more than 2 percent in favor of the company or 3 percent in favor of the consumer, a figure allowing for normal levels of evaporation in wet meters. Under clause 20 of the act, it became legal for inspectors to enter any premises on the request of either consumer or producer and test the meter. This formalized a process under way in London and elsewhere. The Gas Light and Coke Company, for example, had appointed inspectors to monitor abuse of wet meters, while Liverpool had appointed an inspector of meters in 1856.[44]

Reading gas meters was once something other than the epitome of a tacit, mundane practice. Test meters were legalized by the Standards Department of the Board of Trade in 1871, allowing meters to be themselves more easily metered. It was recommended that meters be tested at least every three years since accumulations of dust and dirt in valves could slowly jam the machinery while meters' warmth was inviting for insects. This required specialized testing stations, complete with standardized gas holders, accurate bottles measuring cubic feet, carefully regulated temperature and conversion tables.[45] In the early twentieth century, a quarter of a million meters were being tested annually in London. In large cities like Glasgow, maintenance was a permanent and large-scale endeavor (see figure 4.6).[46] Gas users were advised to consult their meter readout regularly "as so doing they will be able to detect any waste of gas caused by escapes or extravagant consumption," instructions that tacitly admitted the potential unreliability of the system.[47]

By the 1880s, the basic act of meter reading, for consumer and inspector, had been standardized through the familiar interface of three dials, with the outer two dials running clockwise and the center one running counterclockwise. This particular arrangement had become trusted largely through a process of accustomization. "In reality it is the simplest thing

Figure 4.6 Meter-repairing workshop, Glasgow. From *Municipal Glasgow* (1914).

possible," observed Sugg, ignoring the fact that such simplicity was an effect of training, education, and habit. Gas companies, he noted, usually provided customers with cards informing them how to read the dials (the counterclockwise direction of the central pointer sometimes caused confusion), "and the consumer can satisfy himself that he has received value for money."[48] Such cards are another example of the tactics used by inspectors to legitimate their authority over single, delimited, technical dimensions of domestic life. This periodic act of domestic measurement was facilitated by the development of ornamental meters, positioned openly in halls and stairwells, rather than languishing in cellars, unseen, rusting, and riddled with insects.

Meters were also used to check the consumption of gas in streetlights. The "average meter system," for example, was pioneered in Reading in 1863 and further developed in Nottingham in 1868, following the revelation of "great negligence on the part of lamp-lighters, who, especially during the height of summer, lighted the lamps earlier and extinguished them much later than the hours stated in the table."[49] By fixing sturdy meters to the base of a certain percentage of lamps and protecting them against cold and vandalism, the average amount of consumption could be calculated. Idle or bibulous lamplighters were then detected and disciplined. In St. Pancras, the vestry fitted meters on every twelfth lamp in 1874, reportedly saving £3,000 per annum thereby.[50] Were he still alive, Ure would have been horrified to discover that lights still required human lighters—and vindicated by their negligence.

The meter demonstrated the seamless way in which discipline mingled with self-discipline, but in a liberal way. It linked self-interest with corporate or municipal interest, making rational behavior easy without truly coercing: it invited liberal subjects to set their own limits to consumption. Meters were also differentiated quite starkly according to social class, as the rise of the prepayment, or "penny-in-the-slot," meter, popular from 1892, demonstrates. When a coin was deposited in the meter, a valve was released, causing gas to flow. A change wheel allowed the gearing to be swiftly altered if gas rates rose or fell, while meters were protected, in theory, against fraudulent use or theft.[51] When one ran out of money, one ran out of light: thus, thrifty practice and judicious saving were forcibly encouraged. A rational user, who saved coins and routinely monitored the dial, always had light available. The lives of the careless or lazy were periodically punctured by unwanted blackness. The prepayment meter embodied the idea that the poor lacked foresight and needed constantly to be reminded to pay for apparently limitless energy. The rhythm of payment mirrored that of wages: frequent but small. "If they do not have the penny, they go without," mused one commentator, "but due to a peculiar economic law, it is easier to get one penny for each of thirty days than thirty pennies on the thirtieth day."[52]

Prepayment meters were essential to the introduction of gaslight into working-class homes (figure 4.7). "Prior to the invention of the Slot meter," stated one London gas manager in 1899, "so few weekly tenants in London used gas that it may be taken as correct to say that gas was practically unknown in the dwellings of the working classes."[53] In the same year, the Gas Light and Coke Company reputedly collected "75,258,000 pennies" from 114,668 customers.[54] Two years earlier, George Livesey, the chairman of the South London Metropolitan Gas Company, remarked on prepayment meters' prodigious use: "An illustration of the popularity of these machines is afforded by the statement that there has been at times an actual dearth of copper money in south London, and by the additional fact that a single collection from these machines has yielded 10 tons in weight of copper."[55] Landlords liked prepayment meters. Writing about a later electric version, one engineer observed: "It relieves the proprietor of all responsibility as regards the consumption, over which he has practically no control."[56] They also might have broader environmental effects. According to one commentator, gas cookers, paid for with prepay-

Figure 4.7 Collecting coins from a slot meter. From Young, "Lighting London" (1902).

ment meters, were actually contributing to improved urban atmospheres in certain areas of London, by producing less smoke than open grates.[57]

Let us turn now to the burner itself. Candles required regular attention in the form of snuffing and trimming: oil lamps, similarly, were fragile and fickle. Early gas lamps, too, required the operator to perform a convoluted series of operations involving switching the gas on, opening and adjusting valves, and adding weights to regulators.[58] The lamp itself was, at this time, little more than an orifice spewing fire into space. By the 1820s, the opening had been modified: there were now basically two kinds of gas lamp, the flat flame and the argand. Flat-flame burners themselves came in two characteristic types, taking their names from the shapes of their flames, the batswing (1816) and the fishtail (1820). The fishtail design, with two jets providing a thin sheet of fire, was the commonest type of gas burner in Britain in the 1880s. Argand lamps were based on the old oil lamp of the same name, with a glass chimney drawing air into the flame to aid combustion.[59] They were substantially more complicated than flatflame burners, with many more parts, which required cleaning and attention. Sugg, who pioneered a new wave of argand lamps in the 1860s, noted that the addition of governors made them more reliable: "For reading, writing, drawing, painting, &c., it is at this day the most perfectly steady light in the market at the disposal of the gas consumers."[60]

Flat-flame burners were themselves undergoing modification. Research into the candescence of carbon particles in gas demonstrated the combined importance of air supply, pressure, and the shape of the burner's orifice. Excessive pressure could ruin the flame: early flat-flame burners might have to be "stuffed with wool, or pieces of wire gauze"; otherwise, they rapidly clotted with tar, causing the gas to "eddy and swirl" as it issued from the lip. Numerous remedies were developed. The Brönner burner, for example, had a large cavity directly beneath the slit that automatically lowered the pressure of incoming gas: other burners incorporated concave heads or wider openings. The steatite head, pioneered by Sugg in 1868, proved an enduring improvement. Steatite (or soapstone) was ideal because of its "inferior conductive capacity for heat, and its non-liability to corrosion."[61]

The construction, organization, and financing of gas networks necessitated the development of novel forms of engineering and management practice. Perhaps the first "gas engineer" worthy of the name was Samuel Clegg. Many early gasworks were designed and built by specialist engineers, like John Rofe or Thomas Newbigging, whose status grew across the period: by the 1860s, such a figure might make as much as £500 per annum.[62] Their knowledge was not simply technical or material but also managerial and financial: "The gas-engineer or manager of the present day is of a very superior order, possesses more extended views, and has a good knowledge of all that concerns his *métier*."[63] Gas contracting, too, developed rapidly: important early figures included John Gosling, who promoted gasworks construction at numerous places, including Maidstone, Birmingham, and Canterbury, and John Grafton, who built gasworks at Carlisle, Edinburgh, Sheffield, and Wolverhampton.[64] The British Association of Gas Managers was founded in 1863.[65] Finally, the development of gas networks also produced new forms of inspection. The Gas Light and Coke Company appointed Samuel Lay as its first inspector of mains in 1815 and soon after appointed a "mechanical inspector" to examine burners and fittings.[66]

I will not outline in detail the gas system's material expansion: there are many good historical studies of the technological, economic, and organizational dimensions of this process.[67] This growth was regarded by contemporaries as historically unprecedented, indicating Britain's industrial prowess. In 1827, Matthews noted that there were already around two hundred gasworks in Britain: "The astonishingly rapid progress of Gas-Lighting, in the course of only a few years, affords a striking and instructive proof of the great effects which may be produced by the combined exertions of science, ingenuity, and perseverance."[68] Towns like Whitby, Brigg, Tadcaster, and Diss built gasworks in the 1820s and 1830s. Early use of gaslight was largely limited to shops, factories, pubs, and streets, but the general fall in gas prices between 1830 and 1880 allowed its spread, mainly into middle-class housing as well as some rural areas.[69] "It is not an unusual circumstance," observed Newbigging and Fewtrell, "for villages in the United Kingdom, with populations ranging from 700 to 800 persons, to be able to boast their own gas works."[70] Progress was physically measurable by the number and size of gas holders, or "black volcanoes," dotting the landscape.[71] There were forty-seven gas holders in London by 1823, and, by 1865, they were beginning to dwarf St. Paul's Cathedral (figure 4.8). The development of telescopic holders and the abandonment of trussed or girdered roofs "led the way for [their] almost indefinite expansion," and, in 1894, the South Metropolitan reservoir could hold twelve million cubic feet of gas.[72] The languid rise and fall of these rusting pistons became part of the daily rhythm of the industrialized metropolis. Underground, the network of mains expanded so rapidly that precise geographic knowledge of the system became impossible. The *Builder* estimated that there were five thousand miles of mains beneath London in 1875, more than all the city's sewers and water mains combined, and varying in thickness from two feet to six inches. Sometimes, it was rumored, fifteen to twenty such mains would meet at tangled, knotted junctions.[73] Nationally, one 1883 estimate put the number of public lamps in England and Wales at 375,536 and the number of consumers at 2,019,846.[74] Gas was also, by this date, being used by consumers to power a growing range of domestic machines: water heaters, cookers, coffee roasters.

In 1870, the Gas Light and Coke Company opened the largest gasworks on earth, at Beckton in East London. To enable easy access for coal-bearing barges, the works were built alongside the Thames, near the outfall sewer of the northern section of Bazalgette's sewage system at Barking Creek. The gas took an hour to flow through an eight-and-a-half-mile cast iron main to the older works in Brick Lane, a distance that contemporaries found prodigious: "The extent to which gas will have to travel from the point of departure is a remarkable feature in the Beckton project."[75] Commentators hoped that gas production would be banished to London's marshy perimeter: "The public knew nothing of the battle

Figure 4.8 Three-lift gas holder, City of London. From Newbigging and Fewtrell, eds., *King's Treatise* (1878–82).

between gas and air which was going on beneath their feet; and when the city burst into its wanted illumination at night none but the initiated were aware that the gas which burned so brilliantly over a great portion of the civic area had been manufactured in the bleak country on the river side, some eight or nine miles away."[76] "The public," commented the *Engineer* in 1872, "have the satisfaction of seeing the manufacture of gas gradually removed to a great extent out of London," before noting that plans were already under way to double the size of the works.[77] Liberal infrastructure operated through distance: gas became domesticated, deindustrialized, removed from consciousness. It also functioned through translation: gas was translated into whatever kind of use the individual chose. The Beckton gasworks itself remained in operation until 1969. Rusting, decaying, and wrecked by strategic dynamite blasts, it formed an appropriate set for the Hue fight scenes in Stanley Kubrick's 1987 film *Full Metal Jacket*.[78]

Controlling gas networks, like organizing streets, became a formal question of municipal government. Public ownership of gasworks became very common, particularly in the north of England: Manchester's gasworks was run by the Police Commissioners from 1817 to 1843, when it passed to the corporation. Municipal ownership was greatly eased by the passage of the Gas and Water Clauses Act of 1847, which reduced the time and expense of the private statute process.[79] In Glasgow, the 1866 Glasgow Police Act allowed the corporation to tackle the endemic social problems caused by the "want of compulsory and systematic lighting in the city." The corporation was empowered "to erect and maintain lamps and lamp-posts, and other appurtenances for lighting in a suitable manner all public and private streets, courts and common stairs within the city; to light the dial-plates of turret clocks and city timepieces; and to appoint an inspector of lighting to take charge of that work and be responsible for the good conduct of the lamplighters and others appointed by him."[80] Individual mobility and orientation were enhanced by the illumination of clocks and streets: collective self-monitoring could expand, in small strategic pockets, beyond dusk. Municipal control of illumination facilitated nocturnal forms of the various visual regimes on which liberal government relied: oligoptic, supervisory, and inspectoral.

By 1900, there were 222 municipal gasworks in Britain and over twice this number of privately run gasworks, a consequence of the reform of joint-stock company law. Private companies, however, were subject to maximum dividends and the regulation of stock options.[81] In London, private ownership remained the rule throughout the Victorian period. The Gas Light and Coke Company received its act of incorporation in

1810, allowing it to attract sufficient capital to build three gasworks—in Peter Street, Curtain Road, and Brick Lane—and appoint Clegg as principal engineer.[82] From 1830 to 1857, there was frequent and even violent competition between London companies, resulting in chaotic accumulation of mains: "Many of the public thoroughfares were occupied by the mains of as many as four different companies." It was often impossible to tell whose was whose.[83] In 1857, the companies agreed to divide the metropolis into thirteen districts, effectively creating thirteen monopolies, generating sharp rises in prices, which were then regulated by the Metropolis Gas Act of 1860.[84] These companies were frequent objects of reformers' ire. Firth, predictably, fulminated about "the galling domination of this Gasarchy."[85] By the time he wrote, however, the number of companies was declining because of amalgamation (there were only three by 1885), while several parliamentary acts had provided more stringent regulation of their operation.

Calculating Light: The Rise of Photometry

When Matthews derided Accum's explanation of gaslight's brightness before Parliament, he portrayed him as an untrustworthy witness. Accum had proved incapable of using Rumford's photometer and had resorted to less "objective" techniques of measurement. He expressed himself numerically, but his numbers lacked the solidity and authority of those produced through a reliable instrument. As Graeme Gooday has shown, when a nineteenth-century scientist gained the trust of fellow professionals or the wider public, this trust was distributed between the human operator, the machines or instruments he used, and the practical techniques used to produce results.[86] Accum's eyes, measuring stick, and equation of brightness and size failed to amalgamate into a trustworthy whole.

The mere sharing of quantified standards alone, then, is insufficient to win the trust of others. Communities of chemists, engineers, and analysts also shared practices, textual conventions, forms of etiquette, and styles of self-fashioning that, fused with numerical forms of representation, might become consecrated and durable.[87] Shared, replicable, portable forms of measurement were absolutely integral to the management of infrastructures. Telegraphs and railways, for example, demanded data (electric resistance, e.g.) that could be collected quickly and easily and would be broadly perceived as accurate and trustworthy. Instruments, from yard measures to electricity meters, became more publicly visible and available over the century. Imperial pound and yard standards, for example, were from 1855 kept at the Royal Mint, the Royal Observatory, the Royal Society, and the Houses of Parliament. The 1824 Weights and Measures Act obliged local authorities to maintain secondary standards for the use of surveyors, builders, and engineers. These standard measurements should be placed "in the care of corporate bodies in large cities," they should be "accessible for the use of the public," and they should be periodically tested and, if necessary, renewed.[88] The Manchester City Council received a new set of standards, for the measurement of feet, inches, parts of inches, decimal grain weights, fluid ounces, and standard avoirdupois grain weights, in 1879. Local metrological statements about sewers, streets, and food were, thus, connected, theoretically, to a chain stretching from a battered inspector's measuring stick up to official standards preserved in revered sovereign institutions.

As Gooday shows, this kind of practical municipal measurement was usually far removed from the epistemological debates that animated academic physicists. Nowhere is this observation more apposite than in the case of light. Calculating the brightness of a London gas burner bore practically no relation to contemporaneous arguments over whether light was fundamentally electric in nature, for example, or whether ether was required for its transmission.[89] But questions of reliability and trust of measurer and instrument were just as significant for the gas examiner as for Maxwell. Hence, I explore the development of the relation between photometrist and photometer in this context. Alexander Pelham Trotter, for example, commanded the authority and inspired the trust that someone like Accum so patently lacked. To do this, he needed dependable, proved instruments as well as a trustworthy character expressed through personal modesty and care. He also needed established standards and well-tuned, practical, bodily routines.

The photometer is historically a very significant instrument, yet it has often been overlooked by scholars of instrumentation.[90] Its basic function is to measure visible light, the term *radiometry* referring to all instruments gauging levels of invisible light falling beyond the limited parameters of human vision.[91] The invention of the term *photometry* is commonly attributed to the eighteenth-century polymath Johann Lambert, although detailed work on the comparison of light sources had been undertaken earlier in the century by the mathematician Pierre Bouguer, who wrote on the subject in 1729.[92] Much of this early work was astronomical in focus, but, in the later eighteenth century, Antoine Lavoisier and Count Rumford, among others, became interested in using photometers to compute and improve levels of urban illumination. All photometric devices utilized Kepler's inverse-square law, which stated that the brightness of an illuminated area diminishes at a fixed geometric ratio as one moves away from it. A light source producing the same illumination as another, but from twice the distance, is four times as bright.[93] Thus, photometry was fundamentally comparative, aiming to measure a given light against a standard by calculating the distance at which it cast an identical quantity of light. Rumford's photometer, developed in 1794, operated by interposing a stick between the lights and a white screen: the observer would compare the shadows cast until they became identical and then use the inverse-square law to calculate how much stronger or weaker the test lamp was than the standard.

Matthews, then, was palpably disingenuous in dismissing Accum's claim since any attempt to accurately use Rumford's photometer in Pall Mall in 1809 would have been practically impossible. The immobility of streetlights, the effects of the weather, and the need to position the screen at the same height as the lampposts would all have militated against easy comparison. Accum's argument was a reasonable response to adverse conditions of measurement, rather than professional incompetence. Indeed, the later development of the jet photometer, which simply measured the height of the gas flame, sometimes automatically with a pencil and a continuous strip of paper, shows that Accum's intuitions were hardly aberrant.[94] Rumford's method itself fell into disuse across the century, as Dibdin noted in 1889: "The method is one which few practical photometrists of the present day would venture to adopt."[95] By this date, practically all photometric experiments used Robert Bunsen's system, developed in 1843, which utilized a piece of thin paper with a greased spot, situated between two lights. The spot disappeared when the intensity of light hitting the paper from either side was equal, allowing the distance of each light from the paper to be calculated with relative ease. The paper disk was, by Didbin's time, usually mounted in a carriage sliding along a graduated horizontal bar, facilitating quick reading of distances (figure 4.9). The disk also revolved easily, enabling both sides to be compared.

"Never perhaps in the history of mankind has so humble a thing been so universally honoured," effused Dibdin of the Bunsen photometer.[96] The disk could be replaced by a star-shaped attachment, which assisted with the measurement of colored lights. Around bar and disk were barnacled additional devices: sighting boxes, gauges, governors and meters for gas, mirrors to allow both sides of the disk to be simultaneously viewed, and screens to occlude flicker and produce a homogeneous plane of light. A photometer was less a discrete instrument than a ganglionic apparatus, the whole of which required a space purged of light and insulated against vibration in order to operate optimally.[97] Hence the velvet cur-

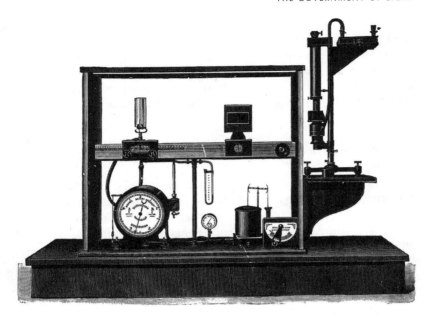

Figure 4.9 Table photometer using the Bunsen system, as used in official London testing places. The standard is on the right, the test burner on the left, and the photometric screen in the center. From Webber, *Town Gas* (1907).

tains, which, as Dibdin instructed, "should be kept free from dust, which is a great reflector of light, and a dreadful tell-tale of the reliance to be placed on an operator's work."[98] Surfaces should also be as dark as possible, to prevent glare and reflection. Trotter cited special recipes for dead black paints, grumbling that even the varnishes sold by photographic equipment dealers failed to fully absorb light.[99]

Dibdin referred to "the exquisite degree of perfection to which a properly-fitted gas-testing station has been brought," before noting that its complex composition "invites the most careful attention to every detail."[100] To produce a reliable reading, necessary for the production of urban knowledge, the photometrist had to read instruments carefully and attentively, measure gas exactly, and align the apparatus perfectly. Yet the quest for reliable knowledge ultimately generated more problems than it solved, and I devote the rest of the chapter to these problems. First, I consider something left unproblematized thus far: the standard itself. If a yard sufficed for distance, what should be the standard unit of light? Second, I examine the difficulties of setting legal standards for municipal and private burners, measured against the standard. Third,

I will outline the physiological dimension to photometry. The eye was itself part of the apparatus, yet it was an unreliable organ that required discipline and correction. Finally, I tackle some of the shortcomings of photometry as I have depicted it thus far, especially its inability to measure light at angles other than the horizontal and its failure to adequately calculate the light falling on plane surfaces rather than points.

The Persistence of the Candle: Victorian Standards of Light

Dibdin's optimism about the "exquisite" nature of late-Victorian photometry was not shared by all his contemporaries. In 1894, the *Electrician* depicted the science as "immature and defective."[101] In the early twentieth century, the American illuminating engineer Louis Bell moaned: "Of all the physical constants none are in so unsatisfactory a state as those pertaining to illumination."[102] There was still no consensus on standards or even nomenclature. Photometric instruments were cumbersome, fragile, and multiple. Numbers varied significantly from reading to reading, while observation was complicated by the wayward economies of the eye.

In 1852, the *candle* became the first parliamentary standard of comparison for gaslight. The term itself was vague, referring to any loosely cylindrical mass of combustible matter (beeswax, tallow, turpentine) with a central wick. These different substances or mixtures of substances "burn[ed] with flames of different colours, and afford[ed] . . . light of varying degrees of intensity."[103] The parliamentary standard, by contrast, required a fixed, stable, and replicable composition. This was defined as a combination of spermaceti (the waxy substance located in the cerebral cavities of sperm whales) and beeswax, weighing 1200 grains avoirdupois (one-sixth of a pound), and burning at a rate of 120 grains per hour.[104] Dibdin advised that such candles should be eight and a half inches long, with a shoulder diameter of eight-tenths of an inch, rising to eight-and-a-half-tenths of an inch at the base. This was, then, a candle with a very specific material form and composition. The candlemakers Miller and Company claimed that the standard candle should "consist exclusively of spermaceti . . . ,pure white and dry, having a melting-point of as nearly as possible 109°, and to which was added just so much air-bleached bee's-wax, having a melting-point of 140°."[105]

Thus, the term *candle* was universally adopted as both the concrete standard and the abstract unit of intrinsic radiance.[106] If the grease spot of a Bunsen disk became invisible when a gas lamp was four times further from the spot than a standard candle, then, by the inverse-square law, the gaslight was producing light of sixteen candles. However, ensuring that all photometric examinations used candles of identical size and composition proved impossible. A committee appointed to investigate photometric standards by the Board of Trade in 1879 reported that candles generally lacked a carefully defined chemical composition, were frequently adulterated with other materials, and had wicks of wildly varying shape and form.[107] The candle, observed exasperated photometrists, was no more scientific a unit than the old "barleycorn."[108] A measurement of sixteen candles lacked validity because standard candles themselves were not identical. In 1883, the Committee on Standards of Light for the British Gas Institute found that the standard candle displayed variations of between 1 and 16 percent. Additionally, even legally perfect candles might be unsatisfactory since new techniques to chemically refine spermaceti were themselves changing the nature of the substance used for the standard. The 1888 report of the Standard of Light Committee of the British Association for the Advancement of Science noted that such improvements meant that standard candles were probably giving less light than formerly since more carboniferous material was being extracted. The sheer number of such reports tells us that standards of light were becoming increasingly significant and that the candle was failing to function as a standard: it was problematic (figure 4.10). In 1890, Dibdin urged fellow photometrists to exercise scrupulous care in selecting and preparing candles and tending to their wicks, to obviate their physical irregularity:

The candle selected for the test should be a straight one, with the wicks central in the longitudinal axis; and it should not be too tapered from end to end. The sloping top is to be cut off at the shoulder; and the candle then equally divided in the centre. The two new ends thus obtained are to be trimmed so as to form new wicks, which, when lighted and burning, are to be turned so that the plane of the curvature of the wick shall be perpendicular to the plane of the curvature of the other wick.[109]

This was a convoluted process, but it guaranteed more reliable results and contributed to the rhetoric of care and cultivation of trust surrounding ideal photometric practice.

Given the efflorescence of novel forms of illumination in the nineteenth century, it might seem perplexing that something as ancient as a candle remained a scientific standard. But ease of manufacture, habit, portability, cost, and the law combined to keep this "antiquated and ill-defined unit" in use, despite numerous attempts to replace it.[110] These alternatives included incandescing platinum wire, developed in France from the 1840s. The Violle platinum standard, first adopted in France

Wick from 1870 candle. Light deviation from 1-candle Pentane flame + 5 per cent.
Consumption of sperm, 121 grains per hour. Melting point of sperm, 110° F.

Wick from 1884 candle. Light deviation from 1-candle Pentane flame + 20 per cent.
Consumption of sperm, 119 grains per hour. Melting point of sperm, 108° F.

Wick from 1892 candle. Light deviation from 1-candle Pentane flame + 16 per cent.
Consumption of sperm, 108 grains per hour. Melting point of sperm, 104° F.

Figure 4.10 Evolution of the wick and the chemical composition of the candle. From Dibdin,
Public Lighting (1902).

at the 1884 Paris Congress of Electricians, was defined thus: "the light emitted by one square centimetre of platinum at a certain temperature, defined by the ratio of two amounts of radiation—one being the whole radiation emitted by the platinum, and the other being that portion which is passed through a certain absorbing medium."[111] The obvious problems with platinum were its expense and the level of laboratory skill required to maintain the incandescing metal at an absolutely constant temperature. German physicists failed to successfully replicate it. The 1888 Committee on the Standards of Light did not consider it a practical standard. Although experiments at the Davy-Faraday laboratory would demonstrate its worth, it was unlikely to be utilized anywhere beyond a national level.[112]

More practicable and successful was the pentane standard, an oil burner developed in 1877 by Augustus Vernon Harcourt, the anaesthetist and Oxford professor of chemistry. The original pentane burner was cumbersome, even when compared with the ideal use of candles as recommended by Dibdin. One had to fill with it with pentane, screw in the box, open the tap, light the lamp, let it warm up, then turn the tap to adjust flow: "It cannot be denied that it is a most tedious tool to work with, as the continual adjustment required takes the observer too much away from the Photometer disc."[113] By the time of the British Association committee report on standards of white light in 1885, an improved pentane standard was the most promising alternative to the candle: "a unit of light which is practical in construction and adjustment, [possessing] extreme accuracy."[114] The Metropolitan Board of Works (MBW) was urged to adopt it in 1887, in order to produce more reliable statistics on London's light levels. However, doubts persisted about the amount of skill required for its operation. In 1895, the Photometric Standards Committee was still pressing for the burner's adoption; the ten-candle pentane standard was legally recognized, alongside the candle, from 1898, but only in London.[115] John Fleming, a professor of electrical engineering at University College, London, claimed in 1901 that candles were "falling into disuse as a practical standard."[116] By 1902, the pentane standard's legal use was extending into the provinces, having been adopted in "Birmingham, Hastings, and elsewhere."[117] In 1909, Charles Stone, a chief gas inspector in New York, described it as "without an opponent."[118]

Municipal Photometry

When Matthews wrote in 1827, photometry was rarely used to produce significant numerical data about the performance of a given town or district's gas lamps. It was an occasional practice, not yet integrated into circuits of urban government. The testing of gas lamps in Marylebone, for example, seemed an institutionalized version of the kinds of practices he vilified elsewhere:

> In the parish of Mary-la-bone the average number of hours for lighting the lamps is twelve; and that a representation of a burner is posted up in the watch-house, that all the patrols may know the proper sized light supplied by five cubic feet of gas per hour, and if they find the flame of any lamp less than the diagram placed in the watch-house, they have orders to enter it in the minute-book of the night, in order that a message may be sent the next morning to require the attendance of some officer of the Gas Company to account for the deficiency, or shew just cause why a fine should not be levied upon them according to their contract.[119]

Watchmen thus monitored and assessed gas lamps purely on the basis of subjective memory of the correct size of the flame. No specialized photometric expertise was brought to bear on the lighting of parishes.

Municipal regulation of light levels developed after 1850. In that year, a law was passed stating that light in the City of London should be supplied at no less than 12 wax candles or 10.3 spermaceti ones. Soon after, the wax standard was dropped, and, in acts of 1860, 1868, and 1869, the level was progressively increased to fourteen or sixteen standard candles for the City, depending on the company supplying the gas.[120] For the rest of London, the level was fixed at twelve candles for common gas (an increase from ten) and twenty for cannel gas (a brighter substance produced from cannel coal, which has a higher hydrogen content than normal coal) in the 1860 Metropolitan Gas Act.[121] By 1907, London gas companies were obliged to deliver illumination of between fourteen and sixteen candlepower.[122] By this date, practically all British companies were obliged to provide light at some kind of minimum level, ranging from ten to twenty-five candles: although the rise of the gas mantle, discussed in the next chapter, would complicate matters, legal standards remained "a real safeguard for the consumer."[123] Photometers were absolutely necessary to measure a product purchased daily by vast numbers of companies and individuals: they were included in the Weights and Measures Acts of 1878 and 1889 as "measuring instruments used in trade."[124] Light was now, in theory, standardizable and commodified. It was essential to produce a standard test burner, with which to measure gas against the standard candle. In 1868, Sugg's London Argand was selected by metropolitan authorities to perform regular tests on London's gas: by the 1880s, its use was widespread throughout Britain and its colonies.

Equally important was the establishment of testing places. Under the 1847 Gasworks Clauses Acts, companies or authorities providing gas were to provide some sort of space for the examination of gas, although, without legal standards, this was toothless legislation. The Gas Light and Coke Company built its first photometer room in 1857.[125] Gas inspection could be a contentious process. In 1858, for example, after a squabble over the quality of streetlighting in St. James's Vestry, London, the vestry refused to provide a testing place, which led to an awkward effort to wheel a photometer into place in the street: "The end of the Photometer could be placed bodily over the street lamp. The disc was fixed at a point 50 inches from the flame; and the candle mounted on a traveling holder on the reverse side of the disc [an unsatisfactory reversal of habitual photometric practice]."[126] The 1860 Metropolitan Gas Act was more specific, stating that testing stations should be provided within a thousand yards of every gasworks in London.

The MBW was given powers in the early 1860s to supervise London's gas industry and to test meters.[127] Aside from the provision of testing stations, the MBW also utilized clauses in the City of London Gas Act of 1868 to appoint gas examiners. London's first gas examiner was appointed in 1869, four more a year later.[128] Each inspector had a metering house, and, in the early twentieth century, there were twenty-two of these dotted across London.[129] Most of these men were chemists, like Dibdin, who was a superintending gas examiner for the MBW before its dissolution and replacement by the London County Council. Experts were, thus, devoting themselves to the monitoring and measuring of urban illumination.[130] When one W. M. Williams wished to compare solar with terrestrial illuminants, he turned, not to optical scientists or physicists, but to London's gas examiners because these were the men who conducted the vast majority of metropolitan photometry.[131]

The gas examiner was expected to perform three tests daily, or four if one test produced results falling below the legal light level. These tests should be scattered over the course of the day since it was acknowledged that candlepower could vary dramatically over the space of a few hours.[132] Each test required attention to detail and very specific bodily practices. According to the 1871 Gasworks Clauses Act, two candles, once selected, should be burned for ten minutes before use, to produce a "normal rate of burning, which is shown when the wick is slightly bent, and the tip glowing," at which point the candles were weighed, then turned so that one flame's edge was perpendicular to the other's face, and a stopwatch began timing the test.[133] The two candles should, between them, consume forty grains of spermaceti in ten minutes (they sat in a balance during the experiment). The observer then took ten measurements at intervals of one minute, ideally turning the disk after five measurements to neutralize any minor unevenness in it. After the test, the measurements were added, divided by ten, and multiplied by two (to calculate the brightness of the gaslight in single candles), and then corrections were made for temperature and pressure.[134] One simply recorded the readings of the barometer and thermometer and consulted tables in the Instructions for Gas Referees. The final figures, which were already averages, were then recorded in a book.

In provincial cities, similar routines were instituted, especially following the Gasworks Clauses Act of 1871, which made all companies supply a testing place and apparatus.[135] Gas testing was undertaken in any available laboratory space: Manchester's Gaythorn gasworks was used for the analysis of the city's gas as well as its food. Laboratories, often rudimentary and threadbare by our standards, were becoming nationally more common following legislation like the 1875 Sale of Food and Drugs Act, the development of bacteriology, and the establishment of universities.[136] In Manchester, John Leigh was producing photometric measurements "almost daily" in 1867.[137] These photometers themselves were subject, by the 1890s, to four annual examinations, after which they were usually adjusted against a mediating standard. These tests, like metropolitan ones, consisted of producing the average from ten readings, which could themselves be averaged over the course of a year to produce a single annual figure. In 1869, City of London tests showed the illuminating power ranging from 13.46 candles to 16.64.[138] In Manchester, this level was higher: 20.32 standard candles in 1870–71, but only 18.53 in 1874–75. In Birmingham, the average figure was 17.21 in 1882.[139] These averages publicly pronounced how light the city was and whether it was getting lighter or darker. Metropolitan figures could be used to check that the gas companies were not defrauding the public (examiners were also expected to test the gas for pressure and possible adulteration). One examination of gas provided by the Metropolitan Gas Company in 1876 revealed a level of only 7.02 candles.[140] Sugg argued that differences could often be explained by looking at the quality of coal or the type of burner used: the Houses of Parliament and Westminster, for example, were supplied with gas from cannel coal, which was invariably of a higher illuminating power (around 20 candles). In Scottish cities, he noted, the coal was of very high illuminating power (25–30 candles), but this was nullified by the poor quality of burners.

Just as municipalities quantified their death rates and levels of crime, so they measured the brightness of their gas. Yet, although producing this constant stream of data was relatively straightforward, it was also acknowledged to mask several serious problems. It was soon realized that testing gas close to its source of production through a standard burner often produced figures bearing little relation to the actual lighting of the district in question. In 1876, the original requirement that metropolitan tests be made within one thousand yards of the gasworks was abandoned, meaning that examiners could now measure gas where it was consumed. This, however, raised the issue of how to use photometers outside the laboratory. Such a delicate apparatus was fragile and fixed and required a degree of environmental calm completely unrealizable in busy streets.[141] Sugg was one of the first to develop a portable photometer, consisting of a small oil lamp, screen, mirrors, and bar; it was "easily taken to pieces, and packed in a box for conveyance; and is as readily set up again for use, with a little practice, in five minutes."[142] Yet, despite protestations of examiners and chemists, readings from portable photometers lacked legal force.

These readings did, however, demonstrate the tendency of laboratory readings to give a distorted picture of local light levels. Here is William Preece reporting on measurements of gaslights to the Streets Committee of the Commissioners of Sewers of the City of London in 1885: "The gas lamps in the City are supposed to give an illuminating power of fourteen candles, when burning five cubic feet of gas per hour, and they do so when burning steadily and regularly in the laboratory, but when placed in the street lamps the supply of gas becomes irregular through age and dirt in the burner; the flame flickers about through imperfect combustion and through drafts, the lanterns become dim and the glass, therefore, obstructive, and the result given is only ten candles instead of fourteen."[143] Crucial here was the dissatisfaction of photometrists with the ability of their measurements to provide an accurate index of urban illumination. Preece implied that the burners, and, perhaps, the quality of the gas, were to blame for the poor reading. Gas manufacturers might have reasonably responded that the instruments, the standards, or the photometrist's eyes were at fault. In other words, if photometric levels were demonstrably falling, this might not mean that the level of illumination was itself declining. Subjective elements of vision were acknowledged to complicate readings, as this report of 1881 suggested: "Two gas examiners of London obtained 16.5 candles and 19 candles, respectively, as the illuminating power of the same sample of gas, using the same photometer, and candles from one packet. This variation is considerable, and shows a difference in treatment . . . far to [sic] great for trained observers working under a common superintendent."[144]

The Errors of Observation: Photometry and the Eyes

"The science of photometry," stated Trotter, "consists of intelligent apprehension of the principles of the subject; the art of photometry lies in skilful avoidance of errors."[145] As historians of science have argued, the act of reading an instrument is never a purely ocular and cerebral act. The body itself is heavily implicated in the exercise, as attempts to replicate experiments have demonstrated.[146] A numerical reading is the product of a potentially precarious alignment between "two *separate* and *spatially separated* instruments, bodily and artefactual, that are not necessarily ever completely merged in a unitary ensemble."[147] Techniques of correction or compensation, in the form of mathematical equations or tables, have been used to overcome such problems, as has automation. As Trotter implied, a third way was to discipline the body, to cultivate sound habits, to concentrate, and to exercise care.

All instrument reading required disciplined visual practices: what was unique to photometry was that the entity being measured, light, physically affected the very organs through which measurement was to be made. Thus, one could not actually measure anything intrinsic, or noumenal, about light, but only its subjective comprehension, its reception within the body: "Photometry is not the measurement of an external dimension or force, but of a sensation. It is difficult to make a quantitative measurement of our sensations."[148] Perception was affected by physical changes in brightness, by flutter and fluctuation. The eye should not become strained or blinded: "If the intensity [of the light] is too strong, the tired eye partially loses its ability to recognise small differences of intensity: if the light is too weak, on the contrary, the eye no longer easily grasps the difference of intensity...and the measurements are similarly less precise."[149] The irreducibly temporal nature of vision, and especially the production of afterimages, was particularly evident. "To look straight at a lamp," argued Trotter, "for examining the height of a flame or the position of a filament, or even to light a pipe, is enough to make accurate work impossible until the eye has recovered, for after spending some time in a darkened room the eye becomes very much more sensitive."[150] Such inattention to one's own eyes explained the "sometimes ludicrous difference of results obtained by independent observers in individual cases."[151]

For Bell, these visual problems meant that photometry was almost intrinsically useless: "A very little experience will convince the experimenter that the results depend upon the *general state of the eye*, the personal equation of observer, practice, preconceived notions of relative intensities, and other factors so variable that the result is little better than guesswork."[152] Yet, although these results may have said little about the nature of light itself, most illuminating engineers considered photometry indispensable. By the early twentieth century, candlepower had been measured routinely by photometry for fifty years, and the data were legally necessary. It was, observed Fleming, "difficult to imagine that anything else can be a substitute for the human eye in testing the relative value of two lights for visual purposes."[153] For most photometrists, the aim was to standardize practice in order to purge measurements of the worst errors and to ensure that a photometrist in Dundee followed the same physical routine as one in Swansea. The first solution, as legal practice indicated, was to produce averages. Carelessness, misbehaving instruments, or ocular eccentricities might produce freak results, which could be neutralized by sheer weight of numbers. The magnitude of personal errors, Trotter concluded, could "under any given set of conditions... only be determined by investigating the deviation of a considerable number of similar observations from their mean."[154] Personal ocular error could also be counterbalanced by photometrists working in pairs and each observer using both eyes.[155]

The observer should also use his body in specific ways and become physically accustomed to certain routines. Speaking before the Institute of Civil Engineers in 1892, Trotter emphasized how photometrists should harmonize the motions of head, eye and arms: "After very little practice, it is not difficult to make the two lights vary until [the difference] is hardly sensible. The muscular sense of moving the lever enables its mean position to be found, or the alternate motion may be continued for a couple of seconds, while the observer looks at the scale and estimates the middle position."[156] William Abney, the editor of the *Photographic Journal*, argued that observers should work quickly and rely more on sensation than perception. Thought was the enemy. "The operation of equalising luminosities must be carried out quickly and without concentrated thought," he noted, "for if an observer stops to think, a fancied equality of brightness may exist, which other properly carried out observations show to be inexact."[157] The understanding of the eye's temporal economy enabled gazes to be quantified. One should not stare too long at the flame: dictating to an assistant might stimulate a blurred cascade of spectral afterimages. In 1908, Kennelly and Whiting conducted experiments on photometrists, concluding that twenty-five observations of about fifteen to twenty-two seconds each was the largest number that could be achieved without ocular fatigue.[158] Meanwhile, the flicker photometer, in which the screen was illuminated and eclipsed in rapid succession, was devised to "exercise the eye at its maximum sensitiveness." The method was particularly useful when comparing lights of different colors at varying levels of brightness since it overcame some of the problems of relative chromatic perception associated with the Purkinje effect.[159]

Elsewhere, scientists were attempting to eradicate the eye from the apparatus altogether, despite warnings that the psychological and perceptual dimensions of vision "may make simple physical measurements simply illusory."[160] In other words, the figures would be very precise, but they would bear only a (possibly calculably) tangential relation to how people actually saw. By 1920, however, the eye was often represented, somewhat paradoxically, as the weakest link in the photometric chain: "The question of the precision of photometric measurements is of peculiar importance in that in this field, more than any other, the precision obtainable is limited by other than physical factors; namely, the ability of the eye to decide when two adjacent areas appear equally bright."[161] The quest was to reduce photometry to a branch of energy measurement. The photoelectric spectrometer (1922) was indicative of the trend.[162] Subjective perception and objective measurement were being forced apart.

From Point to Sphere and Surface: Radial and Illumination Photometry

Measuring light in a laboratory was widely understood to be very different from measuring it in the street. But the differences involved extended well beyond technical questions of pressure and the condition of burners, as Preece indicated in 1885: "To measure the intensity of a source of light is a very simple matter when we have a darkened room and a reliable photometer; but to measure the light in a street, or in a hall, when it emanates from many sources, and when these sources are fixed in many positions, and at various distances, and when it falls at various angles, is a very difficult thing."[163] People almost never used light in the way it was measured in the laboratory. Light invariably hit a surface from many points and angles, yet the traditional photometer, with its bar and disk, measured only one point of light from one angle, the horizontal. The instrumental solution to this was to refine photometers in order to measure light as it was seen.

The radial photometer was designed in the 1880s to measure light falling at angles above and below the horizontal plane. In its basic form, the test lamp was placed on a pivoted bar, enabling it to follow a semicircular orbit around the disk. The standard moved along the bar as usual, and the disk itself could be tilted. This allowed the effect of particular globes and reflectors on the overall distribution of light to be measured: "It is to be hoped that in future all comparative tests of the value of various burners will be so conducted as to show the actual work done by them, not only in one direction, but in all directions."[164] Radial photometry was particularly important for calculating the impact of lighthouse illumination. The holophotometer, devised by Harcourt, was specifically designed to measure the total amount of light emitted by a given beacon. Photometric readings could now be taken at every ten degrees to the horizontal and a distribution chart composed.

The question of the illumination of surfaces (roads, watches, street signs) became more pressing as the sheer volume of urban light increased: "We do not want to know so much the intensity of the light emitted by a lamp, as the intensity of illumination of the surface of the book we are reading, or of the paper on which we are writing, or of the walls upon which we hang our pictures, or of the surface of the streets and of the pavements upon which the busy traffic of cities circulates."[165] Again, traditional photometry merely compared two points of light and rigorously excluded all other light sources. This was completely artificial, argued Preece, since streets were illuminated by the combined effects of numerous lights, some direct, some reflected, including those from houses, as well as rays reflected from walls, street surfaces, and trees. In short, photometrists had been measuring light when they should have been measuring illumination: "The measure of the illumination of a surface, as far as intensity is concerned, is quite independent of the source of light itself."[166] Nobody saw by looking directly at a point of light: the electric arc lamp, being tested in numerous streets, had a high intrinsic radiance, but the light was often blinding, poorly focused, and badly distributed. In 1884, Preece described a new form of photometer designed to compare, not points of light, but surfaces: the science of illumination photometry was under way. As an electrophile, he utilized electricity, but the principle was applicable to gas. He placed a lamp in a small box, the top of which had a white paper screen with a grease spot: "When it was desired to measure the illumination of any space, such as the surface of a street, this box had simply to be put at the place to be measured, and the current had to be regulated until the grease spot disappeared. The current of electricity became the measure of the illumination, and a simple table gave the result in terms of the new standard."[167]

Preece was aware of the rudimentary nature of his device and knew that his results carried little validity beyond his immediate circle of metropolitan electrical engineers. According to Trotter, there were few attempts to replicate Preece's experiments until the early 1890s, perhaps because photometrists already had too much to occupy their time, but also because they had become settled into routines that they might be reluctant to break. Trotter described an appliance of his own design and pointedly distinguished it from regular photometry: "The illumination photometer is a portable instrument for use in all places where light is used—the street, the church, the school, the house, the railway station, the railway carriage. *It is not concerned with the lamps, but with what the lamps do.* . . . By measuring the illumination of a well-lighted bank, knowledge is obtained for prescribing that illumination for another bank, but the photometer does not tell whether the lamps are so arranged that the clerks' eyes are dazzled."[168] Such devices were sometimes called *illuminometers*. Trotter took as the standard the amount of illumination falling from a single candle on a surface at a distance of one foot: the *candle-foot* or *foot-candle*, used from around 1866, being the initial unit of measurement.[169] He argued that, "as the heights of lamp-posts and the widths of streets are measured in feet," it made sense to express illumination itself in feet.[170] Preece disagreed and, at the 1889 Electrical Congress in Paris, argued that the term *lux* should replace the obsolescent *candle-foot* (one lux being the illumination provided by one carcel lamp over one meter). The lux was formally adopted as an international standard of illumination in 1896, but Dibdin was still using the already quaint-sounding *candle-foot* in 1902. He did, however, provide charts for illuminating engineers to convert between the two units of measurement.

The measurement of illumination made it possible to represent light geographically through the distribution curve, which was formed by laboriously plotting contours corresponding to particular levels of illumination, or "equi-luminous lines."[171] Islands of radiance, centered on lampposts, might be revealed floating in pools of murk (figure 4.11). This allowed closer connection and correlation to be forged between the quantity of light and the kind of visual practice to be secured. In 1892, Trotter performed experiments to calculate the level of illumination inside railway carriages on the Metropolitan and District Line (0.3 to 0.9 candle-feet) and even that falling on the president's desk at the Institute of Civil Engineers (0.8 candle-feet). He argued, on the evidence of his measurements, that illumination equivalent to 1 candle-foot was "comfortable."[172] Dibdin concurred, noting: "It is for most people the best illumination for reading, and it is to be found on the most well-lighted dining tables and billiard tables."[173] Thus, it became possible to

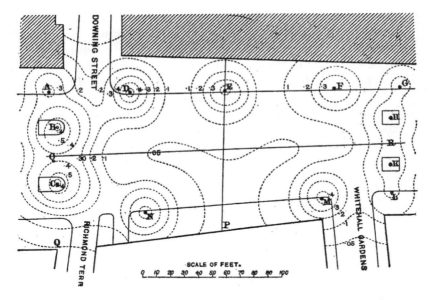

Figure 4.11 Illumination contour diagram for Whitehall, London. From Trotter, *Illumination* (1911).

argue that the president's desk, say, was not sufficiently illuminated for the act of reading and writing or that the light in the billiards room was just sufficient for a gentleman with normal perception to line up a long pot without straining his eyes. Trotter, certainly the most avid illumination photometrist in Britain in the early twentieth century, carried his measurements further by attempting to measure daylight with a blackened viewing tube and a stout book of equations and conversion charts. Daylight illumination photometry would measure all light "which would be produced at the spot in question, if all buildings in the neighbourhood were demolished, and the illumination were produced by light from a uniformly grey sky."[174]

———

The development of gas networks, designed to secure levels of illumination necessary for multiple industrial, commercial, and domestic uses, entailed more than just the construction of gasworks, mains, and burners. It involved maintenance, inspection, meter reading, photometry, and the measurement of impurity, which, in turn, involved legislation, institutions, and training. These practices were continuous and widespread by the later nineteenth century. They demonstrate the extent to which the Victorian "growth of government" was very often caused by the need to regulate large technological systems that, despite the best intentions of engineers, failed to regulate themselves.[175] The everyday regulation of gaslight, in the form of the inspection of gas mains and municipal photometry, was never perfected, acting instead as a kind of perpetual incitement to govern better, produce more data, and measure more accurately. The next chapter follows the light itself into the spaces where it was used and examines the forms of perception that it made possible.

FIVE

Technologies of Illumination, 1870–1910

We can well understand the perplexity of the inquirer who attempts for the first time to ascertain the most suitable degree and kind of illumination for any particular purpose. If the use to which the light is to be put is to show the colours of pictures or decorations, he will prefer a different light to that which he would require if he desired merely to show the way about in dark weather, whilst for the quiet reader a subdued yet sufficient light withal will be necessary. WILLIAM DIBDIN, *PUBLIC LIGHTING BY GAS AND ELECTRICITY* (1902)

Writing in 1902, Dibdin was struck by the immense diversity of available lighting systems and the wide, nuanced range of visual tasks for which they were used. Light "varying in intensity from a fraction of a standard candle to many thousand such candles" could be produced simply by judicious choice of illumination technology.[1] Surgeons scrutinized the inner contours of the human body with delicate bulbs, while armies carried powerful searchlights along with their arsenals. In addition to being industrialized, governed, and quantified, illumination was increasingly functionally differentiated, designed to facilitate perception almost everywhere human beings ventured: from operating theater to battlefield, to railway, to workshop.

It would be tedious—indeed, pointless—to catalog every innovation in the field of illumination technology between 1870 and 1910. Among the most significant developments were acetylene, the regenerative gas lamp, the gas mantle, the Jablochkoff candle, the arc light, and the incandescent electric bulb, not to mention many novel forms of oil lamp.

Many of these technologies were not actually invented in this period: rather, they became practicable, affordable, and relatively widespread, often because of enormous advances in infrastructural systems. There was a proliferation of multiple forms of illumination technology, rather than the rise to dominance of electric light at the expense of other light forms. Only in retrospect, with much historical simplification, as well as a leaden, reductive dose of Whiggery, does electric light seem inevitably poised in 1900 to become the dominant twentieth-century illuminant. The incandescent gas mantle was, in many ways, the most successful illumination technology in the first decade of the twentieth century. It must also be emphasized that this period witnessed, as in the cases of street widening and smoke abatement, large amounts of failure, compromised experiments, and accidents.

This chapter traces some of the tangled ways in which illumination and visual practice coevolved while suggesting critical points of connection between illumination technology and liberal subjectivity. I begin by outlining the various new forms of illumination technology that emerged and spread during this period, before considering how manufactured light was used and experienced. I examine three perceptual capacities that were particularly valued: the ability to distinguish colors, the ability to see at a distance, and the ability to distinguish detail. Following this, I look at certain illumination arrangements designed to produce specific visual effects: attention in the factory, comfortable reading in the library, and tasteful decoration in the home. Finally, I examine the medical, sanitary, sensory, and bodily dimensions of various illuminants, notably their impact on the eyes, their impingement on other sensory systems, and their broader environmental effects.

Mantles, Carbons, and Bulbs: The Proliferation of Illumination Technology

Gaslight thrived in the age of electricity. Despite what is sometimes asserted, it was no less "scientific" than electric light, as William Mordey, the future president of the Institution of Electrical Engineers, observed in 1906: "Gas lighting is just as scientific as electric lighting. Electrical engineers have taught gas engineers something, and gas engineers in return have taught us something."[2] Gas and electrical engineers used similar methods to measure the potency of their lamps, while the mechanical, physical, and chemical sciences were as important to men like Sugg as they were to electrical innovators like Joseph Swan and Sebastian

Ferranti. Conversely, the older, more practical, and less laboratory-based craft tradition was as vital to the development of electric light as it had been in the early days of the gas industry.[3] Improvements to gas illumination, in terms of brightness, focus, efficiency, and manipulability, utilized the resources of both "elite" and "popular" science, something evident in the use of new types of gas: oil gas, albocarbon (gas enriched by naphthalene), acetylene, and even those released by metals.[4] In 1902, Dibdin described acetylene as "winning its way into favour, especially in rural districts, churches, hotels, railway stations, and country mansions, where either electricity or coal-gas cannot be obtained at reasonable prices, or has not yet been introduced."[5] There were increasing numbers of gas lamps designed for specific purposes or trades (bakeries, railways, and markets) and many new techniques of illumination control (bypass pipes to avoid waste in railway stations, safer designs for theaters).

There were also new forms of gas lamp. Self-intensifying lamps, drawing on hot-air engine technology and chimney design to automatically coax more air into the flame, were being "widely used for public lighting in many large towns of the United Kingdom" in 1911.[6] The regenerative lamp, the most enduring form of which was the industrialist Friedrich Siemens's model of 1879, operated on the reverse principle, whereby products of combustion were utilized to heat the gas itself before burning. Like the self-intensifying lamp, this was inspired by earlier technology, in this case Siemens's own 1861 regenerative coke furnace.[7] The major advantages yielded by the regenerative principle were threefold: "First, the particles of carbon are liberated rather earlier in the flame; second, they are raised to a more exalted temperature; and third, they remain for a longer time in the incandescent condition."[8] These results were quantifiable: the photometrist F. W. Hartley concluded that "the light yielded per cubic foot of gas burner per hour was therefore equal to 5.62 that of a standard candle," an ordinary gas burner giving between 3 and 3.5 candles per cubic foot. Some models could also project good light at all angles.[9] Finally, the regenerative burner, by recycling its own waste, lessened the atmospheric damage that had become a marked criticism of gaslight by the 1880s: in 1888, observers of a demonstration at the Royal Institute of British Architects noted that, if placed "sufficiently far above the head," it "prevent[ed] any sensation of radiant heat."[10]

The most radical, and significant, innovation in gas illumination technology was the incandescent gas mantle, invented by the Austrian Carl Auer von Welsbach in 1885–86. Mantles were delicate, conical lattices composed of the oxides of rare earth metals, strengthened with a stiffening agent. This flameless gaslight was described in 1886 as follows: "The light

Figure 5.1 Bray reversible inverted burner with gas adjuster. The latticed mantle is inside the glass globe. From Grafton, *A Handbook of Practical Gas-Fitting* (1907).

emitted is, at a distance, hardly distinguishable from a twenty-candle in-candescent electric lamp, and by a modification of the composition of the impregnating liquid a yellower light is obtained, resembling that of the best gas-lights, but much more brilliant."[11] The light was easily manipulated, especially since different metals generated different colors, and was also substantially more powerful than older gas burners (figure 5.1). The 1898 Welsbach Kern burner could produce between twenty-five and thirty candles per cubic foot of gas. The 1893 model "placed gas in a position unapproachable by its competitors as a cheap illumi-nant for both indoor and outdoor lighting." In 1902, "some millions of these burners are now in use."[12] High-pressure incandescent gas burners could produce light of six hundred candlepower. Glasgow used 392,387 mantles alone in the year ending May 31, 1913.[13] Liverpool, which Dib-din regarded as England's best-lit city, had nearly 9,000 incandescent gas lamps in 1901 and only 152 electric arc lamps. Strategic clustering of mantles produced even, diffuse illumination: "By this means a much more even distribution of light can be obtained than from a single point, as is the case when electricity is used" (figure 5.2).[14]

"The once-doomed gas-burner has, thanks to Welsbach's mantle, in many instances replaced the incandescent electric lamps that were to

Figure 5.2 Gaslight at Victoria Station, London, early twentieth century. The station was lit by four hundred high-pressure gas lamps, supplying light of between 175 and 1,000 candlepower. From Webber, *Town Gas* (1907).

doom it," declared Archibald Williams in his popular 1906 text *How It Works*.[15] Electric light was, clearly, the mantle's main competitor and, as Schivelbusch notes, its model.[16] When new gas lamps were displayed, the comparison was routinely made with electric light as well as with other forms of gaslight. Following a demonstration of the Wenham regenerative burner in 1885, the *British Architect* observed: "The purveyors of the electric [light] will need to do something much better and cheaper to compete with any one of these lamps."[17] The point here is obvious but worth repeating. There was nothing inevitable about electric light's eventual salience: indeed, electric light often seemed simply to have stimulated the gas industry to vastly improve its service. "There can be no doubt," announced the electrical engineer John Slater in 1889, "that we now get a far higher value of gas which we burn than was the case ten years ago."[18]

Let me make some general points about electric light here, although its history has been well documented elsewhere.[19] First, electric forms of illumination had a very patchy history in the 1880s. Some commentators praised electric light for its purity, but others found it unbearable, and many more found it quite unremarkable. In 1881, electric light was derided as "ghastly and unpleasant" in so obviously supportive a publication as the *Electrician*.[20] Its systemic reliability was an even greater issue. Manchester's Victoria Station, for example, tried arc light for a year, abandoning it 1882. "On several occasions," lamented one writer, "the station has been left in darkness when trains have arrived, and the unsteadiness of the light has been very great."[21] In the same year, experiments with street and dock lighting in Liverpool proved "unsatisfactory." The corporation discontinued its operation, and twenty years later the city was overwhelmingly, and successfully, gaslit.[22] Some towns and districts abandoned gaslight but turned instead to oil. In Wimbledon, a dispute between local authorities and gas suppliers led to gaslight's abandonment in 1882. Three years later, it was reported that the new paraffin lamps provided better and cheaper light and that the public was entirely satisfied.[23] Flat wicks and Mitrailleuse burners made the lights apparently less prone to meteorologic influence than electric light, and the reservoirs were designed to cast no shadow.[24] In *How London Lives* (1897), W. J. Gordon considered petroleum the second most important metropolitan illuminant after gas.[25]

Electric light installations became rather more widespread and reliable in the 1890s, not least because of the more favorable economic climate created by the 1888 Electric Lighting Act.[26] Nonetheless, at a meeting of the North British Gas Managers at Perth in 1890, oil was described as a "more active competitor" to gas than electricity.[27] In 1894, the *Engineer* commented that, although the electric light was now "a competing force," this did "not prove that it [would] conquer the whole domain."[28] The permeation of British towns with electric lights remained uneven and idiosyncratic. In 1899, the largest number of electric lights per head (1.899) was in Lynton/Lynmouth; by way of comparison, Manchester had 0.313 and Dublin only 0.087.[29] All the main public streets in Whitehaven were lit electrically by 1902, but this was very much an exception. The distribution of electric light followed no simple pattern, determined as it was by innumerable local technological, economic, political, and topological factors.

A second basic point is that *electric light* was shorthand for several very different systems that transformed electric energy into illumination. Discounting early vapor lamps, there were two main forms of electric light: arc light and incandescent light.[30] The principle of arc light involved passing a battery-generated electric current through two carbons that, when slightly separated, produced a brilliant spark: this was demonstrated in numerous experiments, including those of Humphry Davy, in the first decade of the nineteenth century.[31] Public displays occurred

Figure 5.3
Brush-Vienna electric
arc lamp, with
regulating mechanism
(*top*) and carbons
(*bottom*). From
Maycock, *Electric
Wiring* (1899).

spasmodically thereafter, most notably those of William Staite, who illuminated numerous buildings and streets by arc light between 1847 and
1853.[32] By the 1880s, there were various forms of arc light, all of which
were distinguished by their sheer brightness (figure 5.3 and figure 5.4).
Such raw effulgence was, however, often a hindrance. As Paget Higgs,
the enthusiastic author of 1879's *The Electric Light in Its Practical Applications*, admitted: "Arclight experiments have only succeeded in blinding
the bypassers, and projecting long shadows behind them."[33] Two years
later, the *British Architect* expressed what was becoming a consensus on
arc light: "The light resembles an intense glare, rather than anything
else. It has not even the softness of moonlight to recommend it. Then
again it seems to lack concentration. Evidently there is need of some
alterations [*sic*] before anything like a general adoption of electricity for
public lighting."[34] This was an unwelcome, astringent, and distracting
torrent that was exacerbated by impure or bent carbons, which might
make the light flicker annoyingly or simply burn out. Like candles, arc
lamps disintegrated and needed replacing. Burning splinters fell away as

Figure 5.4 Farmer-Wallace arc lights at Liverpool Street Station, London. The light is very bright, but not particularly well diffused. From Walter Besant, *London in the Nineteenth Century* (1909).

the carbons fragmented. Board of Trade regulations from 1888 insisted that arc lamps "shall be so guarded as to prevent pieces of ignited carbon or broken glass falling from them, and shall not be used in situations where there is any danger of the presence of explosive dust or gas."[35]

The electric, or Jablochkoff, candle was a variation on the arc light principle, "composed of two carbons placed side by side with a slip of insulating substance between them, which burns away with the carbon exactly in the same way as the wax of a wax candle is consumed with the wick."[36] Originally, the insulating substance was composed of plaster of paris or kaolin, but, by 1882, the candle's makers had perfected a combination of baryta (an alkali earth that could withstand high temperatures) and lime. The incandescing insulating substance was itself a major source of illumination. Other electric candles, those of Wyld and Jamin, dispensed with this alkaline filling altogether and deployed electromagnets to separate the carbons, which burned down at an equal rate, avoiding some of the problems bedeviling arc light. The simplicity and brilliance of these candles struck observers forcibly. The photographic expert T. C. Hepworth, viewing them in Paris, noted: "The glitter and the general effect of the spectacle were altogether beyond description."[37] In London, gas shares underwent a "temporary panic," and the candles were installed at several sites, most prominently along the Thames Embankment and Waterloo Bridge, where sixty of them burned from 1878 to 1884.[38]

Thereafter, however, the electric candle's demise was swift. It proved hard to translate basic brilliance into effective and well-distributed illumination, as the electrical engineer James Shoolbred demonstrated in 1879.[39] In the same year, Joseph Bazalgette and T. W. Keates observed that the Embankment was not as effectively illuminated as it might be since "the projection of the light from the Jablochkoff candle is rather upwards."[40] Others found it palpably disagreeable: "There is something irritating in the electric light, and the effect, if it were universally applied, must be . . . to have some disastrous effect on the nerves."[41] Meanwhile, Jablochkoff candles on the Embankment often self-extinguished during the night. In 1883, the Jablochkoff Electric Light and Power Company lost the contract to light the Strand Vestry to the Swan-Edison Company, and by October of that year it was bankrupt.[42] The lights were removed from the Embankment, and, by early 1885, London's premier promenade was, once more, illuminated by gas.

Incandescent electric illumination needs less description since it has become ubiquitous in the West over the twentieth century. Demonstrations of incandescent light, produced by electrically heating a highly resistant wire or filament, appeared occasionally during the nineteenth century. Some early incandescent lights were combustible, like candles and arc lights.[43] By the 1890s, the bulb's form was stabilizing, with a carbon filament inside a glass globe from which air had been mostly evacuated. Bulbs came in various strengths: eight, sixteen, and, occasionally, five candlepower.[44] Numerous inventors contributed to this development, from Swan and Lane Fox to, of course, Edison, who in 1880 described his incandescent bulb as follows:

The [incandescent] light is designed to serve precisely the same purposes in domestic use as gaslight. It requires no shade, no screen of ground glass to modify its intensity, but can be gazed at without dazzling the eyes. The amount of light is equal to that given by the gas-jets in common use; but the light is steadier, and consequently less trying to the eyes. It is also a purer light than gas, being white, while gaslight is yellow. Further, the electric light does not vitiate the surrounding atmosphere by consuming its oxygen, as gaslights do, and discharge into it the products of combustion. The heat emitted by the lamp is found to be only one-fifteenth of that emitted by a gaslight of equal illuminating power: the glass bulb remains cool enough to be handled. Of course, there are no poisonous or inflammable gases to escape, and the danger of fire is reduced to *nil*, with a consequent reduction in the rate of insurance.[45]

In Britain, many were initially suspicious of such panegyric, not least because demonstrations of the incandescent light seldom matched the rhetoric.[46] Although Edison's main contribution was to conceive of electric light as an integrated system, many of the necessary elements (dynamos, transformers, wiring arrangements) were still in an embryonic phase.[47] Hepworth thought that incandescent electric lights "represent a system which, I think, will never lead to any practical result."[48] The light needed shading and in 1900 still gradually coated the inside of its bulb with a sooty deposit: "Constant work under such glare has been the cause of a great deal of eye-disturbance, and almost daily I am meeting with cases directly traceable to this as a cause."[49] Electric light of all kinds would remain more expensive than gas until the development of the National Grid (1926–38).

Historians, then, should be suspicious of Edison's bombast. Edison did, however, effectively delineate all areas where gas and electricity would be compared: brightness, steadiness, color, smoke, heat, safety, and cost. He gestured toward the spaces where light was being used, toward the home, factory, street, and office, as well as toward the eye itself. I follow Edison here and examine the practical interaction between eye, body, and light as it was unsettled and transformed by the introduction of new illumination technologies.

Three Critical Visual Capacities: Color, Distance, and Detail

Comparisons between gas and electric light were routinely made on the basis of their ability to replicate sunlight. The most reliable way of performing this comparison was to use the spectroscope, an instrument that recorded the particular pattern of lines emitted by a given radiating body.[50] By the later nineteenth century, a precise spectroscopic nomenclature for describing lines and units of wavelength was established and the language of color standardized through the development of the tintometer and the color wheel. The scientist and dyemaster Michel Eugène Chevreul's arrangement of 1,440 colors on a circle divided radially and concentrically facilitated the matching and naming of tints (mixtures of pure colors) and tones (tints with added black or white).[51]

These resources were soon adopted by those seeking to demonstrate the superiority of electric over gaslight. Higgs claimed that the spectroscope provided irrefutable evidence for the advantages of the new illuminant. Gaslight, he argued, produced a predominance of fatter, cooler waves and, hence, a spectrum dominated by red, orange, and yellow. Little light was produced from the hotter end of the spectrum, which was the chemical reason for the light's yellowness. It was simply too cool to generate wavelengths of all frequencies necessary to activate the retina in same fashion as sunlight: "It is impossible to add the indigo and violet, and this is the cause of [gaslight's] inferiority. The electric light is more complex."[52] Other scientists concurred: "As far as mere colour is concerned . . . the electric light approaches nearer to the sun than does the gas-flame."[53] Elsewhere, experimenters used color wheels to measure perception of colors by gas and electricity. Weber's chronoptometer was utilized by Herman Kohn in 1880 to make statistical tests of the eye's performance at distinguishing colors by gaslight, electric light, and sunlight. This somewhat rudimentary device, consisting of colored disks pinned sequentially to a black velvet screen, was used to measure the chromatic perception of "fifty eyes": "Electric light always improves the colour perception when compared with gaslight, on an average the perception of red from two to six times, the perception of green from two to seven times, the perception of blue from one and a-half to two times, the perception of yellow from two to five times." For some experiments, "even the longest room in the physiological laboratory in which I carried on my experiments . . . was not long enough to determine the limits of perception." This entailed opening the laboratory doors and using an adjacent corridor, at which point, "owing to the great distances, the conversations had to be carried on in a loud voice."[54] Such unsophisticated techniques produced vital statistical evidence of electric light's ability to approximate sunlight.

Such results merely confirmed something many workers and readers already knew well. Gaslight was yellowish: "We have difficulty by artificial light in illustrating the exact colours, as the gaslight being yellow, it imparts a yellowness to some colours and takes it out of others."[55] This yellowness could be at least partially negated by deploying blue glass in chimneys, shades, or glasses.[56] In light and air cases, if a defendant who regularly worked with color could show that a plaintiff's building deprived him or her of daylight and forced reliance on gas, then "special loss" might be demonstrated.[57] Arc lights were vital weapons in the war on yellowness and, thus, of immense promise for all industries producing colored materials, especially fabrics. This report from the *Warehouseman and Drapers' Trade Journal* of 1882 was typical: "The general employment of electric light for indoor purposes is much to be desired on several accounts. It is a matter of common observation that colours cannot be properly selected by gas or candlelight. Not merely do blues and greens get mixed up, but almost every tint and shade is altered by the yellow of the lamps and candles, and it is one of the great advantages of electric light that it *enables us to see colours as they really are.*[58] Night could be chromatically normalized. Rather than squinting through bilious gloom, workers could now confidently match colors at night or in dark factories, a development that quickly cemented new visual expectations: "We doubt if we could get along now," reported one mill owner, "if we were to return to the old gas lighting."[59]

This was also true for such goods at their point of consumption. Of the Parisian Magasins du Louvre, Hepworth observed: "Here ladies can buy their silks without any regard as to what a tint may look like in daylight; for daylight is here manufactured for them by the aid of Gramme and Jablochkoff."[60] Combined with glass, electricity tossed spangles over gloves and made dresses glisten and shimmer. Cabinets and window displays, argued Bell, with a dash of hyperbole, "developed mainly by the stimulus of electric lighting."[61] By the 1890s, drapers were the most common users of electric light in "small provincial towns."[62] Choosing clothes by gaslight became a faux pas, as demonstrated by Mr. Pooter, the asinine protagonist of George and Weedon Grossmith's satire of suburbia *Diary of a Nobody*: "Bye-the-bye, I will never choose another cloth pattern at night. I ordered a new suit of dittos for the garden at Edwards', and chose the pattern by gaslight, and they seemed to be a quiet pepper-and-salt mixture with white stripes down. They came home this morning, and, to my horror, I found it was quite a flash-looking suit. There was a lot of green with bright yellow-coloured stripes.... I tried on the coat, and was annoyed to find Carrie giggling."[63]

Pooter's folly was less serious than the duping of customers. Butchers, for example, had long been accused of using gaslight as a tool of deception. The veterinarian John Gamgee accused vendors at London's Newgate Market of selling bad meat by gaslight, which imparted tempting color and sheen to rotting sausages or leathery chops. Meat, he noted, "when brought out under a gas illumination on Saturday night does not show its true colours ... [and] finds purchasers in the poor and hard working population."[64] As well as thwarting such mendacity, electric light was useful in chill rooms, where gas flames radiated unwanted heat. Butchers responded to electric light with concern. Following the introduction of electric light into Smithfield Market in the 1890s, butchers fitted gold-leaf reflectors to beguild and revivify their beef and mutton, which now appeared steely and livid: a case of electric light imitating gas, without the heat.[65]

The flight from yellowness, then, was not universally lauded. Most people were accustomed to seeing yellow. This is how normal night appeared: ochreous, cosy, peppery. The whiteness of electric illumination was often an unpleasant shock, registered chromatically as bluish. For those familiar with yellow light, noted Preece, a whiter light "appeared to be blue": "The Americans did not call it blue at all. When they had been accustomed to them the imaginary blueness rapidly disappeared."[66] Speaking at the Institution of Electrical Engineering in 1892, the photometrist Major-General Festing noted: "The apparent blueness of the light of the arc-lamp is due to the effects of contrasts. The effect of an arc-lamp running in full daylight is that its light appears distinctly yellow. On the other hand, if the shutters of a ball-room be opened in the early morning, the daylight admitted will appear of a very ghastly hue to those who have been dancing all night—although, to one who has had a good night's rest, early morning light does not seem so."[67] As the electrical engineer William Ayrton said: "*White light is what you see most of.* Simply that."[68] In other words, *white light* is only a relative term, used to refer to what people have become accustomed to seeing. Its meaning is flexible, and its referent can change over time.

New perceptual habits had to be slowly learned: instantaneous revolution in color perception is, perhaps, physiologically impossible since such perception is always relative and never absolute. And what was true of the eyes was true of the spaces in which they saw. Paint, wallpaper, carpeting, clothing, and cosmetics had often been designed to be seen by gas, oil, and candle. The introduction of electric light generated numerous chromatic problems. In the theater, it disturbed the colors of scenery, costumes, and cosmetics. "The effect of electric light on stage scenery is very far from satisfactory," admitted the *Builder* in 1882.[69] In her memoirs, the actress Ellen Terry contrasted "the thick softness of gaslight" with its "lovely specks and motes" with the "naked trashiness" of electricity.[70] Actors at Chicago's Academy of Music walked out during the first night that the theater used arc light, grumbling that the new illuminant distorted their makeup.[71] Rather than simply replacing one stabilized set of nocturnal chromatic norms with another, electricity simply made decorative calculations more complex. In the early twentieth century, fashion and engineering literature provided lists of how multiple forms of artificial lighting affected chromatic arrangements. George Audsley's *Colour in Dress* (1912) provided a tripartite chart for "ordinary gas-, lamp-, or candle-light" (which, e.g., darkened blue and made brown warmer), "incandescent gas-light" (which brightened crimson and lightened orange), and "electric light" (which turned blue to violet, reddened brown, and darkened green).[72] Finally, introducing electric light into art galleries, often undertaken for atmospheric reasons, made the rather erroneous assumption that all paintings were produced by and to be viewed by daylight. Rembrandt, for one, painted several canvases by candlelight and intended some of them to be seen thus, while sixteenth-century Italian portraits were often composed by candlelight.[73] The glare of electricity may have left them soot free, but it probably robbed them of their intended viewing conditions.[74]

Color perception was, thus, a vital, and contested, area within which debates over illumination technology were forged. Spectroscopy did not simply prove electricity's superiority. Sugg, for one, used the spectroscope to argue the opposite. One should not aim to replicate sunlight, he argued, since noonday sun was too blue and vivid; we draw curtains to protect our eyes from its glare. Rather, the illuminating engineer should imitate morning daylight diffused through clouds, like one experienced when reading the newspaper at the breakfast table.[75] Gaslight, he argued, was better equipped to produce such light. Gaslight, then, often persisted, not because of a blithe disregard for science among civic corporations, but precisely because the subtleties of spectroscopy facilitated the distinction between sunlight and daylight to be mobilized against arc light.

Less delicate, chromatically nuanced powers of illumination were, however, required to traverse larger tracts of space. These were the concentrated focus on a point or area (the searchlight) and the point on which to focus (the signal or beacon). The searchlight was particularly developed in military contexts. For centuries, the primary form of military illumination had been the light ball, simply projected by cannon. It was inaccurate and easily extinguished by enemy troops. Lights fitted with parachutes were developed in the mid-nineteenth century, but these burned out quickly and also revealed the position of the firer.[76] Warfare, like surgery, often stopped at night: it was simply impossible to know with accuracy where, or who, the enemy was.[77] Arc lights promised to make nocturnal warfare more common. They were used first by the French during the Crimean War (1853–56) and then on a larger scale by both sides during the Franco-Prussian War (1870–71), for signaling and as searchlights.[78] There was no need for detailed perception here: one simply needed to identify flags and uniforms and locate masses of troops. The importance of such recognition was highlighted in this report from the Argentine navy in 1882: "We have used the light for signal purposes, first hoisting our flags and throwing the light upon them, and then turning the light on the other ships and observing their flags. Of course the light would be useless for the purpose were we not able to absolutely distinguish colours by it. We are laying six miles off the shore, and at a house two miles inland it is possible to read small print by our light. At first, the people thought it was a new kind of comet, and predicted another revolution."[79]

Searchlights were also important in merchant shipping, icebreaking, and dredging, but the military connection was most significant. Several technological developments made electric searchlights integral to warfare: parabolic reflectors, portable dynamos, gas motors, and transportable towers (figure 5.5).[80] Tests with searchlights in 1892 showed new portable dynamos functioning flawlessly in "ankle-deep" mud.[81] In 1886, the engineer Julius Maier observed that electric light was "indispensable" for naval operations.[82] Its potent beam could illuminate enemy coasts, help avoid collisions, identify torpedo boats, and, when equipped with shutters, project coherent signals.[83] Norms of camouflage were disturbed: in 1894, Germany repainted its entire submarine fleet bluish-gray, considered the least-detectable color by arc. Searchlights could also be used to deliberately blind and dazzle, waging war at the retinal level: "The blinding beam of light rightly handled by the defenders increases the difficulties of attack quite considerably and has a demoralising effect on the attackers. . . . The retina becomes over-excited, and the attacker is incapable of at once seeing clearly again, even when the beam of light has been turned away."[84]

"All civilised nations," observed Major R. L. Hippisley of the Royal Engineers in 1891, "have come to regard the electric light as a necessity in time of war."[85] With the outbreak of war in 1914, London Electrical Engineers Territorials were distributed round the British coast to operate searchlights.[86] By this date, the National Physical Laboratory was engaged in experiments with camouflage design, searchlights, luminous watch dials, and signaling.[87] Signaling, of course, has a long history, from hilltop beacons and church bells to the extensive optical telegraphy systems built during, and after, the Napoleonic Wars (1804–15).[88] Signals transmit a simple sign across space, rather than illuminating an object from a distance. The heliograph allowed signals to be transmitted by day, with limelight, oil, or electricity allowing messages to be sent by Morse code after dark.[89] This technique, combined with telegraphy, was used during British imperial campaigns in Afghanistan and Egypt. Although electric lamps were developed, the army was still using oil lamps, visible at two to four miles, for signaling during the First World War.[90]

Figure 5.5 Siemens-Schuckert transportable electric searchlight and tower. The tower is elevated, and the searchlight is in position. From Nerz, *Searchlights* (1907).

Signaling was particularly vital to transportation networks: land and sea traffic was increasing in volume and speed, so the circulation of simple information, indicating position or providing instructions (stop, go, left, right), was imperative. On railways, the familiar hinged semaphore posts were first used in 1841. These signals were usually placed at a height of eight to ten feet in order to be visible in fog. In the early twentieth century, most signals were still oil lit, with gas or electricity adopted only near large towns.[91] Red danger signals were first adopted on the Liverpool and Manchester Railway in 1834. From 1852, ships were legally bound to use illuminants at night: green lights to mark the starboard bow, red for the port.[92] These colors were chosen on the basis of research into color blindness (the term dates from 1844).[93] Very few people could fail to distinguish between them.[94] Optical scientists estimated that between 3 and 5 percent of the population suffered from some form of color blindness, and they urged employers of engine drivers, sailors, and coast guards to test potential employees for defective vision. Inquests into maritime collisions revealed that color blindness was clearly to blame for several major accidents.[95] The issue was raised in medical, legal, and engineering contexts: "It is self-evident that, if red lamps and green lamps are the same to an engine driver, he will sooner or later—probably sooner—run past a signal and wreck his train. Railway companies are, therefore, very particular that their drivers shall not be colour blind, and to this end they subject the eyesight of the men to tests."[96] Tests should be practical rather than linguistic, based on the capacity to match rather than to name. Sometimes, one color was used as a background and the test color overlaid, while the Holmgren test, involving the use of juxtaposed skeins of colored wool, was used from the mid-1870s.[97] Such tests also raised the specter of malingerers faking color blindness, so extra tests were necessary in order to detect cheats.[98] As with decoration, solutions to perceptual problems generated unpredictable new problems.

Accidents revealed the vital importance of reliable information transmission. The most important visual technology here was the lighthouse. In midcentury, Britain's coastal lights were failing to prevent the annual loss of around one thousand lives and a substantial number of ships. Speaking at the Royal Institution in 1860, Michael Faraday argued that lighthouse design must progress along with the rest of society: "The use of light to guide the mariner as he approaches land, or passes through intricate channels, has, with the advance of society and its ever increasing interests, caused such a necessity for means more and more perfect to tax the utmost powers both of the philosopher and the practical man in the development of the principles concerned, and their practical application."[99] The number of lighthouses, lightships, and buoys indicating positions of coast or rocks increased vastly by the century's end, and the question of the best form of lighting arose early. Lighthouses were traditionally oil lit, and most continued to be. Trinity House, the body responsible for funding lighthouses, opposed the introduction of gas for numerous reasons, including expense, maintenance, and ease of use as well as potency. Tests to ascertain the most powerful form of light simply provided both camps with facts to support their cases, as Roy MacLeod has shown, while agglomerations of oil lamps were regularly used in similar situations, like the illumination of night work.[100]

Gas lamps were introduced at the lighthouses at Howth Bailey, Dublin, and Granton, Edinburgh, in 1865. Gas buoys lit the entrances to the Thames and the Clyde in the early 1880s.[101] Arc light, meanwhile, was deployed at the South Foreland lighthouse, near Dover, in 1858, and, by 1871, photometric analysis at Souter Point showed that its beam had a maximum intensity of 700,000 candles.[102] But arc lights were never widely utilized for lighthouse illumination. Britain had only about six arc-lit lighthouses in the 1890s, not least because of the expense and difficulty of use in contrast to oil lamps.[103] The proliferation of lighthouses and buoys also necessitated greater distinction between lights and techniques to reflect, refract, channel, break, flash, revolve, or color beams to improve navigation of perilous coastlines. In 1893, André Blondel claimed that his flashing lighthouse beam could reach a prodigious eighty-five miles, a figure that he contrasted with the reach of English designs, which, he claimed, revolved too fast, resulting in "gushes" that dazzled rather than securely indicating.[104] This had led to electrical experiments at Dungeness being discontinued in 1878: "The full glare of the electric light at a low elevation was found to be dazzling and bewildering, so much so that it was impossible to judge accurately of the ship's distance from the shore."[105] The light was visible, but its location remained elusive. The solutions to this included taming and channeling light, particularly by the use of holophotal lenses, which gathered light into a relatively homogeneous stream of directed illumination (figure 5.6), as well as telemetric devices to help judge distances: "It is well known that the distance of any object at sea, and more especially the distance of a light, cannot be estimated by the eye with any approach to accuracy."[106] Fallibilities of arc and eye required further mechanical correction.

Fog and smoke threatened clear perception across the century. When fog descended, cities had to be illuminated irrespective of whether it was day or night, which created a double problem, of penetration and storage. Light must be bright enough to cut through thick smog. In 1889, Marcet found metropolitan gaslights to be "actually invisible in dense black fogs."[107] In January 1870, a Manchester fog forced most of the city's transport to stop.[108] The city's gaslights were incapable of indicating the position and distance of objects with sufficient reliability for public safety to be guaranteed. The city was also running out of gas, which could be neither produced nor stored at the requisite rate. On "dark

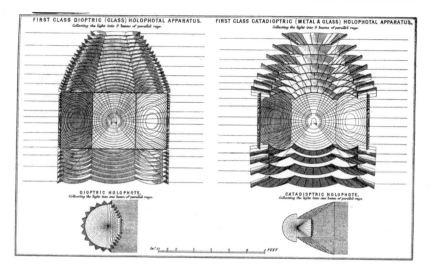

Figure 5.6 Dioptric and catadioptric holophotal apparatus, using glass, and glass and metal, to collect rays into parallel, horizontal beams. From Stevenson, *Lighthouse Illumination* (1871).

and foggy days," the city council heard, "business had to be suspended, and workpeople dismissed, for want of the requisite supply of gas."[109] Four days of thick fog exhausted Manchester's gas reserves in 1904, when the city languished in total darkness.[110] One of the advantages of electric light technology was its ability to cope with such contingencies, whether by using batteries, accumulators, or generators.[111]

Illumination also had a global role in the securing of transportation networks. Lighthouses and buoys were significant technologies of imperial power, funneling trade through reliable channels.[112] Before the development of effective onboard dynamos, traffic through the Suez Canal was forced to halt at night. Lighting ships electrically enabled them "to pass through it at night, instead of coming to anchor, as at present." In 1887, ships traveling through the canal were fitted with electric lights: vessels thus equipped reportedly cut their average time of passage from thirty-eight hours to twenty-two and a half.[113] Powerful, directed street illumination also promised to accelerate the speed of vehicles, particularly early automobiles. Fast but safe nocturnal driving is unthinkable without a secure sense of the distance and position of other entities occupying the road, as Bryant and Hake observed in their 1911 book *Street Lighting*: "With the advent of the automobile and other swiftly moving vehicles, the distinctness with which objects may be seen must be improved, so that the danger may easily be avoided."[114] Ten years later,

Trotter observed that new forms of electric light enabled road traffic to travel "at a pace which a few years ago would have been condemned as reckless and furious."[115] This acceleration of traffic necessitated lighting vehicles at night, to make them mobile beacons: from 1869, an act of Parliament compelled all London cabs to "carry a lighted lamp from sunset to sunrise."[116] The material history of economies of goods and information is often depicted as a process of mobilization, deterritorialization, and speed. Illumination has been an indispensable technical element of this process.[117]

These televisual technologies were essentially macrovisual: they provided basic, large-scale information about location and color, devoid of nuance. A complementary set of microvisual devices aimed to bring small, local areas of the world into clear focus: these were the technologies of detail. Attention to detail, as noted in chapter 1, was a characteristic expected of self-governing individuals.[118] Smiles emphasized: "It is the close observation of little things which is the secret of success in business, in art, and in every pursuit in life."[119] For this ingenuous humanist, close observation was simply cultivated through force of will, but illumination technologies were instrumental in creating the physical conditions of possibility under which attention could be constituted. Under such favorable visual conditions, careless nonobservation could truly become a failure of the self.

This question of detail was routinely addressed by optical scientists and illuminating engineers. Laboratory testing suggested that, by stimulating more retinal cones, whiter light increased visual acuity: electric illumination could, thus, reproduce the physiological experience of daylight. Electricity might, thus, play a role in fulfilling Helmholtz's dictum that "sensory pleasantness" corresponded "to the conditions that are most favourable to perceiving the outer world, that permit the finest discrimination and observation."[120] Hence, scientists spoke of the "distinguishing power" of illumination technologies, equated, for example, with the distance at which reading could comfortably take place.[121] Trotter calculated that street illumination "begins to be useful when it is comparable with moonlight," the value of which he estimated at one-thirty-sixth of a foot-candle.[122] In 1895, Blondel defined the minimum conditions of public lighting as "good sensation of light, so that [observers] are able to read printed matter at the foot of a lamp-post."[123] Here was a double limit: the limit of moonlight, which facilitated distance perception, and that of legibility, which stimulated detail perception.

In major urban streets at least, distance perception alone, the discernment of the general outline and location of other objects, was usually regarded as insufficient. In 1891, for example, the *Electrician* complained that gas lamps in London's Piccadilly failed to provide conditions approaching general legibility. Rather, they merely marked distance and position, serving as "buoys," which was sufficient for coastlines or rural roads but hardly apposite for the center of London: "The illumination *as* illumination, for the purpose of reading a newspaper or finding a fallen sixpence, is of the feeblest kind."[124] To achieve such visual conditions, numerous solutions were gradually emerging. There was a slow and uneven provision of illumination of privileged information relating to location and time. The fixing of street nameplates to public lamps became more common from the 1870s. The St. James and St. John Vestry, Clerkenwell, for example, passed a resolution in 1874 to place street names on lampposts "in embossed ruby letters on ground glass at the back," which would "act as a direction for people in search of streets in neighbourhoods with which they are unacquainted."[125] Graphic inscriptions and information (bus timetables, graffiti, adverts) have wreathed lampposts like ivy ever since.[126] Similarly, the illumination of public clocks was becoming more widespread. When the clock-face of the Houses of Parliament was electrically lit in 1880, the *Electrician* commented: "The contrast with the face looking towards Great George-Street, which is still lit by gas, is very striking. The electric-lit face has a silvery-white appearance, whilst the gas-lit face has a dark dingy red tinge."[127] These little pockets of strategically illuminated detail made the city a murky surface studded with significant information: sufficient for autonomous movement, safety, and punctuality. Early electric lights might also produce generalized legibility in the most important public streets and spaces. When St. Petersburg's Nevskii Prospekt was lit by arc light in 1883, a writer commented: "In every point of Nevskii it was possible to read easily."[128] Conditions of legibility, so vital to autonomous subjectivity, were being fabricated in specific nocturnal spaces.

Aside from facilitating individual motion, urban illumination had long functioned as a technique of social security.[129] According to the most familiar formula, dark streets bred crime and fear as well as signifying the gloomy past that was being transcended by technology. This became a simple, pervasive way in which historical progress was imagined. During George I's reign, argued the *Builder*, householders had been obliged to hang lights outside their homes until 11:00 P.M. only, after which "highwaymen continually rode into the streets... and perpetrated the most open outrages with impunity."[130] The direct connection between policing and lighting persisted well into the nineteenth century. The Metropolitan Police, for example, were instructed in 1832 to ensure that London's parish streetlamps were lit "according to the terms of the contract."[131] Gaslight had itself become regarded as "a powerful auxiliary agent in the prevention and detection of crime" long before this date.[132] When it was suspended, owing to strikes, systemic faults, or the policy of lighting streets only during certain lunar phases, there was often a perceived rise in crime, as in Sunderland in 1875: "The chief constable of the borough informed the magistrates that in consequence of the corporation issuing orders that no lamps were to be lighted during the summer months, the streets and houses were decidedly unsafe, as his limited number of men were unable to prevent burglaries, robberies and assaults, owing to the darkness, and the identification of offenders was impossible."[133]

The use of light to deter criminals operated in many ways. Generalized illumination exposed actions to public vision, legibility, recognition, or shame, extending oligoptic visual and moral norms explored earlier in this book, in certain streets and institutions, into the hours of darkness. Disciplinary illumination, it should be noted, was associated, not merely with the police, but also with an active, attentive public. Lamps positioned outside houses and important buildings created zones of visibility, within which the criminal was made vulnerable to public perception. Fanlights afforded some protection, but they might themselves provide access for burglars. The positioning of small lights outside houses was preferable, and the mode of deterrence took the agency of the criminal into account: "A lamp burning before a house throughout the night constitutes the best safeguard against burglary. The *front* of the house is safe. Burglars, as a rule, do not like the *assistance* afforded by a gaslight. Neither do they like to turn it out, because its absence will probably attract the attention of the police."[134]

Some disagreed: light might enable such men to better assess how to break into houses. Certainly, criminals utilized public illumination, setting in play a new dialectic between inspector and inspected. Caminada recalled capturing a base coiner by the name of "Raggy": "When he arrived opposite the Commercial Hotel, in Hardman Street, I saw him take something from his pocket and examine it under a lamp. Whilst he was thus engaged I walked up and arrested him."[135] Streetlights allowed the discernment of faces, numbers, and signs: they would greatly help the identification of criminals.[136] In 1881, for example, the *Electrician* reported on an accident in which a pedestrian was hit and severely injured by a hansom cab, the driver of which was captured as a result of electric light: "A police constable . . . pursued him for a short distance, and, facing the station, where the electric light brightly illuminates the street, was able to take the number of the offending vehicle. Cabby

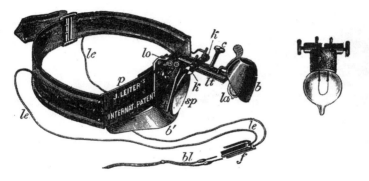

Figure 5.7 Leiter's forehead lamp, producing 2.5–4 candlepower. From Reiniger, Gebbert, and Schall, *Electro-Medical Instruments* (1893).

has consequently been summoned—much to his surprise and much, we hope, to the education of his brethren, whose reckless careers would be speedily checked were electric lighting more general."[137] Night did not, of course, become day, but it was gradually and unevenly permeated by specific networks of inspection. Inspectors were increasingly equipped with torches to enable their work to continue by night or in fog. Omnibus inspectors, for example, were given tiny lamps to check tickets and, hence, reduce the petty defrauding of municipal governments.[138]

Alongside nocturnal inspectability, subsurface realms of machinery, infrastructure, and viscera were also exposed to surgical illumination. Portable lamps or torches were vital to the thankless, routine inspection and maintenance of sewers, gas mains, and subterranean cables. These techniques were, perhaps, most fully developed in the field of medical illumination. Various medical lamps, using paraffin and electricity, were developed over the century: the laryngoscope, lamps for abdominal operations, and the iconic forehead lamp (figure 5.7).[139] Such lamps complemented older techniques (stethoscopy, e.g.) through which the body's secrets might be revealed without ripping it open. The cystoscope, pioneered by the urologist Max Nitze, became the main endoscopic tool, with an electric light positioned at one end of a tube, an eyepiece at the other, and several interposing lenses allowing scrutiny of the bladder, urethra, and larynx. When equipped with pincers, the device could be used for rudimentary operations. The visceral folds and surfaces of bodily organs were revealed to the surgical eye (figure 5.8). The surgeon Edwin Fenwick described the appearance of a normal bladder as follows: "A wave crosses like the motion of a snake beneath a blanket; this is merely the peristalsis of a heavy-laden coil of gut in the recto-vesical

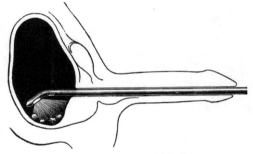

Cystoscope lighting up bladder base.

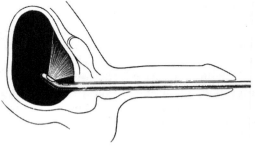

Cystoscope lighting up anterior wall.

Figure 5.8 Cystoscope illuminating bladder. From Fenwick, *Electric-Light Cystoscopy* (1904).

pouch." In chronic cystisis, however, one could observe "polyhedral or rectangular quiltings, the swollen and almost translucent tissue enclosed between the depressed seams or puckers being thrust forward as globose or polypoid bodies, not unlike the vesical myxomata in children."[140]

The proliferation of illumination now promised periodic, but noninvasive, scrutiny of bus tickets, pipes, and inflamed throats: torches and scopes were vital tools of inspectability. Nobody complained about this, apart, perhaps, from fare dodgers or the occasional patient singed by a lamp: it was not insidious or disciplinary in nature. Rather, the technological expansion of inspection was part of the liberal impulse to monitor, know, and maintain critical mechanical and organic systems while attempting to respect their integrity and privacy.

Structures of Illumination: Diffusion and Focus

By the early twentieth century, the illumination of interior space had become a specific kind of engineering: the *Illuminating Engineer* was first published in 1908. But illuminating engineering, or "the art of direct-ing . . . [candle]power for the use and convenience of man," was never simply the routine application of rules and formulas: it was art as well as science, involving pragmatism and subjective judgment.[141] "It is im-possible to define rules governing all conditions of lighting," observed one electrical engineer. "The style and methods to be employed can only be decided by experience."[142] When lighting a home, illuminating en-gineers almost invariably faced a building not designed for electricity or even gas. Rapidly changing building standards meant that they encoun-tered innumerable shapes and heights of room. Finally, of course, the optical economy itself needed to be taken into account. Illumination was increasingly being assembled with the normal eye in mind.

Two basic structural frameworks were used when lighting buildings or spaces: these were *general* and *local* illumination. Local illumination was a technique of concentration, detail, and focus. General illumination was one of diffusion, distance, and volume, modeled on the analogy between cloud strata and ground glass: "When the sky is clouded, the sunlight pierces the clouds as through a ground glass, and the whole sky is like an immense illuminated ceiling, radiating light from every point and in all directions. The objects illuminated diffuse in their turn the light which they receive, so that there is an intercrossing of rays, producing the effect of a *mean amount of light* everywhere. This is *general illumination.*"[143]

Engineers used numerous strategies to replicate this arrangement in-side buildings. A common technique was to interpose a ground glass ceiling between overhead lamps and the space to be lit, producing, ac-cording to Dibdin, "a thoroughly uniform illumination . . . in all parts of the room." The gaslighting of the Houses of Parliament was organized on this principle. This kind of diffusion was also produced by enfolding each lamp (gas or electric) in a ground, opalescent, or ribbed globe, to fracture and homogenize rays, something especially necessary with arc light and high-pressure incandescent gaslighting.[144] Arcs might also be placed in "big funnel-shaped covers" to disperse light across the ceiling.[145] The ratio of individual lamps to ceiling space was calculated (figure 5.9). For ordinary illumination, "where close inspection of materials, etc., is not necessary," one sixteen-candlepower lamp per hundred square feet was recommended (shops and offices might bunch lamps a bit closer).[146] Engineers also recommended the whitewashing of walls, white walls re-flecting more light. In outdoor spaces, floodlighting was perhaps the most common form of general illumination.[147] Electric floodlights were first used at a British sporting event on October 14, 1878, when a football match between two Sheffield teams was watched by a crowd of twenty

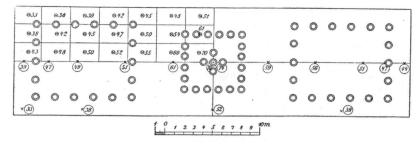

×⓷⓷ Measured values of the horizontal illumination.
℮⓸⓷ Predetermined values of the horizontal illumination.
◎ Positions of the Nernst lamps.
Maximum horizontal illumination, 76 lux (7·05 f.c.).
Minimum ,, ,, 33 ,, (3·1 ,,).
Mean ,, ,, 56 ,, (5·2 ,,).

Figure 5.9 Calculable illumination. The figures—for a restaurant lit with Nernst incandescent lamps—show illumination greatest in the center of the room and gradually declining toward the corners. From Bloch, *Science of Illumination* (1912).

thousand, most of whom came to admire the technology as much as the game itself.[148] The impact of these illuminating strategies was to produce a relatively homogeneous volume of light, midway between the beacon, on the one hand, and the surgical lamp, on the other.

Local illumination, by contrast, aimed at the concentrated lighting of a small space. The inspector's torch and the surgeon's cystoscope are examples of completely mobile local illumination technologies. Fixed forms of local illumination included banks of lamps for lighting billiard tables and oil, gas, and electric lights fitted in railway carriages.[149] The white glass "Christiana" gas burner was described by Sugg as "reflecting the light downwards and diffusing it all around, so as to make it agreeable and soft to the eye, and yet powerful near its work." The fixed and the mobile were combined in various contrivances by which lamps could be swiveled, raised, and twisted, of which the reading lamp is the archetype (figure 5.10): "The object of this moveable burner is to concentrate a strong light on the book or newspaper; thus avoiding the necessity of employing so large a burner as would otherwise be required to produce on the page the same amount of light from a greater height."[150] The universal swivel, allowing a lamp to be moved into almost any position, was permeating offices and factories in the early twentieth century.[151]

Environments could, thus, be structured to engineer the possibility of focus and easy attention. The psychologist Edward Titchener described attention as an internal state identical to "sensible clearness."[152] But attention was clearly not simply an interiorized, psychological condition,

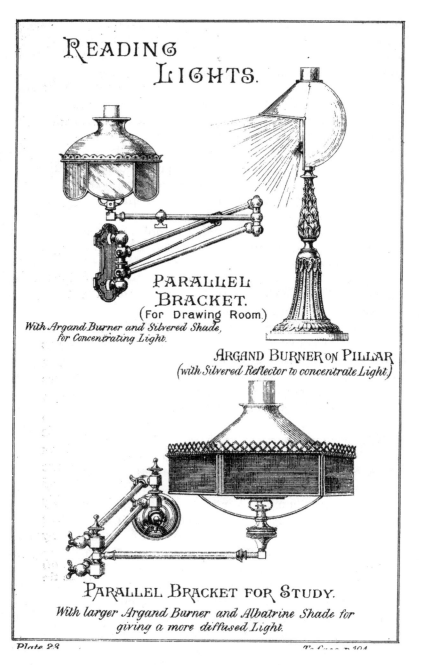

READING LIGHTS.

PARALLEL
BRACKET.
(For Drawing Room.)
With Argand Burner and Silvered Shade,
for Concentrating Light.

ARGAND BURNER ON PILLAR
(with Silvered Reflector to concentrate Light.)

PARALLEL BRACKET FOR STUDY.
With larger Argand Burner and Albatrine Shade for
giving a more diffused Light.

Plate 23

Figure 5.10 Reading lamps with shades and brackets. From Sugg, *Domestic Uses of Coal Gas* (1884).

bounded by the body, but a transient state of optical stability produced by the interaction of psychological, physiological, and environmental systems. Lighting arrangements could, thus, be calculated to create focus and fixity. This meant, not a gratuitous flooding of light, but a careful consideration of surfaces and space to be lit, a sensibility toward glare and respect for shadow: "In a shadowless space we have to depend upon binocular vision to locate points in three dimensions, and the strain upon the attention is severe and quickly felt."[153] Such strain was not felt merely within the eye and its muscles. Another psychologist, Walter Pillsbury, observed that attention was produced by "the attitude of the man's body, the direction of his eyes, and other bodily signs.... As a whole then it seems that the physiological effects of the attention are as widespread as they well could be." Brighter and better-distributed light could avoid the hunched, squinting nocturnal deportment that generated headaches, eyestrain, and backache. Attention was, therefore, something linking eyes and body with their total visual and technological environment: "Nearly every trade and profession has given its members a mind adapted to receiving impressions that would not be received by the great mass of men."[154]

Technologies of illumination, then, shaped the visual experience of numerous forms of labor. They unquestionably made night work a far more pervasive human experience than it had ever been before.[155] In 1832, Matthews noted that gaslight "has given a most important impulse to several branches of our national manufacturers."[156] Later in the century, the *Electrician* argued: "Rather more work is done at night than in the daytime, because of the lower temperature, and because the workmen are not so distracted by passing sights and sounds."[157] Edison dreamed that electric light might permanently destroy night, eradicating the need for sleep, and allowing everyone to work all the time.[158] Of course, this most Protestant of fantasies was absurd, but the alliance between artificial illumination and work was strong. Mining, quarrying, and fishing all benefited from powerful oil, gas, and arc lamps.

Once revered as "peculiarly adapted for the useful occupations," gaslight was by the 1880s the subject of serious critique as a mode of workshop lighting.[159] Burners were regarded as wasteful, while the light itself distorted colors and gave "false ideas of size and shape."[160] Eyes became hot, tired, and droopy. The whiter light produced by electricity promised better color perception, and, for this reason alone, electric light was adopted in numerous factories. According to the *Halifax Courier* in 1880, a worker said that he preferred electric light "because it enabled him to see work better that required his close attention."[161] Within a reasonably short period—a few weeks or months—eyes became so accustomed to viewing by the new light that, if gas was reintroduced, work might become painful or difficult. Benjamin Dobson, a textile machinery manufacturer, described the return to gaslight from electricity following a mechanical breakdown in 1893: "The result was a deputation to the manager on the part of the workmen to know what they were to do, as they could not see how to perform their work by gaslight; and on one or two occasions since then, when the light has failed through one cause or another, the workpeople have declined to work with the gas, stating that they preferred to wait until the electric light was on again, and then they could pick up the time lost."[162] The importance of electric light to the clothing industry was emphasized during 1886 parliamentary debates over revisions to the 1882 Electric Lighting Act. Shoolbred observed that, in "cotton, wool and silk weaving, and corn and dye works, and lace and hosiery," the new light technology had improved efficiency and visual acuity.[163]

Illuminating the workplace usually involved the judicious combination of general and local illumination: "The most economical scheme of illumination is to furnish general illumination in moderate amount, and to re-inforce it, in points where brilliant light is needed, by extra lights at these places."[164] The shadows cast by general illumination, Dobson argued, should be like those produced by sunshine, "so natural that the eye has no difficulty in following detail in any visible part."[165] In the early twentieth century, the precise amount of light necessary for particular spaces and practices was calculated. In 1908, the American illuminating engineer William Barrows argued that "assembly rooms, corridors [and] public spaces" needed 0.5–1.5 foot-candles, "bookkeeping and clerical work" 3–5 foot-candles, and drafting and engraving between 5 and 10 foot-candles.[166] Trotter recommended 4–5 foot-candles for weaving colored goods, 10–15 for surgery, and 15–20 for watchmaking.[167] Like contemporary research into ocular reaction time and fatigue, Trotter's figures attempted to define standardized and normative technological environments within which productive, efficient perceptual practice could occur (figure 5.11).[168] Illuminating engineering and photometry promised to give visual work a more scientific foundation. In 1913, a large survey of factory illumination was undertaken, involving four thousand measurements in fifty-seven factories, resulting in an extensive report published in 1915 and a set of recommendations to be implemented by engineers and monitored by inspectors.[169]

Similar technological developments, albeit on a smaller scale, were evident in spaces for private reading. Childhood reading habits, of course,

Figure 5.11 Gaslit printing machine room. "The arrangement of the lights in industrial buildings should be governed by the character of the work carried on and the needs of individual workers." From Webber, *Town Gas* (1907), 142 (quote).

generated multiple anxieties, often centering on illumination: "Children should not be allowed to read or work by flickering or dull light."[170] At home, localized illumination was usually sufficient, but in public libraries, however, a mixture of general and local illumination similar to that of the workshop was required. Experiments with gas and electricity were often successful in producing an environment conducive to scholarly concentration. The British Museum employed arc lights in its reading room from October 1879, enabling scholars to work until six in the evening. Despite early problems with carbons, including hiss, flicker, and fragmentation, the trustees of the museum agreed to their permanent adoption in February 1880. The following winter, the library and its adjoining rooms opened until seven, while the rest of the museum closed three hours earlier. New arc lights had been fitted, making the lucubrations of students easier: "There is now none of that disagreeable hissing and painful blinking that detracted from the service of the former lamps." Something of the experience of these perceptual conditions can be gleaned from this report: "The silent manner in which the light seems to glide into existence in a moment, and illuminate the prevailing dusk as if by magic, is very beautiful, and convinces even the most sceptical that for libraries, museums and picture galleries, or, indeed, any place where the silence should be undisturbed in any way, the electric light is that to be preferred."[171] In 1890, most of the museum's rooms were open until ten at night: an inching forward of hours available for research and self-improvement. This was not a simple triumph of electricity, however: other libraries were contemporaneously adopting regenerative or incandescent gaslight.[172] By the early twentieth century, library manuals were recommending specific, calculable levels of illuminating power for reading rooms: James Dugald Stewart recommended at least three, and up to six, foot-candles.[173]

Domestically, the process of spatial differentiation within the nineteenth-century home was accompanied, in middle- and upper-class housing at least, by a parallel process of illuminatory specialization, whereby specific rooms adopted light sources, lamp arrangements, and chromatic effects deemed suitable to their function. This differentiation was visible in suggestions for domestic illumination: the hall should be bright enough that visitors could read the house number when approaching the door, staircases should be bright to avoid accidents, dining rooms should have warm radiation, drawing rooms should be brilliant by night, and so on.[174] Sugg described light fittings and arrangements appropriate for different spaces: passages and cellars should have plain brackets and governors designed to limit consumption of gas to three feet per hour, stables need wire guards to ensure that the gaslight does not ignite straw, while upper passages should adopt governors maintaining consumption of four feet per hour and be "sufficiently ornamental" since they are in regular view.[175] Domestic lighting brought much local illumination into play, for aesthetic and physiological purposes. Bell warned against the use of large-scale general illumination in the home as it "deadens shadows [and] blurs contrasts."[176] In more private domestic spaces, precise arrangements of illumination were developed: "Pendants can be made to slide up and down so as to alter their height from the ground, and are nearly always used in bedrooms, and other places where no ornamental effect is required."[177] Sugg outlined the specific arrangement of gaslight for those preparing for balls or parties: "The object of the two lights is to enable ladies, when dressing, to see both sides of the head. But, as a rule, this kind of lighting does not give them a true idea of their appearance in the dining or drawing room, or at public assemblies. The effect of light coming from above the head is necessary to enable a lady to form an idea as to how she will appear in a ball or dining room lighted in this manner."[178]

One could be secure in one's vanity with such accurate simulation, but the effect would be chromatically accurate only if the ball or party was lit by an identical illuminant. Hence, electricity disturbed chromatic and decorative norms themselves undergoing a long, complex, and culturally mediated process of mutation. But electric light was often advanced for domestic purposes because of its flamelessness, which gave it potential versatility, as the electrical engineer Percy Scrutton boasted in 1898: "We can put it into positions where gas would burn down the house in no time, such as close to curtains, among bed hangings, and so on. We can screen it with effective silk and muslin shades, or with any other delicate material. We can put it into cabinets to light up collections of different objects. We can even immerse the electric lamp in water, a means of decoration which has proved exceedingly effective on dinner tables." Electricity promised to open enticing new vistas of tastelessness. "We are accustomed to so much more light in our rooms nowadays to [sic] what we were formerly content with," Scrutton continued, "that to give us satisfaction we should have to use such a number of candles that the trouble and cost would put them out of count for general use."[179] Of course, Scrutton was exaggerating, and candles were long used routinely, and not just by the poor, spiritualists, or crepuscular decadents like Huysmann's Des Esseintes.[180] Dibdin argued that they were excellent for reading since they were portable and provided soft light and reading required no color discernment.[181] Candles were still recommended for bedside reading by Trotter in 1921.[182] We are speaking, again, of a proliferation of light sources, and of their functional differentiation, rather than the simple replacement of one illuminatory technology by another.

Incorporating Illumination: Eyes, Lungs, and Light

Men like Dibdin and Trotter were not simply engineers: they were engineer-physiologists. They did not see their responsibilities terminating where their lamps physically finished, for the radiance they produced penetrated the pupil and lens and stimulated the retina, optic nerve, and brain: it was incorporated. Webber's *Town Gas*, for example, included discussions of the "physiological standard" of illumination for the eye and the "degree of brilliancy" a "normal retina" could visually tolerate.[183] The physiology of vision was not confined to medical and optical textbooks, as this quote from Dibdin, which I repeat from the introduction, shows:

Now-a-days many people spend a considerable proportion of their lives, especially during the winter months, working by artificial light. Every time the direct rays from a light source impinge upon the retina, the iris, or "pupil" of the eye rapidly closes until the intensity of the rays passing through it is reduced to bearable limits. As soon as the direct rays cease to enter the eye the pupil expands in order that sufficient light from a less illuminated object can act upon the optic nerve, otherwise the object viewed would be invisible, or nearly so. The constant action of the pupil, or guardian angel of the eye, as it might be termed, combined with that on the optic nerve, and the crystalline lens, becomes most fatiguing, and in time unquestionably affects the power of vision.

Illuminating engineers sought to harmonize the mechanisms of electricity and gas with the organic systems of pupil, ciliary muscles, and retina, to produce a single, operative system. Two factors, in particular, were emphasized. The first, as Dibdin suggested, was that of steadiness; the second was that of brightness. In the earlier nineteenth century, gas was often represented as producing a less gyrating, fluttery flame than candles: "Persons accustomed to read or work long by ... [candle] lights have their eyes injured; and this injury is not so much from the light itself as from its fluctuation."[184] By the 1880s, the stable point of incandescent electric illumination was now contrasted with the erratic gas flame, which, according to the electrical engineer Robert Hammond, flickered "every tenth part of a second," nearly the speed at which the illusion of constant motion was produced: "The constant fluctuation of a gas-flame keeps the retina of the eye in a constant varying vibration." By contrast: "The uniform volume of light received from an electric lamp makes it more comfortable and less dangerous to the eyes than any other form of illuminant known."[185] Hammond certainly cannot have been referring to arc light, which frequently resisted such stabilization: "An unsteady or flickering light has an ill effect on the sight, and electric lighting on the *arc* system is open to this objection."[186] This irregularity was "irritating to the worker," and elaborate clockwork counterbalances were invented to combat it, further encrusting an already baroque technology.[187]

Similar problems were produced when workers were provided with an inappropriate amount of light. Eyestrain was regarded as particularly inimical to accurate, industrious seeing; below certain calculable thresholds of illumination, the eye reached its limits of dilation and lost focus. It blinked rapidly and became strained and ungovernable. When eyes were fatigued, the self-control so integral to autonomous visual perception began to disintegrate: "It is curious how little control there is over the absolute direction of the eye when the light has almost disappeared."[188] Under such conditions, workers squinted or rubbed their eyes rather than focusing on their work. This optical pressure, noted Clarence Clewell in 1913's *Factory Lighting*, was "largely avoided if the entire working surface is liberally illuminated." This meant a calculated provision of general illumination, shaded and diffused, rather than a crude agglomeration of bright arc lights. "We must adapt the illumination to the requirements of vision," noted Clewell, "and not compel vision to accommodate itself to unsuitable illumination."[189] Capturing perception, mobilizing it productively, involved careful avoidance of excessive brightness. During the early days of electric light, men employed in trimming arc lamp carbons developed symptoms analogous to those of snow blindness.[190] The *Lancet* concurred: "The vibratile impulse of the electric force is obviously stronger than the delicate terminal elements of the optic nerve in the retina can bear without injury. . . . The electric light is too hard; it needs to be softened."[191] Hence the use of shades to protect perception.

Illumination technology did not, however, impinge on the eyes alone. By sensibly affecting the atmosphere within which perception occurred, its impact extended to the entire sensorium as well as other bodily systems. Different forms of gas burner produced characteristic sounds, from the "singing noise" of argand burners to the "squeaking and roaring" emitted by fishtail burners when pressure varied.[192] This noise was magnified when adjusting the light at the stopcock, which sometimes caused flames to flare and whistle "most distractingly," drawing attention to the light itself rather than the spaces and surfaces it illuminated.[193] Moreover, the unpredictable effects of vapor and heat on air currents could produce a heavy atmosphere, with curious sonic effects. "In electricity," wrote Slater, "the architect would probably find a valuable acoustic ally," useful for constructing concert halls.[194] The silence so craved by men like Carlyle, then, might be produced by electric light. However, early arcs proved frustratingly noisy. If electrodes were positioned too closely, diminished in potential, or ran on weak current, they produced "a disagreeable hissing sound."[195] Carbons also burned at unequal rates, so their position needed regulating, often by electromagnets.[196] The arc, it appeared, had little leeway: if it was too short (under a tenth of an inch), it invariably hissed; if it was too tall (over five millimeters), it produced a "flaring arc, which consumes the upper carbon very quickly."[197]

Combustible light forms were notoriously malodorous: tallow and paraffin reeked, while the products of gas combustion, like the stench from drains or middens, aroused increasing concern as thresholds of aromatic tolerance waned. Dobson moaned about the quality of atmosphere in his factory when it was gaslit: "In spite of the best ventilation by Blackman propellers; even at six o'clock in the morning, on entering the room, the smell was most objectionable."[198] The effluvia, he claimed, led to diarrhea (electric light, moreover, allegedly stiffened the stools of his employees). Electric light, in turn, was accused of olfactory irritation: the Gas Light and Coke Company referred to its "offensive smell" in 1878. This indicates the extent to which the entire sensorium was brought into phenomenological and discursive play when artificial illuminants were being compared.[199]

Heat was, perhaps, the most significant by-product of gas flames. Early accounts of electric light often emphasized its coolness rather than its brightness or whiteness. This report, from the 1885 meeting of the Royal Institute of British Architects, was typical: "The large gas-burner, which usually makes the room almost unbearable, was removed, and a ring of thirty-six Swan incandescent lamps was arranged round the base of the dome. . . . The effect was very pleasing, and the difference in temperature from that which ordinarily prevails was most marked."[200] Gaslight's heat "dries the eyes, the lids, the forehead, and temples," an effect that might be precluded by saucers of water or tempered by mopping the brow with a damp cloth.[201] Opticians sometimes recommended tinted spectacles to obviate "the scorching heat of the gas light."[202] Parched corneas triggered "pain and headache."[203] Electric light preserved the liquid economy of the eyeball: eyes were freed to weep. Even arc lamp carbons, which released some heat, did not "sensibly affect the surrounding air," producing cooler spaces more conducive to the chilled and sober emotions.[204] Hot, clammy air, long associated with factory work, was sybaritic or simply soporific: heat from gas produced "lassitude among the workpeople."[205] Again, electric illumination could function as a liberal technology, able to secure the kinds of visual practices so integral to liberal subjectivity without simultaneously undermining physiological well-being.

Concerns about the effects of gas on the atmosphere of the workplace were as old as gaslighting itself. At the same 1809 parliamentary committee where Accum struggled to describe the size of gas flames, Lee, of Phillips and Lee's gaslit Manchester factory, stated that he had "not seen the least alteration in the health of the workers" since the adoption of gaslight.[206] It reportedly produced less atmospheric organic matter than oil.[207] But, by the 1880s, the chemical constitution of gas, and its products of consumption, was under far closer examination, as were its effects on large groups of people. Schoolchildren, in particular, were threatened by its fumes: "At this time of year towards 3.30 we must light up for an hour; the air is already vitiated, and brains begin to be fagged as a consequence. All at once seventy gas jets at least are alight, and each at a low computation consumes as much oxygen as six persons;

the equivalent of *420 more* people is crowded in whose breath—well, we will not say what they breathe out, chemists will tell us that."[208]

Atmospheric norms had been established by chemists earlier in the century.[209] Carbonic acid, which was "very hurtful to animal life... a narcotic poison," was present in air at levels of 0.04 percent: if its level exceeded 0.1 percent, then air was understood to be "impure."[210] The chemist Charles Meymott Tidy estimated that a batswing burner disgorged forty-three hundred cubic inches of carbonic acid per hour. Another carbon compound, carbonic oxide, was produced by imperfect combustion. Tests in 1879 showed that "inferior coal-gas commonly contains as much as twenty per cent of carbonic oxide. The result of the inhalation of this gas is to render the blood corpuscles useless."[211] This point raised the question of the initial purity of gas: legal thresholds had been established by earlier gas industry regulation, but this did not prevent sickly vapors like sulphureted hydrogen (hydrogen sulphide) escaping from gas pipes and burners. Urban journalists discerned the effects on the faces of the desensitized poor: "The brown, earth-like complexion of some, and their sunken eyes, with the dark areolae around them, tell you that the sulphuretted hydrogen of the atmosphere in which they live has been absorbed into the blood."[212]

More tangible evidence of this chemical and sanitary impurity could be demonstrated by its effects on physical environments. In 1859, a commission reported on the impact of gaslight on the delicate pigments of the nation's paintings: sulphureted hydrogen, it was found, blackened paints with a white lead base, while acids slowly corroded surfaces and frames. Unless perfect ventilation could be achieved, gaslight was inadmissible in such spaces of silent contemplation. In 1887, the impact of gas on books at the Birmingham Free Library was assessed: "Leather exposed to the foul air in which gas had been burning for 1,077 hours was seriously deteriorated, for the extent to which it would stretch was reduced from ten per cent to five per cent, while the strain it would bear was reduced in the ratio of 35 to 17, or about two to one."[213] Electric light in libraries, it was argued, promised to save both eyes and books.

Nonetheless, we should be cautious here. These reports might reflect shifting thresholds of tolerance, but they also often formed a key strategy of the embryonic electricity industry seeking to market a product as sanitary: the very fact that this was a, if not *the*, logical connection to forge is indicative of the wider environmental context of illumination technology. New forms of gaslight (regenerative, incandescent) were marketed identically. By the 1880s, many forms of gaslight existed that simply released the products of combustion via a chimney (figure 5.12). These

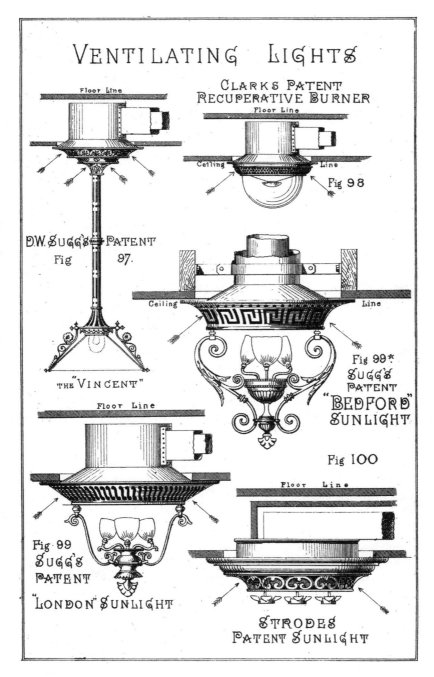

Figure 5.12 Ventilating gaslights. The products of combustion, and vitiated air, are removed by the lamp. From Sugg, *Domestic Uses of Coal Gas* (1884).

were explicitly sanitary technologies and appeared as such in sanitary manuals.[214] Gaslights were also enlisted to stimulate and sustain air currents within buildings: they were part of a broader atmospheric economy of which light was a single dimension. In theaters, for example, sun burners or chandeliers were regularly used as much for ventilation as illumination.[215] Simply replacing gaslight with electric light could lead to rising temperatures. Following the adoption of electric light, the Savoy Theatre was reported to "frequently [stand] at the tropical figure of 84 degrees" because existing air currents had been unsettled.[216] Parkes concluded that "the growing use of the electric light will necessarily modify the arrangements for ventilation," rather than simply improving them.[217] Additionally, coolness was not necessarily a universal virtue. The warmth generated by gaslight was often appreciated by workers, tailors, and printers in unheated mills and workshops.[218] Electric light, particularly that produced by arcs, was also shown to produce noxious chemicals (notably nitrous oxide), and there were claims that it committed chromatic heresy, by visibly yellowing paper.[219]

Illumination produced a delicate, unpredictable set of interacting bodily and environmental effects: there were sufficient variables for any new light form to be promoted as potentially perceptually, physiologically, or physically beneficial. But the marketers of electric light did have one substantial advantage over their rivals. The human body itself, or at least some of its salient subsystems, was increasingly viewed as electric in nature. Experiments on animals and criminals by Luigi Galvani in the later eighteenth century established the contested existence of animal electricity, and, by the 1840s, there were flourishing experiments into the manifold electric qualities of the human body. Medical electricity came into fashion: clinics or sections of hospitals were devoted to electric bathing and massage or the extraction of sparks from bodies with a conductor.[220] Electricity was being used to treat, among other maladies, aneurysm, neuralgia, hangover, writer's cramp, facial palsy, hemiplegia, cataract, gout, locomotor ataxy, and tooth decay.[221] Electricity could retard the formation of cream, positively stimulate yeast cultures, and incubate eggs. From at least 1845, currents transmitted through soils reputedly enhanced the growth of crops.[222] Electricity's white light could be adopted to grow fruit, vegetables, and flowers: tests in 1880 showed: "Electric light was clearly sufficiently powerful to form chlorophyll and its derivatives. . . . Fruit, excelling both in sweetness and aroma, and flowers of great brightness, may be grown without solar aid."[223] Lettuce could be grown in two-thirds the usual time if exposed to electric light.[224] Other experiments produced conflicting results, including a batch of

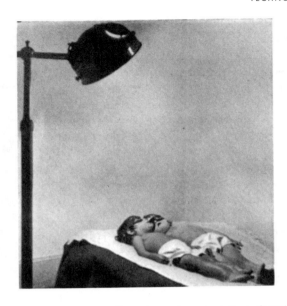

Figure 5.13 Mercury vapor lamp used to treat rachitic children. From Gamgee, *Artificial Light Treatment* (1927).

unpalatable Scandinavian celery, but that electric light was a more salubrious, vital form of radiance was not contradicted.[225] What stimulated vegetable bodies, it followed, would also stimulate human bodies. Artificial sunlamps were developed from the 1890s, and, in the 1920s, arc and vapor lamps were being used in several artificial light treatment centers, for example, in Hull (figure 5.13).[226]

Electricity could, thus, function as a metaphor for, or model of, the body, or the relation between the two might be seen as one of pure identity. This was the case for electric theories of light and vision. Faraday and James Clerk Maxwell concluded that light was electric in essence.[227] "Light," declared Heinrich Hertz in 1889, "is in its essence an electrical phenomenon," and, by implication, "the domain of electricity extends over the whole of nature."[228] If the radiation hitting the eye was intrinsically electric, then the physiological processes developing therein could be seen similarly. The physicist Oliver Lodge argued that retinal rods and cones were tiny circuits transmitting current to the brain when stimulated by electromagnetic radiation.[229] The brain was a storage battery and the eye an electric converter plugged into the boundless electric sky.[230]

This all proved useful for advocates of electric light, like Preece, who, writing on the "sanitary aspects" of electric lighting, asserted confidently, if somewhat oxymoronically, that electricity was "naturally the proper source of artificial illumination," precisely because of these various analogies and homologies. This could be expressed optically, medically, or, more generally, environmentally or sanitarily. Electric light was good for eye and body because these entities were themselves, however enigmatically or vaguely, electric. Preece, for example, emphasized the impact of electric light on the whole body: "That the electric light is a powerful element of health is evidenced by the fact that those who use it not only feel all the better for its introduction, but their appetite increases, and their sleep improves, and the visits of the doctor are reduced in frequency."[231]

Preece strategically ignored several serious public health and safety issues, to which I turn in the next chapter. He also avoided reference to the possibilities of extracting extra labor from workers toiling in electrically lit offices and factories. Electric light could allow employers to bypass overtime laws, which limited the space within which workers could operate at night: the 1901 Factories and Workshops Act stated that, during overtime, four hundred cubic feet per worker constituted overcrowding, a volume that could be increased by the secretary of state if lit by anything other than electric light.[232] Addressing the Association of Municipal and Sanitary Engineers and Surveyors in 1891, Preece broadened his discussion from the medical to the sanitary. Municipal engineers, he urged, must electrify their cities. Electric light was, here, the natural counterpart of the sewer and the drain: a technology of salubrity and cleanliness. Its advocates and engineers were the children of Chadwick rather than Edison: "The chief duties of municipal engineers have been to improve sanitary matters, to remove vegetables and decaying matter from water, and by a proper system of drainage to remove all dangers from health [sic]. But the duties of a municipal engineer will not be complete until he takes in hand the electric light.... Gas burnt in a bedchamber is as bad as bad water and bad smells."[233]

––––––––

This chapter has explored numerous aspects of illumination technology in the period 1870–1910: the proliferation of technological forms, the visual capacities they aimed to provide, the art and science of illuminating engineering, and, finally, the medical and sanitary implications. The pattern of development of illumination technology was incredibly complex, and I have tried to preserve this complexity while highlighting a number of substantial trends. The first point to be made is the simple increase in both number and type of light sources. A second point relates to the nature of perceptual shifts. Along the axes of color, distance, and detail, there were many, slow transformations of perceptual practice that did not produce "better" vision: rather, the definition of *better* was revised, and then challenged, and multiple unpredictable consequences (the problem of color blindness, the revision of decorative norms) set in motion. A third point is that these perceptual shifts were spatially localized, limited to city centers, coastlines, battlefields, operating theaters, drawing rooms, and factories. There were never indiscriminate floods of light, and deliberate floodlighting was hardly representative of contemporary lighting. The overall pattern is much more variegated, nuanced, and eclectic. A fourth point is that all such innovation in illumination had to work through, rather than against, the eyes: raw force and sheer brightness were never particularly useful attributes, even for lighthouse illumination. A fifth point is that, like all municipal engineering, illuminating engineering was broadly sanitary at root, in its consideration of the total environmental experience of a given society or city: air, heat, and smoke were as important as light itself.

A final point relates to the pattern of expansion. Innovations in gas technology tended to permeate urban space more rapidly because the infrastructure itself was already in place. Purer gas could flow through the same network of mains, while an incandescent mantle could simply be fitted where an old burner had been. At the same time, older forms of light, especially gaslight, were finally reaching working-class zones in large numbers, especially after the development of the prepayment meter. Electric light, in contrast, remained limited in the 1880s to isolated, local plants, which sprung up haphazardly like mushrooms: one in the garden of an enthusiast, another in Godalming, Surrey. Electricity was not yet systemic, still less a uniform, stratified, and dominant system.

SIX

Securing Perception: Assembling Electricity Networks

Nothing that has yet been done in the way of lighting streets or isolated establishments by means of machinery on or near the premises can claim to amount to a distribution of electricity, which to be practical must be effected from distant stations, and with the same facility as gas or water. CHARLES GANTON, "THE DISTRIBUTION OF ELECTRICAL ENERGY BY SECONDARY GENERATORS" (1885)

When Charles Ganton, the president of the Society of Engineers, spoke on the subject on April 13, 1885, there was no electricity infrastructure in Britain.[1] London's first electricity station, built at Holborn Viaduct in 1882, had already been abandoned. In 1888, London had one institution worthy of the name, the Grosvenor Gallery station, which lit around thirty-four thousand electric lamps via overhead cables. Such early installations were the antithesis of durable infrastructure; they were flimsy and impermanent, prone to failure, with exposed cables and noisy, unreliable generators. These installations hardly fitted well with the environments they were supposed to sustain. Electricity networks were supposed to be built in general accordance with the kinds of urban perceptual norms discussed earlier in this book. They were, ideally, quiet, safe, clean, and smokeless. They should not interfere with the bodily freedoms or sensibility of the subject. Mains and wires should, ideally, be invisible to the public yet accessible for inspectors. Finally, the behavior of the networks had to be predictable, reliable, and durable. As the *Electrical Review* indicated in 1893, assembling such a network was necessarily complicated, laborious, and time-consuming: "The system of mains must be designed for permanency. If a wrong system is adopted and carried out, the mistake becomes a very serious one indeed; it is no trifling matter to remodel a network of electric light supply mains, and it is next to impossible to patch up a defective scheme."[2]

This chapter is not an exhaustive account of early British electricity networks. I largely ignore the system's developing legal framework, while several physical elements are mentioned only briefly, notably the dynamo, motor, and transformer, all of which were absolutely integral to the development of electricity infrastructure.[3] Instead, it examines this process of assembling networks that were simultaneously durable, inconspicuous, and inspectable. I look at mains, streetlights, and domestic wiring and the various anxieties that surrounded the issue of electricity's impalpability and deadliness, before concluding with a brief history of the electrification of the City of London.

Making Networks: Mains, Conduits, and Manholes

The revised Electric Lighting Act of 1888, together with a gradual economic upturn, provided an improved climate for municipal undertakings after the rather unpromising days of the mid-1880s. By 1900, most British towns of reasonable size had built electricity stations, which were usually under municipal control. Early plants were usually steam powered: steam engines drove dynamos, which produced electricity, which entered the mains system via switchboards, which monitored and adjusted supply.[4] These systems, however, developed without national standardization, the result being a multiplicity of local networks, sitting alongside older, larger, more established gas systems.

When Edison lit downtown Manhattan from his Pearl Street plant, he used direct current, which was ideal for densely populated urban areas. But direct current required relatively thick wires, which raised costs for larger areas of supply to prohibitive levels. John Slater estimated that the maximum area of direct-current supply from one station was one square mile. Using alternating current, by contrast, allowed finer wires to be used by lowering the current and raising the voltage with transformers, a technology by no means stabilized by 1890. This promised to transmit electricity across greater distances, with palpable environmental advantages: "As regards city life, the generation of power at large central stations will enable a more efficient and rigid system of smoke abatement to be enforced."[5] Early British systems using alternating current—for example, Cardiff (1891) and Scarborough (1893)—experienced many technological problems, especially surrounding the issue of synchronizing alternators in parallel.[6] Transformer stations, where dangerous, high-pressure electricity was "stepped down" and made safe for domestic consumption, became integral elements of the system. Such stations were "usually grouped together in suitable positions in a district, oftentimes in chambers built under the pavements and roads, and the low pressure currents coming from them are led into the mains which supply the houses."[7]

Mains were the most vital, tenuous, and troublesome part of these developing systems: "The devising of a perfect system of mains is the most pressing problem in the electric light and power industries; so long as the isolated electric plants were in vogue, the interior wiring was all that one had to consider; but now we must take into account the wiring between the central station and the consumer."[8] They had a threefold structure: conductor, insulation, and protective covering (often made of steel tape and yarn).[9] Conductors themselves should be no less than 98 percent copper, which, after silver, was long established as the best conductor of electricity. Gauges, and their corresponding resistances, ranged from 12.7 to 0.0254 millimeters (with the corresponding classifications running from 0,000,000 to 50): this was clearly a predictive and scientific practice.[10] Insulation came in equally diverse forms: vitreous or lithic (e.g., porcelain), oily or organic (wax, gutta-percha), and even gaseous (air itself).[11] At his ambitious, doomed Deptford plant, Sebastian Ferranti adopted a combination of brown paper and shellac.[12] Vulcanized rubber was, perhaps, the commonest insulator, but it was accepted that "no insulated wire has yet been produced which in itself can successfully resist the deteriorating influences present in mortar or plaster."[13] Board of Trade regulations recognized this, by permitting a maximum leakage of one-thousandth of the highest current borne by the wire. Leakage was integral to the system, something to be measured, limited, and governed rather than ignored.

A second problem with the construction of mains was where and how to lay them. The easiest places to do this, along railway cuttings or bridges, were largely absent in crowded city centers. Some public installations ran lines across the street, while early electric cables in the House of Commons were simply laid across the floor. Most often, however, cables were suspended aerially, an arrangement that soon fell victim to shifting sensibilities. Electrical engineers were, ultimately, forced beneath the streets, where they inevitably encountered other networks: mains, sewers, pipes, telegraph wires, and pneumatic dispatch tubes. Haussmann's

HOLBORN VIADUCT SUBWAYS—INTERIOR VIEW.

HOLBORN VIADUCT SUBWAYS—INTERIOR VIEW.

HOLBORN VIADUCT SUBWAYS—INTERIOR VIEW.

Figure 6.1 Holborn viaduct subways. From Newbigging and Fewtrell, eds., *King's Treatise* (1878–82).

Paris had been equipped with wide underground subways for such infrastructure: Hygeia was honeycombed with them.[14] The argument for subways was simple: "Pipes would be inspected, the joints made good, and service connexions opened, without inconvenience either to the workmen themselves, or to the traffic in the streets overhead."[15] Proposals to build subways for London's burgeoning subsurface systems dated from 1817, but they were routinely thwarted by gas and water companies objecting to their expense.[16] Subway building began in the 1860s: those constructed as part of the Holborn Viaduct (1863–69) were lit by glass lenses and gaslights (figure 6.1).[17] Such subways were often criticized for being too small, unlike their capacious Parisian counterparts. In 1897, the vestry surveyor of St. Martin-in-the-Fields, Charles Mason, grumbled that the Charing Cross Road subway was only eleven feet, six inches, by six feet, six inches, and "will not conveniently hold the number of pipes that should be laid therein." There were also insufficient lateral openings for distribution cables or pipes.[18]

Rather than sharing airy subways, electricity mains were laid in conduits: "While the other features may be accessories thereto, the employment of smooth, strong conduits, accessible ducts for distribution of the current, first-class insulation, constant expert attention to details, and thorough organisation in every department, are essential to the production of a successful underground system."[19] The basic aim here was protection against multiple enemies: water, rats, pickaxes, steamrollers. In 1873, when the first steamroller crawled down a London street, in the vestry of St. Mary Abbots in Kensington, it slowly crushed gas pipes as it went.[20] The answer here, usually, was simply to situate conduits at greater depth. With organic enemies, the solutions were generally chemical. Bitumen, for example, repelled condensation and "disagree [d] with a rat's digestive machinery."[21] There were many different conduit systems, using iron tubing, concrete, or wood, or simply laying copper mains on a "bare strip" system straight into the ground (figure 6.2). Technological norms were as yet unestablished, as the *Electrician* observed in 1890: "There may never be quite the same uniformity in the methods of distributing electricity that has come to be the case with water and gas."[22]

Durability was critical. So too was inspectability, as the Board of Trade's 1888 regulations on electric lighting made explicit: "The value of frequently testing and inspecting the apparatus and circuits cannot be too strongly urged as a protection against fire. Records are to be kept of all tests, so that any gradual deterioration of the system may be detected."[23] This entailed making systems inspectable as well as training inspectors and providing them with recording techniques. The electrical inspector, his satchel bursting with rasps, drills, and dusting brushes, did not emerge unlauded: "The appearance of this new functionary will be like the dawn of a new species."[24] Subsurface electricity systems, however, physically militated against easy inspection: "Underground conductors are the things which need the closest observation and scrutiny in all their details, though, unfortunately, they are by no means easy to observe and scrutinise."[25] Aside from the provision of portable inspection lanterns, inspectability was achieved through two techniques: the manhole and fault localization.

Manholes had a double function. They were sometimes used to lay cables, through the "drawing-in system." More generally, they provided permanent points of visual, manual, and instrumental access to the subsurface cable network. Generally, manholes were situated at all junction points, where several cables converged, and then at regular intervals (between 70 and 150 feet) along the street "so that taps may be made from the mains for services in the buildings along which they run."[26]

Figure 6.2 The Compton system of bare-strip copper mains. Air and glass act as insulators. From Scrutton, *Electricity* (1898).

Manholes had to be sturdy and protected against depredations of vagrants and vandals: they also required regular ventilation to prevent accumulation of gases from leaking pipes. This entailed overlaying the apparatus of inspectability with another network of inspection, which might be olfactory or visual. "Careful inspection of manhole boxes is, at present, the only sure preventive of explosions," complained the *Electrician* in 1894. The Notting Hill Electrical Company, for example, paid men to sniff manholes for evidence of volatile gas: "An instrument might easily be constructed to measure the percentage of hydrogen present in the manhole, and this without lifting the cover."[27] Such an instrument was soon invented.[28] Manhole technology was, finally, incorporated into the drive toward a smooth, silent street. If a manhole was built in a wood pavement, or a wood pavement built over a manhole, the cover would be slightly lower than the street surface to allow for wear.[29] Wheels, of course, might clatter if passing from manhole lid to asphalt, so soundless designs were developed by the 1890s.[30]

Manholes allowed surgical inspection at critical points, where cables joined or forked. Increasing lengths of mute wire, however, languished between these strategic access points, which made the location of faults along them progressively more difficult.[31] Faults were most commonly caused by earthing of the circuit following the decomposition of insulation (especially around joints) or water leaking into conduits. They usually emerged extremely slowly as a result of street vibration or the glacial, corrosive action of subterranean gases or liquids. Faults called for pragmatic responses depending on the material configuration of the particular system: "In some circumstances it is necessary to disconnect the network bit by bit until the faulty section is arrived at by a process of elimination, while in others the consumers who rely directly on the bad section of main for their supply will assist by calling attention to the lowness of the voltage at their lamps."[32]

Numerous other techniques were devised, notably the loop method, which involved a circuit of cable running from a testing station, through the faulty section, and back. Multiple loops allowed different portions of the mains to be tested and the exact position of the fault calculated. Other systems involved portable equipment, among the most notable of which was a triangle of coiled wire, sometimes requiring several men to carry it, that generated an induced current audible by telephone (figure 6.3): "On placing the search coil over any part of the cable on the station side of the fault, a sound is heard in the telephone. The tester can thus start from the station and walk along the cable with his ear to the telephone, and note where the sound stops."[33] This was more successful on bare cables than sheathed ones, where the sound only slowly disappeared. Ears and nose, again, were vital to electrical engineering and inspection, as were formulas and tables, listed in pocketbooks.[34]

Such were the manifold methods through which engineers attempted to make electricity networks durable, predictable, and governable. Let us now follow the current itself, first to streetlamps, then into the home itself.

Lighting the Street

"Little or no effort has yet been made to improve the *lighting* up of London's streets by night," moaned the *Builder* in 1874.[35] Occasional rooms

Figure 6.3 Franz Probst and triangular fault localization coil, Vienna. From Raphael, *Localisation of Faults* (1903).

and buildings were well lit, but streets themselves appeared stubbornly gloomy. Trotter found taking photometric readings almost impossible in some streets, color and detail being discernible in them only when the inspector was very close to lampposts, sometimes almost touching them. Some readings fell below one-hundredth of a foot-candle. This was merely "beacon lighting": "In such a street one can see carriages as dark masses of shadow rather than as illuminated objects."[36] Such perception was resolutely scotopic, to adopt today's parlance.[37] But grayish, mottled gloom was not the only enemy of perception here: street illumination might also be excessive or unfocused. One could not dazzle, strain, or confuse the eye: "It is difficult to see beyond a brilliant light, and when the eye is exposed to the intense glare of an open arc it does not recover promptly enough in passing on to get the full value of the relatively feeble light at a distance from the lamp."[38] The eye simply adjusted too slowly for such naked, unfocused intrinsic radiance. A sensitivity for the eye's physiological limitations was combined with a more general mistrust of schemes to illuminate entire cities. The brief American vogue for tower lighting, for example, was never replicated in Britain, although

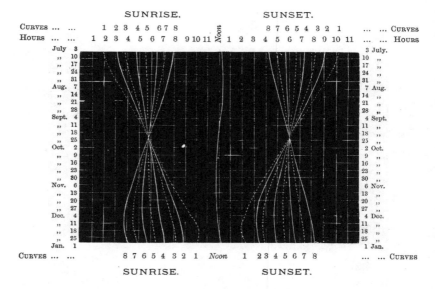

Figure 6.4 Lighting-up chart, showing week-by-week curves from July to January. There are eight curves, from number 1 (sixty-four degrees north: north of St. Petersburg and Stockholm) to number 8 (forty-nine 49 degrees south: Tasmania and New Zealand). London is line 3. From Dibdin, *Public Lighting* (1902).

some lamps were placed at significant heights.[39] Tower lighting did not produce total transparency: it tended to produce unsatisfying, concentric rings of glare and shadow. But it did, perhaps, symbolize a desire for a totally illuminated city. British cities, by contrast, were usually lit so as to attempt to balance the freedom and security of public motion and observation, on the one hand, and the privacy of the subject, on the other.

A central question relating to this was when exactly to light and extinguish streetlights. This might seem a relatively straightforward question since the hours of official daylight were acknowledged to be "one of the most regular laws of nature . . . voluntarily obeyed by hundreds of different individuals" (figure 6.4).[40] Gas engineers might consult a nautical almanac, listing "the number of hours of darkness under various circumstances of time and place, and enabl[ing] him to construct, on an accurate basis, a lighting table suitable to the requirements of the locality."[41] But these local circumstances also included less predictable factors of custom, economy, and level of nocturnal activity, which made hours of streetlighting vary significantly from place to place. In London, the institution of the Metropolitan Board of Works (MBW) did not homogenize lighting-up times since vestries and district boards were individually empowered to light their streets and "so to continue during such time as they shall think fit."[42] The idea of lighting streets all night remained alien to rural districts and many towns, and what little lighting these places had was often extinguished after midnight or during full moons. Moonlit nights, however, could be extremely dark, meaning that "a town is often left in complete darkness for nights together."[43] Practices of streetlighting, in other words, had hardly changed from early modern times, when the "parish lantern" was the primary source of street illumination.[44] Moonlight was used with particular frequency in country districts. Lamps might commonly be lit "one hour before the moon sets and extinguished one hour after the moon rises." Towns followed a set of idiosyncratic practices: at midnight, lamps might be dimmed or alternate ones extinguished.[45]

In the early twentieth century, gas lamps were still hand lit. Traditional lamplighting practice, whereby a stick was inserted into the lantern to turn a cock, routinely damaged lamps and was slowly being replaced by a more delicate arrangement whereby a torch was slipped through a hinged door. This was used in early-twentieth-century Liverpool, following failed experiments with pilot lights and ladders, a system silently enforced by a disciplined system of inspection: "An inspector has been appointed, whose duty it is to visit each of the lamps two or three times a week, under a carefully-organised route table, during the lighting hours, and to report daily upon the condition of the lamps and as to requisite renewals, which are ultimately carried out by an experienced and independent fitter."[46] Public arc lights, too, required daily attention. Carbons needed replacing (or "trimming"), without making lamps inviting targets for drunks or miscreants. Trimmers sometimes used ladders, which were cumbersome and particularly impractical in high winds. Solutions included winches being incorporated into the lamp's base or portable climbing apparatus. Captain C. E. Webber, the engineer who built the City of London's electric light system, designed a system with "alternate side holes in the shaft, which are not conspicuous, and . . . six or eight portable steps of light steel, which can be securely hooked into the holes he ascends, and taken out by the trimmer as he comes down the pole."[47]

Not all lamps were situated on posts: in the early twentieth century, the wall bracket remained common in narrow streets. Some lampposts were ornamental. Special designs might be found on important streets (like the Embankment), outside certain buildings (Cambridge colleges, e.g.), or in particular districts (Fulham's lamps were emblazoned with the borough's coat of arms) or built to commemorate national events (the

Diamond Jubilee of 1897, e.g.).[48] But functionality and durability were far more important requisites. Lampposts are vital urban hardware: they have to resist weather, vibration, and collision. In 1872, the *Builder* referred to the "spectral looks" of lamp standards on Blackfriars Bridge after they had been grazed by traffic.[49] According to the City Commissioners of Sewers in 1877, "hundreds were broken every year."[50] Sturdy materials (iron in particular) and calculated positioning (fifteen to eighteen inches from the gutter) could avoid such waste: paints might be chosen to resist flaking and scratching.[51] Gas mantles were sometimes equipped with antivibrators (conical pieces of metal with rubber or asbestos diaphragms) to protect them from street rattle.[52] These mundane artifacts made illumination more secure and predictable at its point of delivery.

Lamps were themselves globed and hooded to capture illumination and deflect it onto pavements and walls. The usual kind of reflector was a white or blue cover of enameled sheet iron: silvered glass deteriorated when exposed to the arc's heat, while other metals rapidly oxidized. Often known as catoptric lights, these reflectors produced "a continuous ribbon of light."[53] Nonreflecting hoods were also adopted on urban streets to prevent rays escaping into upstairs bedrooms (domestic space should be protected from potentially intrusive public illumination). Specially manufactured glass globes were devised to break, dampen, and direct beams. Opal globes, for example, softened shadow, while ground glass deflected rays onto the ground at the direct foot of the lamp, which often remained penumbral, as the carbon apparatus blocked light rays heading vertically downward. Ribbed holophane globes diffused and softened often-astringent arc light (figure 6.5): "By properly shaping these reflecting and refracting surfaces any desired distribution of light can be obtained."[54]

The final question facing engineers was how to arrange lampposts for optimum perception. This, of course, depended on what exactly *optimum perception* was and to what extent streetlighting could secure it. Blondel suggested that the ideal arrangement consisted of cones of light projected from standards enabling "a passenger, or an object that moves along a line joining two lanterns [to] remain illuminated in the most consistent manner possible."[55] This line was, as Bell indicated, "not in the plane of the pavement, but above it," since pedestrians generally watch other pedestrians, traffic, and signposts.[56] Lighting a main thoroughfare was also completely different from illuminating a suburban drive. Electric light or bright mantles were often regarded as inappropriate for narrow and outlying streets: when Cockermouth was electrically lit in 1881, most streets remained illuminated by oil lamps because their

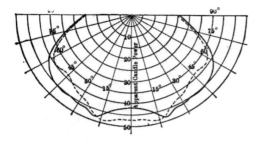

Clear Lamp
Partly Frosted
DIFFUSING PRISMATIC REFLECTOR

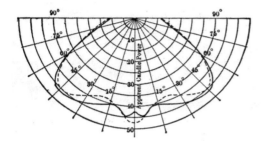

Clear Lamp
Tip Frosted
COATED PRISMATIC REFLECTOR

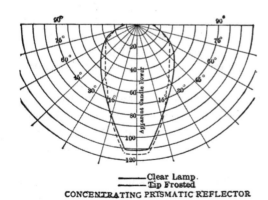

Clear Lamp.
Tip Frosted
CONCENTRATING PRISMATIC REFLECTOR

Figure 6.5 Light distribution diagrams for three different holophane globes, using a forty-watt tungsten lamp. From Barrows, *Electrical Illuminating Engineering* (1908).

Figure 6.6 Testing wagon for street photometry. Bloch, *Science of Illumination* (1912).

narrowness and irregularity made that method technically, and socially, more appropriate.[57] Indeed, when describing how a city should have a hierarchy of light zones, Bryant and Hake argued that suburbs needed calculated dimness in keeping with their paraurban nature.[58] "There are also many cases," concluded Bell, "in which there is no real need of a brilliant illumination, but merely enough light is desired to make the way reasonably clear."[59]

Photometers were wheeled into the streets and put to work (figure 6.6). There were obvious problems, not least surrounding the visual plane it-self, which, as a floating figure of indeterminate size and shape, remained impossible to capture and perfectly calculate. The most detailed measurements were those undertaken by Trotter from the early 1890s. Trotter measured illumination in both horizontal and visual planes as well as accounting for the illumination from multiple lampposts. Urban perception was being subjected to a genuinely sophisticated calculus: "The illumination on the wall of a house facing a lamp, and at a distance measured by the angle of incidence of a ray striking its foot, follows the square of the cosine on that part which is level with the light (that is, inversely as the square of the distance), and falls off to the product of the sine into the square of the cosine at the foot of the wall."[60] For cities, Trotter concluded in 1921, the minimal streetlighting should be between 0.01 and 0.25 foot-candles, to facilitate free and safe motion. He suggested: "Modern lamps should be placed not lower than eleven feet six inches from the ground in order to put them above the range of vision and reduce glare."[61] Too high, of course, and illumination would be lost—and also absorbed by periodic strata of smoke and haze. Forty years before Trotter wrote, Newbigging and Fewtrell complained: "Lamp-columns are generally made ridiculously tall as though they were intended to light the clouds instead of the public ways."[62] Following Parisian experiments, Blondel recommended positioning standards along the street's axial line, at a distance apart of not less than the width of the street. If there were no chains across the street or no central reservation (both seldom seen in British cities), then a zigzag arrangement had to be adopted, which Blondel regarded as considerably inferior.[63] Trotter provided tables of the necessary candlepower to give illumination of 0.1 foot-candle at a point on the ground midway between two lamps.[64] Engineers could then decide, on the grounds of aesthetics or economy, whether to use many short, or fewer tall, lampposts.

This, of course, meant that Trotter's calculus was only a loose and general guide. Street surfaces and shapes, the height of walls, reflected light, foliage, and illumination leaking into the street through windows and doors made every space idiosyncratic and never fully quantifiable. Illuminating engineers had to discover in practice what was feasible and what was not. Cities were to be lit, not identically, but according to their own patterns of streets and buildings, their degree of municipal parsimony, and their own accumulated habits and practices: municipal illuminating engineering was very much rooted in the locality. As with wiring systems and lighting-up times, the physical arrangement of streetlighting varied greatly across the country. Between 1856 and 1874, for example, London's Strand Vestry installed no new streetlamps, while Kensington erected 2,275.[65] In 1902, public lights might be as low as 8 feet from the ground (the gas mantles of Leeds) and as high as 64 feet (Bournemouth's arc lights). Spacing was equally variable: incandescent electric lights in Bristol were 17 yards apart, while Aylesbury's flat-flame gas burners had 130 yards between them. Some areas adopted arrangements that seemed to defy logic: "The most notable feature of Chelsea is the irregularity of the distance between the lamps. Were the lighting of the Borough re-organised and lamp-posts fitted at proper intervals a great improvement in lighting would ensue."[66] The fact that an assortment of different lighting systems was in use meant that, even in relatively well-lit cities like Glasgow, street illumination was typified by variety and unevenness (figure 6.7).

The very idea of building shimmering cubes of light, of making the street into a seductive "interior," was, thus, alien to the practice of municipal illuminating engineering.[67] Indeed, several engineers explicitly argued that perfectly even illumination was undesirable. Webber, for example, found it "less convenient than occasional well-lighted spaces separated by darker portions."[68] Blondel went further, arguing that uniform illumination would mislead pedestrians, who were accustomed to using streetlights as indicators of distance as well as stimuli of detail: "We should then have sources of light which would appear more brilliant at a distance than close to. This fact would completely upset the public, who would no longer be able to appreciate distance or understand the enfeeblement of the globe when looked at closely."[69] This was a consequence of irradiation, an automatic physiological response of the eye that makes bright objects appear larger against a dark background owing to the stimulation of an increased surface of nerve endings.[70] Elsewhere, ocular physiology was used positively to stimulate the perception of light. Dioptric reflectors, which collected light rays into an intense, smooth horizontal beam, had the effect of spreading illumination over a wider area. Such light, argued Preece, appeared brighter than it actually was: "Hence it is that when the eye alone estimates the value of the dioptric lantern it is deceived."[71]

Deception in this case appeared to exist only in respect to Preece's own complicated photometric readings. There was always, again, an elusive, productive gap between subjective perception and machinic measurement, a gap that was not ignored by illuminating engineers but actively exploited by them. Trotter admitted that his contours did not isomorphically replicate actual perception. "The small and gradual changes of illumination," he said, "could not be estimated or even detected by the eye." His machines produced more nuanced readings than his own eyes. He referred here to Weber's law: "The intensity of visual sensation is not directly proportional to the luminous stimulus." Rather, the ratio of intensity varies with the intensity of the stimulus itself.[72] Trotter cited the physiologist Michael Foster and referred to Fechner's work. He engineered with ocular physiology in mind, catering to general norms rather than individual eyes, assuming that total control of subjective perception was impossible, undesirable, and unnecessary. The more vision was understood and calculated, the more absolutely indeterminate it became.[73] "The eye," he concluded, was "the ultimate judge."[74]

Great Western Road.   Low-pressure Incandescent Gas Lighting.

Bath Street.   High-pressure Gas Lighting.

Bath Street.   High-pressure Gas Lighting.

Figure 6.7 Variety and unevenness in streetlights. Gaslighting in Glasgow, early twentieth century. From *Municipal Glasgow* (1914).

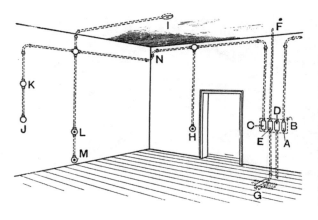

Figure 6.8 Distribution diagram for armored insulating tube system. "A," "C," "D," and "E" are distribution boxes on a subdistribution board ("B"). "H," "J," and "L" are switches, "I" a light, "K" a wall lamp, and "M" a wall socket. From Maycock, *Electric Wiring* (1899).

Intimate Networks: Wiring the Home

In many respects, internal wiring was simply a smaller-scale, more intricate version of city wiring. Copper wires were often arranged in a so-called tree system of mains and feeders, with boxes functioning like tiny manholes through which wire was drawn.[75] There was a similar proliferation of insulation systems, and the system should be tested: one should prepare a detailed plan of wiring in advance, ensure that electric pressure was relatively equalized throughout the building, and compute the overall loss of electric energy through the system. Fitting wires always involved a delicate balance between inspectability, safety, and aesthetics. Wires might be laid on the outside of walls and then painted the same color.[76] Such an arrangement provided accessibility but compromised durability: it left wiring open to "inquisitive, uninformed and malicious persons."[77] More commonly, then, wires were buried in walls, sufficiently deep to avoid being punctured by nails, "protected against injury and their position clearly indicated."[78] Wooden casing might be easily removed to provide access. Frederick Bathurst, who introduced the armored insulating tube system of electrical wiring (figure 6.8) from the United States in 1892–93, stated: "All good wire should be accessible . . . [and] always open to the inspection of owner or wireman." This meant that "the inspector can use his eyes, and have no need to trust to his ears for inspection."[79] This arrangement mirrored that of the street: open to inspection, protected from interference, unnoticed by habitual users. Those responsible for engineering this delicate and invisible system were expected to be conscientious and attentive: "The jointer should be a man with deft, pliable fingers, cleanly habits, a good stock of patience, and a conscience."[80] Fitters should subject wires to mechanical tests before burying them in walls, and, once complete, the system should be tested in its entirety at twice its regular voltage.

There were three other fundamental aspects of the electrification of buildings: safety, control, and measurement. The basic technology of safety in electric circuits is, of course, the fuse. Analogies were often drawn with older systems: "The fuse wire is to an electric lighting system or plant what the safety valve is to a system of steam generation and supply."[81] The fuse was a point of calculated perishability, designed to preserve the durability of the circuit it protected. Fuses (made of metals like lead and tin) were always required where conductors changed in size, most importantly on the threshold between street and home. As with other systemic elements, an array of injunctions surrounded the fitting of the fuse, including, as always, recommendations about accessibility: "[Fuses] should never be placed under floors, inside roofs, or behind wainscoting, or skirting-boards, or in wood cupboards."[82] It was common to fit two sets of fuses here, one inside and one outside the house. These protected the home, physically and psychologically, from the threat of mysterious, unpredictable surges of energy.

The domestic control of electricity was shaped by several new devices: switches, sockets, and plugs.[83] Switches, commonly made of porcelain by the late 1880s, were simply "devices introduced into a circuit to enable the latter to be 'closed' or 'opened' at will."[84] This promised instantaneous control over illumination: "The moment the [electric] light is required, it can be had."[85] Switches could be positioned wherever needed, at great distance from the light if necessary. They made illumination tame and easy. Preece could leave the electric lighting of his daughter's dollhouse "effectively under the child's control" so that she could gain a head start on other girls in mimicking the arts of domestic economy.[86] One could fit switches that activated automatically when someone entered a room (useful for nocturnal lavatory visits, e.g.). They silently embedded the primacy of the visual, by precluding the need for reliance on the tactile: "There need be no groping in the dark for the means of lighting up."[87] The self-control of perception could be furthered by such arrangements, as the following quote suggests:

In ordinary rooms one switch may be placed so that the light may be turned on or off when entering or leaving the room; and the other fixed at the opposite side of the room—say, near the fireplace. In bedrooms, one switch may be placed near the door and the other by the bed, so that the light may be put on when entering, and operated at will while in bed. On staircases, a light on the ground floor may be turned on by a switch on that floor, and turned off at a second switch on the first floor; at the same time a lamp or lamps between the first and second floors may be put on the first floor, and turned off at the second floor. If it is necessary to descend the stairs during the night-time, the advantage of being able to turn the light on in advance will be apparent.[88]

Switches were designed so as not to occupy any intermediate position: light was either on or off, and the Institute of Electrical Engineers was explicit about this.[89] This was, obviously, profoundly "unnatural," in that the human eye had developed to slowly adjust to gradual changes in tone, brightness, and shadow. Some early switches had allowed for the tuning of the eye. In 1883, the electrical engineer James Gordon described a system whereby "a slow screw" was used "so that it takes at least a minute to make a change of light equal to one candle-power."[90] We have traded such delicacy for speed, although the dimmer switch does partly preserve the principle.

Switches allowed one to control one's light: they also potentially enabled the control over the light of others or the formation of techno-logically mediated hierarchies of perceptual control. Illumination tech-nology could, thus, reify social relations. In disciplinary institutions, switches (like gas taps) were never placed in cells, for example, since the prisoner had no right to choose when he or she was able to see: "In pris-ons, asylums, etc. the wires, lamps and appliances must be inaccessible for the inmates, but the switches and cut-outs must be located in a place easy of access for those in charge."[91] This facilitated the uniform impo-sition of "lights out" at the end of the prison day. Subtler asymmetries of power might be reinforced by using wires and switches to control the amount of illumination available to servants. A centralized distribution box with switches controlling the delivery of electricity to the whole building meant that "the master of the house can cut off the supply of current to these rooms, and prevent any waste of current going on during the night."[92] A similar principle was regularly applied in model lodging houses: in London's Peabody dwellings, the gas was extinguished every evening at eleven o'clock.[93]

Some early electric lighting systems promised centralized institutional control over illumination. In 1879, the *Electrician* reported enthusiasti-cally on a San Francisco scheme developed by the Spanish engineers E. J. Molera and J. C. Cebrian. Like many engineers, they were strug-gling with the ongoing problem of dividing the electric current so as to make it power several individual lamps. They argued that it was not the electric current itself but the light that should be manipulated. By

Figure 6.9 Molera and Cebrian's system of piped illumination. A powerful arc
("A") produces light, which is distributed about a mine. From *Mining and
Scientific Press*, July 19, 1879.

ensconcing the "most powerful source of light" available in a prismatic
chamber, the light could be piped "in several beams of parallel rays,
without the smallest stray ray of light being lost." Reflectors could then
be positioned to intercept beams from the central arc light and channel
them into specific positions (figure 6.9). Mains and distributors could
serve whole cities: "A net of properly branched out pipes will put into
communication every room of the building with the service pipe."[94] A
similar scheme was undertaken by the Russian electrical entrepreneur
Chikolev at a gunpowder factory near St. Petersburg a few years earlier.[95]

Four giant arc lights encased in glass could illuminate the whole of San Francisco. They used a single four-thousand-candlepower arc light with a twenty-four-inch Fresnel lens to light one building: the illumination of color in particular was lauded by observers.[96] In 1880, the pair were overseeing further experiments in Barcelona, according to the *Crónica de Catalana*. A private house was lit with pipes and prisms and an electricity company formed to implement their invention.[97]

Thereafter, at least in the Anglophone engineering press, there was silence, although Harlan did refer briefly to a system of light "conveyed by tubes" in 1887.[98] Barcelona and San Francisco both adopted electric light without piping it beneath their streets, and no British engineers attempted to replicate their system. The material problems with the scheme were obvious and crippling; it was enormously expensive and could operate only over limited distances. Yet it also failed because of its centralized, inquisitorial, and almost panoptic nature: "For all large institutions like colleges, libraries, hospitals and asylums, factories, barracks, prisons, and other public establishments, our system, besides totally preventing fires, has the great advantage that the light can never be meddled with by the inmates of the institution, but *it is under the direct control of only one supervising officer*."[99] And again: "the dispersing lenses and reflectors are arranged inside the building so as to illuminate every part, without any obscure corners."[100] This kind of visual architectonics was certainly possible, but it was unsuitable for anywhere but purely carceral institutions. The ability to regulate one's own light and to withdraw into obscurity as and when one chose was a vital liberal right and need. A single supervising officer controlling all the lights in an institution, or even a city, and able to illuminate and benight at will epitomized the kind of centralized observation or state interference quite incompatible with liberty, privacy, and self-government.

Electricity was also controlled and made versatile through the slow miniaturization of motors, which meant that it could be applied to numerous domestic tasks: ironing, heating, cooking.[101] Very early American appliances were simply screwed into light fittings, which led to awkward and dangerous configurations of wire dangling from ceilings (the plug predated the multipurpose socket by around ten years).[102] Ground-level sockets were appearing sporadically in British buildings by the early 1890s, allowing greater mobility of appliances, but also generating novel safety concerns, especially the potential for inquisitive children to electrocute themselves by inserting fingers into them: hence, small shutters might be fitted. The plug and socket system, however, really developed only in the twentieth century. Finally, the structure of lightbulb fittings stabilized relatively quickly. The bayonet and screw caps were soon established as the two main types, although there were other, more evanescent forms.[103]

The final technology of control was, of course, the meter. Building a reliable, durable, legible electricity meter was a pressing problem for the industry, as the physicist Charles Vernon Boys noted in 1883.[104] The technical problems were compounded by the lack of fully standardized nomenclature and uncertainty over the exact object of measurement. As Gooday has argued, one might measure the quantity of electricity (amps multiplied by time), or energy (amps multiplied by volts multiplied by time), or even the amount of illumination itself.[105] Most very early meters (e.g., Edison's) measured quantity by weighing the metallic deposit left as electricity crossed electrodes. This necessitated removing the meter to private laboratories, a cumbersome arrangement that left companies open to the accusation of cheating their customers.[106] By the early 1890s, the energy meter, usually equipped with a dial modeled on the gas meter's, was becoming standard, and it would become common practice to fit the meter between five and six feet high on walls, to further aid clear, comfortable, and habitual reading.[107]

Despite ongoing problems with reliability and precision, such meters facilitated the same mixture of foresight and frugality to which domestic gas consumers had become accustomed. The introduction of two-rate meters encouraged the use of electricity at periods when stations traditionally produced less energy: customers were invited to harmonize their use with these periods and so contribute to overall municipal efficiency. The meter was routinely protected against fraudulent use: circuits were constructed to prevent the siphoning of current, while some meters were made to occasionally run backward, thus thwarting those who applied magnets to retard the machine's motion.[108] Separate meters might also be fitted in servants' rooms, again reinforcing asymmetries of domestic power.

Network Risks: Leaks, Explosions, Electrocution

If the nineteenth century witnessed the slow development of a "networked society," a social system held together by communication, energy, and sanitary networks operative over greater and greater distances, this process was not uncontested or regarded as an unambiguous good: indeed, it inspired much anxiety.[109] Many of these networks were subterranean: by the 1890s, these included pneumatic dispatch tubes, railways, and telephone cables in addition to the more familiar drains, sewers, gas mains, and electric wires, and the earth beneath the streets was becoming crowded. In 1892, the *Electrician* argued that this meshwork was imperceptibly sinking through a soggy morass: London was composed of "busy and comparatively narrow streets, a soil of clay, and rubbish perforated with gas, water, telegraph, drain and other pipes."[110] This sense of terrestrial instability was compounded by uncertainties regarding the precise location of physical networks themselves: "Supply companies use the Ordnance maps as a rule, but the position of the conductors can only be approximately given on them, on account of their small scale and the numerous alterations which have been made since the date of the survey.... We should be glad to know if sketches are regularly made of the bowels of the streets either by the contractors or by the local authorities."[111] The multiplication of systems and the absence of secure knowledge about them combined to make the underground a catalyst for fantasy and fear. The journalist John Hollingshead, in his 1862 *Underground London*, mused: "Imagination generally loves to run wild about underground London, or the sub-ways of any great city."[112]

The experience of gas breakdowns was episodic: periods of smooth functioning were interspersed with minor breakdowns, routine repair, and the occasional major accident. The fully autonomous, self-correcting network remained an engineering chimera. This was particularly manifest in the case of leakage. Gas escaped from reservoirs, pipes, joints, and fittings: it was "apt to escape by leakage, in rooms and in cellars, into water pipes whilst empty, and thence into dwellings, and into sewers, and thence through gratings into the streets."[113] Nuisance cases were brought against the Gas Light and Coke Company for leaks as early as 1814.[114] In 1858, it was estimated that up to one-fifth of all manufactured gas seeped out of the network.[115] Iron pipes were themselves permeated by gas, owing to what the *Journal of Gas Lighting* enigmatically referred to as "osmotic action," but most leaks resulted from improperly extinguished lamps or, most commonly, from the fractures caused by the vibration of traffic or pickax blows.[116] Pipes were unsettled and shattered by the frequent fitting of new road surfaces. Soil visibly blackened around gas mains; if one inserted a metallic cylinder into this ground, gas might escape through it, and, in some areas, "a match applied to the hole will light a . . . flame."[117] This image of a deteriorating network was regularly exploited by the electricity industry: "The day will probably arrive when, before this very Association, a paper will be read on the iniquity of the nineteenth century in permitting the distribution, by constantly leaking pipes, under the public streets, of such a frightfully noxious thing as coal gas."[118]

Such leaks made people nauseous, and they also killed. In one week in 1873, the *Builder* reported that, in separate incidents, seven people had died as a consequence of gas escapes.[119] Houses unconnected to the network were also affected: in 1880, a sleeping Glasgow woman asphyxiated after a main running beneath her cellar cracked and disgorged gas into her home. One did not need to inhabit a machinated house to have one's life irrevocably modified, or even terminated, by the networks securing perception. Today, crumbling and abandoned gasworks and cokeworks are among the worst sources of polluted soil and groundwater.[120]

Strategies of prevention and detection multiplied. Mains might be laid more deeply and be more thoroughly tested (or "proved") before being interred, while fitters might be better trained and consumers educated.[121] Merriman called for "the more general diffusion amongst gas consumers of a knowledge of the principles of combustion, and of the simple precautions to be taken and conditions to be fulfilled in the employment of gas-burners."[122] The lower classes, in particular, were often seen as irresponsible users of illumination: fires caused by oil lamps or candles in the 1890s were still being blamed on their lack of "care and intelligence."[123] Such carelessness implied lack of concern for the visual, in turn reinforcing the kinds of perceptual differences outlined earlier in this book. Gas, of course, smelled strongly. Its aroma was distinctive, and, theoretically, detection became the responsibility of any alert, attentive subject with a well-formed sensorium. Humphry Davy, speaking before a select committee on gaslight in 1823, argued rather casually that gas leaks should be detectable by smell: "unless the delicacy of smell, the sensibility of the olfactory nerves, were blunted by habit; which, however, I suppose is not the case."[124] The implication here was clear: the desensitized were more likely to perish from gas leaks. This did not imply, of course, that detecting gas leaks was the sole responsibility of the individual. Not all leaks were perceptible, and, cumulatively, they could add a significant amount to a family's budget. By the 1880s, a variety of leak detectors were available, some equipped with dials for easy reading, or owners might simply use more rudimentary methods, like applying soap and water to pendants, joints, and seams to see if bubbles appeared.[125]

Public anxiety was most evident over explosions. Early gasworks routinely caught fire and blew up: Clegg recalled an 1813 explosion at the Peter Street Station of the London and Westminster Gas Company that "burnt all the skin of my face, and the hair off my head."[126] This did not prevent him from stating that he would happily sleep on top of a gas holder if asked, a position not shared by others connected to the industry. Wood recalled a time "when the horrified lamplighters of

Westminster Bridge refused to light the lamps."[127] The explosion resulting from searching for a gas leak by candle flame became a routine
affair, wryly reported in the building press.[128] Street explosions were
common, ranging from relatively minor incidents, like that responsible
for wrecking the first London traffic light in 1868, to more disturbing
and seemingly inexplicable ones.[129] In 1873, an iron pillar supporting
four gas lamps in Retford was "completely wrenched from its hold, and
flung down with such force that the surrounding buildings were completely shaken to their foundations."[130] Construction and maintenance
work sometimes triggered explosions. When mains were relaid along
London's Tottenham Court Road in 1880, a workman lit a match to
check for gas, causing an explosion that ruptured one hundred yards of
roadway, catapulting paving stones into houses, and generating six more
explosions at various points along the main. "The aspect of the houses,"
commented one observer, "could only be likened to that which would
follow a bombardment."[131]

The most notorious gas explosion in Victorian Britain was at the Nine
Elms Gasworks in southeast London on October 31, 1865. An explosion
in the meter house punctured the skin of the tank, forcing out a jet of
gas that caught fire. Nine men died, and the charred, skeletal frames of
the gas holders became an emblem for those wishing to banish gasworks
from cities. Thomas Bartlett Simpson, the former owner of Cremorne
Gardens, quickly published a pamphlet condemning not just the company in question but the whole practice of situating gasworks within
cities: "We now find ourselves encircled by about twenty of these dreadful magazines of discomfort, sickness and peril; converting thousands
of tons of coal into coke and gas every day, necessarily accompanied
by poisonous emanations, and an 'unavoidable accident,' at any one of
which may, in the busy hours of day, or in the stillness of night, lay a
neighbourhood in ruins, and bury its inhabitants beneath them." *The
Times* described every gas holder as a potential powder keg, while the
Telegraph opined that the worst aspect of the disaster was its complete un-
predictability. Its effects, argued the strident Simpson, were akin to those
of a "hurricane," and the disaster was "as unexpected as it was awful."[132]

The gas network's final form of risk was the blackout. Blackouts might
be a result of basic systemic failure: gasworks might simply run out of gas
(during times of heavy fog), a subsection of the main system might rupture entirely, or a fundamental part of the gasworks might break or need
repairing. They could also be the result of more nefarious actions, which,
as Schivelbusch has noted, were now targeted at the point of production,
rather than, as in earlier times, at individual lanterns.[133] During Chartist

demonstrations in 1848, the gasworks at Horseferry Road in London was surrounded by a mob: there were fears that the works would be overtaken and the Houses of Parliament plunged into blackness.[134] Numerous Fenian plots to blow up gasworks were uncovered later in the century.[135] Such concerns might lead to important institutions establishing their own energy supply systems. Scotland Yard, for example, fitted its own plant for electric light by 1891, to guard against strike, accident, or attack. Blackouts of gas and electricity followed different rhythms. A gas blackout was felt slowly: there was usually sufficient gas in the system to last for a few days, so there was a gradual descent into darkness as the gas slowly ebbed through the network, followed by an equally slow return to light. Electricity blackouts are felt instantaneously, as "power cuts," and, since electricity's powers so much more than just light and heat, the ensuing social paralysis is much deeper.[136]

The manifold risks of the gas network generated many inquiries. In 1814, the Royal Society carried out a report recommending the distancing of gasworks from populous places and some rudimentary security measures (walls or mounds between gas holders, e.g.). Gasworks were, as noted in chapter 4, rapidly subjected to inspection: an 1816 act facilitated inspection of the Chartered Gasworks of London, and the first metropolitan inspector of gasworks was William Congreve. An explosives expert who included among his many inventions a rocket device for killing whales and unforgeable banknotes, Congreve conducted experiments on the force of explosions and concluded that the potential damage from explosions was significant. He reported to an 1823 select committee of the House of Commons, where Clegg and Davy claimed he exaggerated the risk posed by gasworks and threatened to overregulate them.[137] The chemist William Hyde Wollaston, speaking before the same committee, argued against regulation: "It appears to me, that the Government may rely upon the interest of the parties to take all the precaution they can."[138] The committee agreed and decided not to impose regulations, although inspection continued. Gas companies also routinely employed chemists to monitor their own systems—and especially the leaks from mains.

Legal regulation of the quality of gas really emerged only after 1850, and mains and stations continued to leak and burn routinely. The advent of electricity, some thought, might liberate society from such endemic dangers. "Properly and scientifically employed," said one electrician in 1882, "electricity afforded a means of communication which could be rendered absolutely and perfectly safe."[139] But electricity simply introduced a new set of anxieties alongside those generated by gas, anxieties centering on electricity's impalpability. The presence of gas was tangibly felt in nostrils and eyes, while electricity, invisible and inodorous, carried no such forewarning. "We are endowed with no special senses, by which we can detect the presence of electricity or magnetism," observed the *Builder*; "there is, then, no cause for wonderment that most approach the subject as one of a shadowy nature, if not altogether shrouded in mystery."[140] This was often portrayed, again, in class terms. Hammond reported that, after his house had been wired, his servant was unable to sleep for fear of explosion.[141] Preece argued that, in the absence of general technical education, it was not unexpected that people felt disturbed by having "a natural force of enormous intensity" circulating around their homes.[142] The *Lancet* went further, suggesting that, irrespective of one's knowledge of electricity, there was "at least some grounds for the uneasiness which one of its critics has expressed at the idea of '2,500 volts grumbling in the cellar.'"[143] The consumer of electricity, according to the *Journal of Gas Lighting*, was vulnerable to shock in "the utter absence of any intimation of the deadly power lying in ambush."[144] Gas accidents were caused by carelessness, electrical ones by the surreptitious nature of the energy itself.

Many of the environmental and safety problems associated with gas networks were replicated by electricity networks. Early plants were often extremely noisy and smoky. The noise of London's Gloucester Crescent station was described in 1885 as "almost unbearable" for "those living in the immediate locality."[145] The system was unreliable and prone to breakdown. All the transformers at Ferranti's Deptford station burned out in 1890, leaving the supply disconnected for three months. More typical were minor, irritating interruptions, as in the City in 1881, when the supply simply stopped, without warning or explanation, for several hours.[146] Electricity mains themselves mysteriously exploded with such frequency that the Royal Society conducted an inquiry into the phenomenon in 1895, blaming it on decomposed salts "chiefly derived from the soil," which produced small residues of alkali metals, which in turn generated sparks, igniting any coal gas that had escaped gas mains and entered conduits.[147] This was the result of the interaction of underground networks with the earth. If conduits were not perfectly watertight or securely bolted together, fires or explosions from electrolysis or arcs formed between the conductors and water might result.[148] These occurrences remained rather perplexing, which reflected the imperfect knowledge of subsurface systems, as the engineer Rankin Kennedy noted in 1902: "The subject of explosions in underground electric systems has never been properly investigated. Many of the cases have been somewhat mysterious. What actually did explode and what fired the explosive has in most cases only been a matter of conjecture."[149] It was impossible to fully predict how networks would behave when interacting with variable environments. Risk had to be managed rather than altogether eliminated.

A primary area where such management was apparent was in the comprehension of how electricity interacted with the human body. Contact with powerful electric currents could kill, as deaths through lightning had long demonstrated. Electricity was, thus, a death force as well as a life force: the threshold dividing death and severe pain from therapy needed to be established. Benjamin Richardson, for example, applied the great induction coil at the Royal Polytechnic to numerous animals in 1869, ostensibly to find civilized methods of slaughtering them: "Sheep and other animals were killed by the statical discharge, and the cause of death was shown to be the expansion of gaseous parts of the blood and tissues by which organic lesions of the most extensive kind were induced." The electrophysiologist Jacques D'Arsonval considered five hundred volts as an approximate threshold, beyond which death, and below which stunning alone, would result.[150] Alternating current, it was soon established, caused far more discomfort than direct current at comparable voltages. These experiments thus formed part of wider projects investigating the human body's electric capacities, its resistance and conductivity, as well as its ability to interact with other machines.[151]

"It is certain," observed Henry Lewis Jones, the medical officer in charge of the Electrical Department at St. Bartholomew's Hospital in 1895, "that electrical discharges can kill a man as dead as a doornail." This death resulted, he noted, from heart failure rather than, as D'Arsonval had maintained, respiratory breakdown.[152] That systems being installed in British towns had to be secured against accidental electrocution had long been clear. In 1880, *Engineering* noted: "There can therefore be no doubt that the electric current feeding an ordinarily powerful electric lamp of the Jablochkoff type . . . is quite capable of causing death to any person who is unfortunate enough to come into contact with it so as to 'shunt' the current through any of his vital organs."[153] To be so literally plugged into the system was to risk death. Reports of accidental deaths appeared, including those of Lord Salisbury's gardener and a man at the 1885 Health Exhibition who, rather ironically, died after touching a dynamo.[154] It was especially important to protect workers at central stations or transformer substations. Various methods emerged, including lining walls with wood, making sea boots out of india rubber, coating with ebonite, wearing thick gloves, developing tubes that glowed at high voltage, or at least printing first aid directions on a wall in all stations.[155]

The development of the electric chair was of great rhetorical importance to these debates.[156] In 1888, New York State voted to replace hanging with the electric chair. During deliberations in the assembly, commissioners had spoken coolly of electricity as being "the most potent agent for destroying human life."[157] The "death chair" was designed by, among others, Arthur Kennelly, the former director of Edison's team of electricians. William Kemmler, a murderer, was the first man executed by electrocution, in 1890. British journalists made much of the spectacle. It was reported in *Engineering* as follows: "The voltage is variously stated at 800 and 1800, but it is clear that it was not sufficient to produce instant death. . . . It recalls the barbarous executions of the middle ages." Electric death was supposedly predictable, clinical, and instantaneous, but it took at least eight disturbing minutes. "Finally," the report continued, "his body began to char, and the sickening odour of burning flesh was added to the horror of the scene." Following this, Kemmler's blackened corpse was immediately handed over to scientists.[158]

Such graphic descriptions certainly captured headlines, but they were also used within the electricity industry itself, notably by Edison, to attack the use of alternating current, which had killed Kemmler. The public, argued the *Engineering Magazine* in 1891, was "possessed with a horror of high-tension circuits."[159] In Manhattan, this was most manifest in a crusade against overhead wires, which began shortly before Kemmler's execution in 1890 and lasted until the final such cable was removed in 1905. In one week in January 1890, 154 miles were felled while watching crowds applauded.[160] In Britain, this crusade coincided with increasing concern about the danger, inconvenience, and ugliness of overhead wires. In 1883, the *Electrician* spoke of the "inordinate multiplication of overhead wires," particularly telephone cables.[161] Such accumulation, it was claimed, interfered with transmission: Preece reported in 1886 that high-tension electric wires disturbed telephones "within a distance of three thousand feet."[162] Private companies were held to blame for this "nuisance [which] must very soon become unbearable. . . . The wires and cables are now rapidly shutting out daylight and destroying what little beauty is still left."[163] Despite a select committee reporting that the danger was "greatly exaggerated," accidents resulting from falling wires proliferated.[164] During snowstorms or gales, poles were regularly blown down, leaving a dangerous tangle of wires. Following events in New York, some municipalities in Britain acted swiftly. Glasgow Corporation, for example, decided to dismantle its three and a half miles of overhead wire.

Although overhead wires were not outlawed, their growth was tamed. Rules for electrical fitting and wiring were composed by numerous bodies (the Board of Trade and the Phoenix Fire Office, e.g.) in the 1880s, while a specific calculus relating to the erection of telegraph poles was developed, the main variables being tension and sag.[165] More general legislation followed. The 1890 Public Health Act included clauses empowering authorities to make regulations to protect the public from such wires. Suspended cables could be constructed only with the Board of Trade's permission, and, to prevent the sometimes slapdash workmanship of earlier erections, precise codes were produced to govern the positioning and construction of wires. The 1891 London Overhead Wires Act defined the maximum distance between poles and the angle at which wires could cross streets and insisted that all posts be marked so inspectors could easily distinguish the ownership of particular wires. These were extended and nationalized with the Board of Trade's 1896 Regulations for Ensuring the Safety of the Public. These various acts and regulations provided very detailed information on physical structure, especially regarding the aerial interface between wire and building: "Service lines from aërial lines shall be led as directly as possible to insulators firmly attached to some portion of the consumer's premises which is not accessible to any person without the use of a ladder or other special appliance, and from this point of attachment they shall be enclosed and protected in accordance with the subsequent regulations as to electric lines on the consumer's premises. Every portion of any service line which is outside a building but is within 7 feet from the building shall be completely enclosed in stout rubber tubing."[166] This was a drive toward durability, security, and limited inspectability, by experts only. By this date, underground wiring was becoming more common. Urban electricity networks were, slowly, becoming stabilized and, occasionally, expansive. The final part of this chapter examines this process as it unfolded in the City of London.

A Modest Radiance: Illuminating the City, 1878–1900

It is impossible to walk anywhere in London City without treading on pipes or cables embedded in soil, concrete, or fibre. In fact, so congested is this tract of London's underworld that, sometimes, when they are putting up a new street-lamp, they cannot find a space big enough to dig a hole where the lamp should stand. So they dig as near as possible to the spot, and use a lamp-standard with a twisted root.

F. L. STEVENS, *UNDER LONDON* (1939)

The City of London, 673 acres in size and long the financial hub of the empire, was an administrative anomaly even by the capital's arcane standards, with the City Corporation still protected by a mass of charters, some of which had been in force since medieval times, when the City had been lit, sporadically, by candles.[167] In Elizabeth I's reign, fear of Spanish invasion generated security concerns, and "an Order in Council commanded every householder to do his part in the lighting of the City on pain of death by the common hangman." In 1684, Edward Heming was licensed by the Corporation of London to illuminate streets with patent oil lamps outside every tenth door along the street, between 6:00 P.M. and midnight, on moonless nights between Michaelmas and Lady Day. This was simple beacon lighting: Horace Walpole, we are told, "could not stir a mile from his own house without having one or two servants armed with blunderbusses to protect him."[168] Despite the institution of the MBW, government of the City's sanitation, gas, paving, and lighting remained in the hands of the City Commissioners of Sewers and various committees until 1897, when those duties were divided between the Court of Common Council and the newly formed Public Health Department. By this time, the City was no longer a largely residential space, as it had been in Heming's day. Charles Booth, for example, found that, of 301,384 people working in the City by day, only 37,964 stayed at night.[169] Consequently, its lighting needs were unique. Illumination was required largely for banks, offices, and streets, by day as well as by night, owing to the "not infrequent, and sometimes quite unexpected, fogs which envelop the City in darkness, even at mid-day."[170]

The City operated as something of a laboratory for electric lighting. It was first illuminated electrically in late 1878, when Jablochkoff candles were displayed at Billingsgate Market, Holborn Viaduct, the Embankment, and the office of *The Times*, while a larger-scale experiment took place with arc light in 1880.[171] The Edison Company illuminated the same viaduct with ninety-two filament bulbs in 1882. These installations were experimental and ephemeral: gaslights were routinely relit to prevent chaos. In 1889, there were experiments with oil lamps. Technological problems were compounded by morphological ones. The City's crowded, narrow, winding streets and irregular buildings tended "to make the task of lighting with a medium giving intense and sharply-defined shadows one of considerable difficulty."[172] Preece argued that "he knew of no branch of engineering where more forethought was necessary in order to calculate all the contingencies likely to be met with than with a system designed to illuminate a large city like London."[173] The City would be lit, not by the simple application of an existing system, but by a combination of pragmatism and imagination. The City's complicated political and administrative structure added another layer of complexity. In 1887, for example, the Brush Company's scheme provoked the hostility of the vestry of St. George the Martyr, which objected to mains being run beneath fifty yards of its roads. The antiquated administration of the City, commented the *Electrician*, "appeared to be struggling against the introduction of the electric light into their sacred precincts, as though it were some pestilence."[174]

After thirty-nine separate reports spanning the 1880s, the authorities took steps to invite tenders for electric lighting following the 1888 Electric Lighting Act. The city was split into three provisional areas, and two companies, the Brush and the Laing, Wharton and Down (who soon united as the City of London Electric Light Company), were entrusted with the supply. A specially inscribed commemorative junction box was built into the side of the Mansion House on January 5, 1891, and the first systematic scheme to light the City was under way. The engineering press argued that a radical change in urban illumination was taking place. By June, twenty-five arc lights were lighting Queen Victoria Street, supplied from the temporary station at Bankside, and, in August, *Engineering* declared that, in the City, unlike in the West End, "not a single gas lamp will remain in the street."[175] The same publication noted rather smugly: "After being for many years behind every other important city in Europe and in America in the matter of electric lighting, London has suddenly distanced all its Continental rivals, and is within a measurable space of surpassing New York."[176] Despite one frugal City administrator arguing that electric light was "equivalent to using perfumed water to turn a mill," by the end of 1892 the majority of the main streets were lit by arc light delivered by alternating current via transformer stations.[177] A year later, *Engineering*'s claims seemed to be coming true: the Commissioners of Sewers resolved that, since "electric light is now the light of the City," gas standards would be dismantled.[178] Thirteen miles of main thoroughfares were lit electrically by March 1895, by which time the company was supplying nearly five thousand customers as well as institutions as diverse as St. Paul's Cathedral and the local infirmary. By 1899, there was one electric lamp for every City inhabitant: "The rate at which the City of London has forced its way ahead in the matter of connections is remarkable."[179] A telephone exchange linked the plant with its twenty-three substations.

Like with the Deansgate improvement in Manchester, however, such hyperbole ignored and smoothed the convoluted history of the project. Laying mains, for example, necessitated great interruption to traffic (for

Figure 6.10 Main laying in the Strand, Westminster. From Scrutton, *Electricity* (1898).

a contemporary image from the Strand, see figure 6.10). In January 1892, at a meeting of the Commissioners of Sewers, "attention was called to the disgraceful state of the streets of the City during the past few days," one of the main causes of which was the laying of electric cables, at a rate of around one mile per week.[180] According to the sympathetic *Electrician*: "It was impossible for the work to be done any quicker, or the whole traffic of the City would be stopped."[181] Movement was greatly hampered "by having 100 yards of footway 'up' in nearly thirty places at one time."[182]

Building was also compromised by the sheer density and complexity of the subsurface world of the City.[183] Webber admitted afterward that planning the precise position of the conduits was barely possible because "outside the Postal Telegraph and the City Engineer's office there was no one who could form any accurate idea of the condition of things below the surface of the City streets."[184] Once digging began, he found himself frustrated by the lack of space available for mains: "The occupation of the subsoil by sewers and subways, by the systems of gas, water and telegraphs, is entirely exceptional."[185] While subways did exist in certain parts of the City, they proved unusable, owing to the absence of suitable openings for running cables into houses, meaning that conduits were necessary. Webber originally intended using iron piping from the generating stations to the transformers and bitumen and concrete casing for the lower-current circuits. Eventually, he was forced to make several pragmatic choices owing to the logistic horrors beneath the streets. Although formally bound to maintain a minimum distance of six inches between his pipes and preexisting ones, he found himself compelled to bend iron and bitumen tubes into complicated shapes to thread them between the masses of telegraph tubes, gas pipes, and water mains: "In many cases, such as in King William-street, so contracted was the space that several ways which were actually necessary had to be omitted and taken another way, and occasionally a deviation under the roadway was obligatory."[186] Between December 1891 and July 1892, 271 miles of mains were laid in this tortuous fashion, a quantity that threatened to exhaust the supply of iron pipes.

Similarly, manholes and transformer stations were usually located wherever space permitted: "These boxes were made as large as possible, which is not saying much; I aimed at making them large and deep enough for a man to get entirely inside, but this was rarely possible. Owing to the obstruction underground, they are of every conceivable size and shape."[187] The original scheme involved twenty-two transformer stations receiving high-pressure current from two generating stations, at Bankside and Wool Quay (the latter being eventually dismantled).[188] Finding suitable open spaces, particularly at affordable prices, sometimes proved impossible: the company was forced, for example, to apply to the vestry of St. Benetfink for permission to build a station in a churchyard. The vestry concurred, but only after an annual fee toward church services was extracted in return. "In excavating," Webber recalled, "large quantities of human bones were met with, which had all to be transported carefully for extra-mural re-burial."[189]

Positioning streetlamps was also a question of pragmatism rather than system. Early arcs were often mounted on scaffolding, to enable engineers to experiment to find the best height: Trotter made photometric recordings in the City in 1891.[190] They were described as "eye-sores, with their peg-like excrescences," and many were later removed.[191] Less ugly standards, with bases of "ornamental character," were erected later, with holes cut into the sides to enable men to ascend and trim carbons.[192] Again, fixing the posts proved troublesome, owing to the impossibility of predicting subterranean conditions. Wall brackets were employed in the narrower streets. In busier streets, special round standards were used to save space.[193] After a series of experiments involving ground, opal, and ribbed glass, the latter was adopted as best suited to deflect light downward for the benefit of pedestrians and horses. In smaller streets, engineers had intended to simply adopt the old gas standards as they were abandoned, but the persistence of stagnant gas within them rendered them dangerous, and there were many reports of them being lit, either by "mischievous people passing on omnibuses" or by pedestrians concerned about an electricity cutout.[194] By 1894, these standards were being demolished as obstructions, under the dangerous structure clauses of the Metropolitan Building Act.[195]

The light itself was neither reliable nor entirely safe during its early years. During these years, supply itself was hardly continuous, and the *City Press* in particular found it "intermittent and unreliable," drawing attention to the mayor's dissatisfaction with the distribution to the Mansion House.[196] In his 1893 report on City arc lights, Colonel Haywood found that, of 479 arc lamps, 301 had failed at some point, the majority for short periods.[197] In 1899, the electric lights went out during a meeting of the City Corporation, and candles had to be used.[198] There were problems with generators and insulation. Impatience mounted: "It does not require any great stretch of imagination to conjure up in one's mind such little affairs as possible bank robberies, &c., when these establishments have suddenly to fall back upon candles and ginger-beer bottles to contain the composite dips."[199] In 1897, gaslights were used to illuminate Mansion House to commemorate the Queen's Golden Jubilee: two years later, plans were afoot to illuminate Fetter Lane with gas mantles.[200] Several City streets, as well as the Mansion House itself, were lit with high-pressure gas mantles in 1907 (figure 6.11).[201]

The electricity system was neither secure nor stable. On the contrary, it was dangerous and fragile. In November 1894, a horse drawing a Brougham carriage was traversing a portion of wood paving on Cannon Street when it received a severe electric shock through the pavement. The driver, while watching the horse groaning, twitching, and dying, himself received a shock, immediately after which there were two simultaneous explosions in junction boxes, ripping up the road, and throwing several curbstones through the air. Several other pedestrians received shocks. None of them were killed, but the testimony of a postman, Charles Joseph Willy, gives some indication of the pain and fright of this kind of street accident:

He felt as though sparks were going through every part of his body. He had one foot on the plate of the manhole and another on the pavement. He was drawn towards the pavement, but not thrown down. He had two shocks, one on the lid and one on

Figure 6.11 High-pressure incandescent gaslighting at the Mansion House, City of London, 1907. From Webber, *Town Gas* (1907).

the flagstone, as his feet went down one after the other. When he got both feet clear of the cover he did not feel any further shock. He got away and went on to deliver a parcel at Walbrook. He was just getting on to the pavement outside witness's shop, and a gentleman was asking him if he was hurt, when the street box at the corner of Walbrook blew up, knocking him across the road and against the "Cannon" public house. A piece of stone struck him on the finger, and a piece of iron stuff came down on his toe. His mouth was full of concrete. He went off then—he had had enough of it. . . . He was still sensible of ill effects from the shock.

The accident, it transpired, was the result of a short circuit forming when wires were crossed. In his report, Major Philip Cardew, the first electrical adviser to the Board of Trade, found the City of London Electric Lighting Company guilty of fitting inappropriate fuses and accused one company member of "blow[ing]) up the public."[202]

Infrastructure should be silent and hidden, patrolled by workmen, and its statistics recorded in municipal archives. It should not injure, frighten, or distract. The response to such minor technological calamity was entirely predictable: there should be a coordination and centralization of expertise and an amplification of the play of inspection across the system. Twelve men were immediately employed to check street boxes. A City electrical inspector, A. A. Voysey, had been appointed in 1893:

his duties included the periodic inspection of electric lines and works and the certification and examination of meters. In 1895, he received a new office at the Commission of Sewers, where knowledge of the whole system would be centralized. The *Electrician* noted the importance of the appointment: "The electric inspector has important and multifarious functions to perform, and in the performance of them he must require no little tact and vigilance."[203]

To repeat: infrastructure is supposed to be ignored by its users and forgotten by historians. It should form an unintrusive frame or background to our existence. I have unearthed it here to demonstrate how much work and experiment it took to provide reasonably secure perception. But what of this perception itself? As argued in the previous chapter, electric light did not radically transform the way people saw. Rather, it produced slight reconfigurations of perception: little adjustments to color, shadow, depth, and form. The City did not suddenly emerge radiant from centuries of smoky blackness; eyes were not freed at last to see anew and correctly. Approaching Queen Victoria Street from Cheapside shortly after the inauguration of the light in July 1891, a reporter from the *Electrician* described the effulgence as "very good, but not particularly remarkable."[204] In 1894, the *Electrical Review* grumbled about the "the under-incandesced state of the glow lamp filaments when there is any special demand for extra lamps."[205] A reasonably extensive set of quite fragile networks, slowly being engineered, inspected, and made durable, produced the slenderest of perceptual shifts. It is in the multiple iterations of such projects that these perceptual shifts themselves become durable and, cumulatively, amount to historically significant change.

In *Networks of Power*, Thomas Hughes argues that London electrified at a slower rate, and in more piecemeal fashion, than Chicago or Berlin because of political factors, notably the lack of centralized administration, parochialism, and local management of technological systems.[206] There is much to commend this position, of course, and Hughes's argument is far richer and more nuanced than I am presenting it as being. But there is a tendency in such a position to take American or German technological development as normative and then search for what "retards" normal growth in other contexts. Hence, without vested interests, parishes, archaic bodies, antiquated attitudes, bumbledom, and conservative legislation, the abstract and rational logic of technology would diffuse freely and without restraint. Without human interests to slow it down, technology would expand boundlessly through space.[207] But, if we have learned anything from Foucault, it is that power is never absent in any social, or sociotechnical, relationship and that relations of, and ideas about, power always positively shape human, and technological, practice. The very factors that Hughes identifies—absence of centralized control, commitment to local autonomy, and aversion to system—are, of course, some of the defining characteristics of British liberalism, and the small-scale, variegated nature of electricity networks was not a result of politics "getting in the way" of technology but a consequence of politics positively shaping technology.

Alongside politics, the historian must always consider the role of space and architecture. The City of London's narrow, irregular nature produced a specific set of difficulties making it incomparable with the lighting of downtown Manhattan. Engineers like Webber were sensitive to such local idiosyncrasy. In an article in the *Electrical Engineer* in 1893, another engineer, Arthur Guy, made the point clearly: "When going into the matter of lighting a town by means of arc lamps, the plan of the town must be very carefully analysed and studied, and it is no use trying to find a precedent: each town must be considered and judged from its own advantages and disadvantages. No two are alike: each has its own local peculiarities."[208] Sensitivity to the local, and the resistance to system, was not restricted to liberal politicians: it clearly extended to the world of electrical engineering. Hughes argues that "conservative political interests" produced electrical "backwardness" in London.[209] A more accurate conclusion would be that liberalism, in its distinctively British form, shaped a corresponding form of electrical development, characterized by cautiousness, eclecticism, and localism.

Conclusion:
Patterns of Perception

For the historian, there are no banal things.

SIEGFRIED GIEDION, *MECHANIZATION TAKES COMMAND*

Without the countless objects that ensured their durability as well as their solidity, the traditional objects of social theory—empires, classes, professions, organisations, states—become so many mysteries. BRUNO LATOUR, *WE HAVE NEVER BEEN MODERN*

My aim throughout this book has been to remain close to material artifacts and networks as well as to the eyes and bodies of those interacting with them. I have focused on a set of artifacts and infrastructures that slowly, unevenly, and unspectacularly became absorbed into routine visual practices in nineteenth-century Britain. The objects I have studied (spectacles, plate glass, asphalt paving, gas meters, photometers, mantles, switches, manholes) might have been mundane and eminently forgettable, but they were, as the epigraph from Giedion firmly maintains, anything but banal. As the epigraph from Latour suggests, treating such everyday artifacts as material agents provides us with substantial clues as to how big, overarching things like "modern society" have been organized, maintained, and reproduced. I have endeavored to study the patterns formed, and the practices generated, around these artifacts. My analysis has revolved around three major themes: vision, power, and technology. Here, I make a few concluding remarks about all three.

Patterns of Perception

Scholars of visual culture have, for at least two decades, been excessively transfixed by the twin conceptual idioms of the panopticon and the flâneur. I began by suggesting that these concepts were largely useless when analyzing dominant visual formations in nineteenth-century Britain (and, very probably, elsewhere in the "modern West"). The demolition of concepts is, however, far easier than the painful and laborious creation of new ones. The bulk of this book has explored a series of perceptual patterns or routines embedded through technological networks. These, when repeated in multiple sites, slowly became pervasive, tacit, and often subconscious. Plate glass made pure vision without smell or sound materially possible, while agglomeration, accessibility, legibility, and portability made urban inspection a substantially more routine activity. Judicious construction and management of illumination technologies might allow accurate replication of daytime colors by night, while the introduction of light switches allowed the individual to control her own light with minimal effort.

The particular patterns studied are in no way intended to represent totally the vast and often mysterious range of visual practices in nineteenth-century Britain, and any conclusions I reach must be made with this caveat. These patterns have themselves been revealed partly because of the particular sources I have used. I have had nothing to say about the manifold forms of aesthetic perception, especially those relating to art, photography, and film, and I have altogether ignored those modes of vision belonging to the immaterial and spiritual worlds, from religious visions to extrasensory perception.[1] Of scientific techniques of observation, I have offered only crumbs and fragments, although my brief study of photometry does hint at the messy and embodied nature of scientific perception. My book has had rather more to say about class than gender or race.[2] This is not to suggest that I think such topics unimportant or historically irrelevant: rather, they are too significant to be treated cursorily. My research has also focused more heavily on urban than rural sites, and English sources have featured more heavily than Welsh or Scottish ones. I have gestured toward developments in other nations and colonial sites, but these gestures have been tiny.[3] What my study has lost in breadth, it has, I hope, gained in depth and density.

The perceptual patterns that I have studied offer us, I think, a very particular archaeology of social perception in very particular places (Manchester, London, Glasgow, Leeds). This archaeology has revealed a teeming, muddled multiplicity of visual practices that cannot be reduced to any hegemonic modality (panopticism, spectacle, the gaze, flânerie). Yet historians must be prepared to make generalizations, and I have attempted throughout the book to show that certain pervasive perceptual patterns can be discerned amid the rather turbid jumble of the Victorian city. I will now distinguish some of the patterns I have uncovered. These patterns can be broken into three broad groups of visual arrangements: collective, individual, and productive. These can be further decomposed into subpatterns, and, for the sake of symmetry, I have subdivided them into three further modes or networks. These networks were materially heterogeneous, including architectural and engineering elements, on the one hand, and bodily and visual elements, on the other.

The first cluster of patterns relates to collective perception, or social monitoring. First, we have the *oligoptic*, or the self-regulating visual economy. Oligoptic ensembles, whereby small groups of free, mobile, and potentially self-aware individuals monitor one another, were characteristic of numerous public and private institutions in nineteenth-century cities (libraries, museums, reading rooms, art galleries) as well as less enclosed spaces like gardens, squares, and even streets. The second pattern was the *supervisory*, or fixed oversight. Oligoptic spaces were almost always overlaid with a privileged grid of visual supervision radiating from a special node, usually at an entrance point or a position of relative centrality from which sight lines were maximized without ever being quite panoptic. In some cases, as in Ure's fantasy factories, supervision might be primarily exercised over machines. Third were the networks of mobile *inspection*, usually focusing on technical or vital systems (sanitary networks, gas pipes, food, and water), which were themselves increasingly built to facilitate inspection.

The next three patterns refer to private forms of vision. First was *self-inspection* or *introspection*, encouraged within spaces that were completely private and enclosed and often equipped with visual technologies like lamps and mirrors. Of course, one might choose to obscure oneself from oneself in such a space, but this does not undermine the basic point: there were spaces from which one could entirely escape the play of oligoptic, supervisory, or inspecting gazes circulating in more public domains. Second, and complementarily, we have a set of spaces largely occluded from all eyes save Mill's "peculiar and narrow classes" of inspectors, slaughtermen, or surgeons: the mortuary and the abattoir are, perhaps, most typical here. This very *calculated invisibility* was at least as integral to the visuality of modern urban space as the constitution of glittering and seductive spectacle.[4] Third is modern *voyeurism*, which is absolutely predicated on the tantalizing possibility of transgressing these structures of privacy and occlusion. The voyeur is this society's visual parasite, made possible by its own systems of obfuscation.[5]

A final set of patterns or modes refers to specific configurations designed to accentuate visual capacities relating to motility, reading, production, and concentration. The first visual element here is that of *distant and simple signification*. Archetypes here include the flag and the beacon. Most streetlights, and many lights positioned within halls and corridors, for example, were built simply to demonstrate distance and direction, while traffic lights similarly transmitted elementary information across a reasonable distance. Second, and related to this, are the little nuclei of *proximate and complex signification* provided by everyday objects like street signs, public notices, bulletin boards, and gas meters. These signs were themselves often illuminated or positioned to be inspectable, for example, by being positioned on lampposts. Finally, there are the *technologies of visual detail* (torches, reading and surgical lamps, microscopes), which opened specific spaces, surfaces, and entities (cellars, books, ears, meat) to careful, temporally delimited perception. These technologies often secured the attention to detail necessary for private reading and for many very specific forms of labor.

This taxonomy is already straining and blurring, and I terminate its proliferation here. We have, then, nine basic, recurring patterns of visual perception: oligoptic oversight, supervision, inspection, privacy, obfuscation, voyeurism, distant and simple signification, proximate and complex signification, and detail. These modes cut across day and night, expanding, albeit slowly and unevenly, the temporal reach of practices as diverse as reading, fighting, traveling, working, and performing surgery. This did not amount, of course, to an obliteration, colonization, or disenchantment of the night. Night was permeated, punctuated, and slightly displaced by numerous networks, but these networks were loosely and locally reticulated, even in cities.

These nine patterns were also emphatically nondiscrete. None operated as entirely self-enclosed perceptual paradigms or scopic regimes, especially since they invariably overlaid one another, producing complicated, unpredictable configurations of visual experience. A library, for example, often contained a group of oligoptic reading rooms, a supervisory librarian's desk, private water closets, and technologies of proximate and complex signification (card catalogs, indicators, bulletin boards, magazine racks).[6] Hospitals were not simply the site of the clinical gaze of doctors but also the very detailed gaze of the surgeon, the supervisory gazes of nurses, and the self-monitoring gazes of patients, not to mention the forms of privacy created by curtains, lavatories, and so on. Housing increasingly contained relatively large amounts of private space but also quite intimate but oligoptic spaces (living rooms, kitchens, parlors) and would be periodically accessed by inspectors for the purposes of sanitary and infrastructural monitoring. In more open, public spaces, visual networks of an entirely different topology and scale become increasingly predominant, notably those of distant and (relatively) simple signification, the visual mode par excellence of today's motorist (at least if not driving on Los Angeles freeways or the London Orbital). Henri Lefebvre saw Haussmann's Paris as a concrete and large-scale realization of this kind of perception. The Parisian driver, he comments, "perceives only his route, which has been materialised, mechanised and technicised, and he sees it from one angle only—that of its functionality: speed, readability, facility. . . . Space is defined in this context in terms of the perception of an *abstract* subject, such as the driver of a motor vehicle, equipped with a collective common sense, namely the capacity to read the symbols of the highway code, and with a sole organ—the eye—placed in the service of his movement within the visual field."[7]

Lefebvre's observations are acute. He suggests the hegemony of a kind of perception that is both monosensual ("a sole organ—the eye") and monomodal (the subject is blithely poised within a single visual formation). But lived perceptual practice never became so flattened, functional, and controlled. No single pattern becomes absolutely hegemonic or discernible, either in the nineteenth century or today. The vision networks I have outlined never assumed, individually or in combination, the kind of suffocating dominance that Lefebvre suggests, for three reasons. First, and most obviously, privacy, or freedom from the perception of others, was deliberately engineered into the structure of houses, institutions, and even the public sphere. Second, although some of these networks were impressively long and quite durable, there were significant areas largely untouched by any of them. This was the case with inspection, but also with technological infrastructures like gas- and electric light. It would, of course, be possible to identify certain elements of these regimes operative in Cornish hamlets, along the banks of Cumbrian rivers, or in market towns in Lincolnshire, but one would be ill advised to try to analytically stretch or twist them to fit every conceivable space in nineteenth-century Britain. They were at their densest in urban, industrial, metropolitan areas: indeed, they were part of what made such spaces urban, industrial, and metropolitan. Third, even where these modalities were dense, embedded, and functional, they never absolutely overdetermined certain visual practices. They were fragile, depending on unreliable technologies and even more erratic humans. Lazy inspectors, daydreaming supervisors, sleeping guards, myopic pedestrians without spectacles, shameless bathers, and color blind railway employees all compromised the smooth functioning of whichever visual network they were temporarily bound up with. These visual modalities required active, attentive agents rather than the abject, docile products of the panopticon. Their fundamentally nondisciplinary, nontotalizing, and nonpanoptic morphology, and their conscious reliance on subjective agency, defines them as liberal.

Visual Agency and Liberal Government

I have tried to write a *political* history of vision and its attendant technologies; in other words, ways in which embodied perceptual practices and the material systems shaping or stimulating them materialized, reinforced, or reproduced certain relations of power. By rejecting the panopticon and the flâneur as hegemonic visual modalities, we have also jettisoned the historical (and historiographic) fantasy of total transparency and ocular omniscience. I have, instead, suggested that the multiple, superimposed vision networks of nineteenth-century Britain can be seen as broadly liberal, in that they were invariably designed with certain aspects of human freedom in mind. Freedom to move, as well as freedom to look, within certain limits, was presupposed by oligoptic arrangements, while privacy and freedom from aggressive intrusion were basically respected in systems of inspection. But these particular forms of visual freedom did not crystallize in the absence of power relations or government: far from it, these were deeply organized practices, as physical arrangements for supervision, the management of electricity networks, and the manuals used by inspectors amply demonstrate.

The study of a particular, vital set of everyday bodily practices has, then, allowed us to make some conclusions about the operation of liberal government. As suggested in the introduction, liberalism is, perhaps, best approached as something very broad, multiple, and eclectic. It was both a political philosophy (formal or inchoate) and a mode of sociotechnical organization. It was possible to govern in a liberal way, and it was also possible to engineer in a liberal way: the engineering, sanitary, and building texts that have been the basis of my research are clear testimony to this. This truly eclectic nature of liberalism is something often lost in traditional histories of party, intellectual histories of liberalism, as well the governmentality literature: liberalism emerges as either radiating from great, central brains and institutions or so dispersed as to be meaninglessly omnipresent. I have endeavored to combine elements from these historiographic traditions with the perspective offered by engineering and technology. Thus, a very traditional notion of government, centred on law and institutions, has been central to this study. Almost every aspect of the systems I have discussed, from the construction of electricity networks to the quality of gas, was legally regulated during the nineteenth century. The government of material systems and infrastructures involved much parliamentary time and was itself a stimulus to the delegation of power to municipal and local government. This was an eminently liberal mode of organizing technological systems since it magnified local autonomy, but it still functioned through the traditional means of the state, law, and formal, elected institutions.

But this macrological, statist, and legalistic approach to liberalism is but one perspective on a complex process of technological government. I have concentrated throughout on very local, technological textures of visual practice. At this microscale, the operation of power is often better captured through the idioms of norm and capacity, which I have utilized freely. We are speaking here of the agency made possible by technological networks. The numerous, interlaced vision networks or patterns stimulated and sustained a panoply of individual visual norms and capacities: productive attention, sensory awareness, urban motility, social observation, private reading. These technological networks, then, when meshed with the acting body, actualized optical capacities that made normal and durable the autonomous, rational, judging, distant practices of the liberal subject. The actualization of the "phenomenological vector" did not amount to a determination since these networks were not omnipresent and the individuals within them were not obliged to behave accordingly, most of these practices (being attentive, solitary reading) being normative rather than legally obligatory.[8]

Visual power, as is evident, operates through regimes of both law and norm. If we look at, say, the library, we can see how patterns of perception, superimposed and spatially specific, allowed a recursive or iterative performance of organized freedom. The spatial and visual morphology of the library created a visual environment characterized by oligoptic reciprocity, supervision, pockets of privacy, and proximate, complex signification. One was often under the scrutiny of others, be they fellow readers or superintendents. One could also scrutinize readers or superintendents as well as withdraw to the privacy of the bathroom or one's own home, as and when one chose. Many glances and gazes fell on books, catalogs, and notices. There were laws regulating conduct in the library, but routine conduct almost entirely functioned without recourse to the law. This entire configuration presupposed, in a tacit and routine way, a subject with rights, duties, and freedoms. A dirty, loud, drunk reader, with a dog and a cigarette, would be ejected, perhaps through the combined force of pious frowns and grimaces, or, if persistent, by a librarian, supervisor, or policeman. There is nothing intrinsically liberal about a library, but a library organized spatially, visually, and practically as a partly self-governing, partly overseen institution can be described as liberal in that it expresses the particular kind of organized freedom associated with nineteenth-century British liberalism.

Liberal subjectivity was, and is, of course, an abstract universal, and its exclusionary dimensions have been discussed at length elsewhere.[9] As suggested in chapter 1, there were substantial perceptual elements to such exclusion, and, as hinted thereafter, these perceptual elements were themselves reinforced by the social distribution of vision networks. The spread of gas and electric light followed a very clear social logic: most of the technologies of illuminatory individuation discussed in the last couple of chapters were confined to the homes of the wealthy or respectable. Similarly, absolute privacy and freedom from the gaze of society remained unevenly distributed: lodging houses and tenements at the very end of the century still routinely had shared facilities and were more regularly inspected than suburban dwellings. This suggests that visual networks themselves materially reinforced, reproduced, or reified the perceptual imaginary dividing society and that differences in sensibility or perceptual capacity were less a question of anything innate (character, biology) than of palpable environmental and technological inequality. Society remained, and remains, divided along sensory lines, even if today's middle classes have rather different thresholds of tolerance to those of the nineteenth century.

Superimposition and Displacement: Patterns of Technological Change

Many of the technologies described at length in this book were nineteenth-century inventions. But the historical novelty of, say, the steatite burner head, the access pipe, or asphalt pavement is of far less significance than their relatively pervasive spread.[10] I have examined how a whole swathe of objects became socially embedded and involved in everyday perceptual practice. Some artifacts, of course, were more long-lived than others. Wood pavements and electric candles, for example, proved rather less durable than asphalt and electric incandescent lamps, which themselves have a far briefer history than dirt tracks and wax candles. New technologies never fully or immediately replace old ones: there is never a point of rupture dividing, say, the "age of electricity" from "the age of gas" or "the age of tallow." Instead, technologies, as they become embedded and integrated into everyday practice, become superimposed over, and slightly displace, older artifacts. Illumination is a splendid illustration of this process. The nineteenth century is the history, not of the rise of electricity or even of gaslight, but of the proliferation, concatenation, and spatial juxtaposition of multiple light forms. In 1900, one might routinely have encountered electric, gas, and oil lamps as well as candles over the course of a day.

The process of superimposition and displacement is a far more appropriate characterization of the nature of technological change than more traditional idioms of apotheosis, rupture, and replacement. Superimposition and displacement can, moreover, be operative over very long swathes of time. Illumination is, again, a good example of this. Electric light has certainly become hegemonic in the West, obviously, but has itself never assumed a single form (witness the rise of fluorescent and halogen lamps, e.g.).[11] It is also quite clearly not the only technology of illumination that Westerners use. Candles have remained popular for numerous purposes, notably the religious and the romantic, while numerous British pubs and houses, as well as the Park Estate in Nottingham, retain or have actively installed gaslighting.[12] The last public gas lamp in Manchester, in Aden Street, Ardwick, was extinguished in 1964 (a small ceremony marked its demise). Technological closure, then, is a protracted and potentially interminable process, as older technologies are displaced, remaining in areas untouched by newer objects, or finding new uses in older spaces. Electric light first appeared in certain metropolitan public spaces, country houses, textile factories, and railway stations. It did not replace older networks but sat over and alongside them: even in the City of London, it took many years for the gas streetlights to finally be disconnected. Technologies like gas and electric light could often not simply be transferred into older spaces: witness Sugg's complaints about fitting old houses with gaslight. Technological change, then, is best described as a process of complex, uneven sedimentation in space, rather than abrupt, jarring rupture in time.

These insights can be applied to the relation between technology, government, and society. As mentioned in the introduction, the genealogy of government forms in Europe since the early modern period can be depicted in terms of three superimposed and overlapping processes, relating to territory, population, and technology. Although these processes do have substantially different temporalities, they do not form discrete periods or epochs. Technological government became particularly salient in the nineteenth century, but it clearly existed before this time, just as concerns with territory and population did not wane. Nonetheless, it is this salience that is particularly relevant here. Late-nineteenth-century Britain was increasingly traversed and knitted together by technologies that, if they failed, would wreak social, economic, and medical calamity. Recall Dibdin's assertion that, without gas and electric light, "the whole scheme of present day society would at once fall to the ground" or Armstrong's observation that "modern society" was impossible without engineers.[13] This "modern society" was typified by urbanization, industrialization, capitalization, mobility, and increasing protection from endemic and epidemic disease.[14] All these processes were substantially, although not entirely, technological, in the sense that complex material systems, governed by varying degrees of scientific expertise, were necessary for their function. One does not need actor-network theory to see how inseparable technology and society had become.[15]

What were the characteristics of nineteenth-century infrastructure? Infrastructures themselves were increasingly engineered so as to be durable, inspectable, self-correcting, and disattendable. Building an electricity network thus involved materials science, networks of inspectors, self-regulating devices, and techniques allowing unobtrusive but clinical access (manholes, painted wires, fault localization methods). The perfect infrastructure—everlasting, self-operating, easily viewed by inspectors, yet utterly hidden from the public—remained, of course, necessarily elusive, which, as is suggested in chapter 2, was itself an omnipresent, powerful motor for government action. Failure, accident, and breakdown remained typical as the spectrum of "unintended consequences" of complex technological systems broadened.[16] Among these were various atmospheric, ecological, and environmental consequences that would become more pronounced and politically salient in the twentieth century.[17] These manifold, unpredictable events and trends generated more government activity in the form of the reports, commissions, regulations, and inspectorates that have appeared throughout this book.

Through processes of superimposition, displacement, and sedimentation, infrastructures could slowly restructure certain environmental aspects of daily life (heat, light, air quality, water) within which the bodily self-control and dynamism of liberal subjectivity could germinate. Perhaps there was an ecology of liberalism. This relation, of course, was usually anything but deterministic or direct: we are speaking of the constitution of material a prioris or conditions of possibility, not of technological determinism. One was not obliged to use gas in consecrated ways: one could choose to sit in darkness or to use gas as a means of suicide. Nonetheless, most users of gas, especially when appropriately disciplined through the prepayment meter, judiciously avoided darkness and asphyxiation. It would be inappropriate to characterize this as *disciplinary* in the Foucauldian sense, however. The home was not a coin-operated panopticon, after all, and individuals were free to use illumination in manifold private or oligoptic ways. It is these complex relations between embodied visual practices, spatial organization, and broader assumptions about rights and freedom that are, in significant ways, liberal.

——

I have explored a set of suggestive entanglements, connections, and affinities between certain networks of visual practice (from the oligoptic to the scrutiny of detail), particular concerns of liberal government (the delegation of power, indirect rule, concern for privacy, practical and dynamic agency), and technological devices and infrastructures (spectacles, asphalt, gaslight). Most of these practices, technologies, and connections are still in existence, although they have, in many cases, been superimposed and displaced by more familiar technologies of vision. This superimposition has, however, only intensified the liberal trends palpable in the nineteenth century. Rather than living in a "superpanopticon," we find that our visual experience is still characterized by fragmented, scattered patterns of mutual observation, supervision, inspection, privacy, voyeurism, individual motility, and scrutiny of detail.[18] Our politics and culture are saturated with obsessive, if technologically mutated, forms of these ultimately liberal visual modes. This book has attempted to recover some of their multiple points of emergence.

Notes

INTRODUCTION

1. William Dibdin, *Public Lighting by Gas and Electricity* (London:
    The Sanitary Publishing Co., 1902), 18.
2. For histories of illumination technology, see William T.
    O'Dea, *The Social History of Lighting* (London: Routledge
    & Kegan Paul, 1958); Brian Bowers, *Lengthening the Day:
    A History of Lighting Technology* (Oxford: Oxford Univer-
    sity Press, 1998); and Wolfgang Schivelbusch, *Disenchanted
    Night: The Industrialisation of Light in the Nineteenth Cen-
    tury*, trans. Angela Davis (New York: Berg, 1988). On illumi-
    nation as a "sign of modernity," see John A. Jakle's excel-
    lent *City Lights: Illuminating the American Night* (Baltimore:
    Johns Hopkins University Press, 2001), 15, 33; and Scott
    McQuire's "Dream Cities: The Uncanny Powers of Electric
    Light," *SCAN: Journal of Media Arts Culture*, 1, no. 2 (2004),
    http://scan.net.au/scan/journal/display.php?journal_id=31.
3. Schivelbusch, *Disenchanted Night*, 50, 55.
4. Vanessa Schwartz, *Spectacular Realities: Early Mass Culture in
    Fin-de-Siècle Paris* (Berkeley and Los Angeles: University of
    California Press, 1998), 21; Rosalind Williams, *Dream Worlds:
    Mass Consumption in Late Nineteenth-Century France* (Berkeley
    and Los Angeles: University of California Press, 1982), 85. See
    also Le Corbusier, *The Radiant City: Elements of a Doctrine of
    Urbanism to Be Used as the Basis of Our Machine-Age Civilisation*
    (New York: Orion, 1967); and Paul Virilio, "The Overexposed
    City," in *Rethinking Architecture: A Reader in Cultural Theory*, ed.
    Neal Leach (London: Routledge, 1997), 381–90. The equation
    of electrification and "dazzling cityscapes" is made in Stephen
    Graham and Simon Marvin's splendid synthesis, *Splintering
    Urbanism: Networked Infrastructures, Technological Mobilities
    and the Urban Condition* (London: Routledge, 2001), 45–47.

5. These themes can be found, in one form or another, in much scholarship. See, e.g., Schivelbusch, *Disenchanted Night*; Jose Amato, *Dust: A History of the Small and the Invisible* (Berkeley and Los Angeles: University of California Press, 2000), 84–85; Maureen Dillon, *Artificial Sunshine: A Social History of Domestic Lighting* (London: National Trust, 2002), 21; Murray Melbin, *Night as Frontier: Colonising the World after Dark* (New York: Free Press, 1987), 10; and A. Roger Ekirch, *At Day's Close: Night in Times Past* (New York: Norton, 2005), xxvii.

6. Philip Waller, "Introduction: The English Urban Landscape: Yesterday, To-day, and Tomorrow," in *The English Urban Landscape*, ed. Philip Waller (Oxford: Oxford University Press, 2000), 12–13.

7. The key texts here are Michel Foucault, *The Birth of the Clinic: An Archaeology of Medical Perception*, trans. Alan Sheridan (New York: Vintage, 1994), and *Discipline and Punish: The Birth of the Prison*, trans. Alan Sheridan (London: Penguin, 1991); and Walter Benjamin, *Illuminations*, ed. Hannah Arendt, trans. Harry Zohn (London: Fontana, 1991), and *The Arcades Project*, trans. Howard Eiland and Kevin McLaughlin (Cambridge, MA: Belknap, 2002).

8. Michel Foucault, "The Eye of Power," in *Power/Knowledge: Selected Interviews and Other Writings, 1972–1977*, ed. Colin Gordon (New York: Pantheon, 1980), 153–54. For a nuanced discussion of the desire for transparency in Enlightenment and early-nineteenth-century France, see Sharon Marcus, *Apartment Stories: City and Home in Nineteenth-Century Paris and London* (Berkeley and Los Angeles: University of California Press, 1999), 17–50.

9. A useful outline, with diagrams, is provided in Thomas Markus, *Buildings and Power: Freedom and Control in the Origin of Modern Building Types* (London: Routledge, 1993), 123–27.

10. Jeremy Bentham, *Panopticon; or, The Inspection-House: Containing the Idea of a New Principle of Construction Applicable to Any Sort of Establishment, in Which Persons of Any Description Are to Be Kept under Inspection; and in Particular to Penitentiary-Houses, Prisons, Houses of Industry, Work-Houses, Poor-Houses, Manufactories, Mad-Houses, Lazarettos, Hospitals, and Schools*, in Jeremy Bentham, *The Panopticon Writings*, ed. Miran Božovič (London: Verso, 1995), 34, 100, 35–36, 36.

11. Foucault, *Discipline and Punish*, 203.

12. Bentham, *Panopticon*, 93.

13. Jacques-Alain Miller, "La despotisme de l'utile: La machine panoptique de Jeremy Bentham," *Ornicar?* 3 (1975): 4. Miller describes the panopticon here as "le dispositif polyvalent de la surveillance, la machine optique universelle des concentrations humaines." On the "diagram of power," see Gilles Deleuze, *Foucault*, trans. and ed. Séan Hand (Minneapolis: University of Minnesota Press, 1988), 73; and Hubert Dreyfus and Paul Rabinow, *Michel Foucault: Beyond Structuralism and Hermeneutics*, 2nd ed. (Chicago: University of Chicago Press, 1988), 190–93. On the "urban panopticon," see Suren

Lalvani, *Photography, Vision and the Production of Modern Bodies* (Albany: State University of New York Press, 1996), 185. For the "panopticisation of society," see Stuart Elden, "Plague, Panopticon, Police," *Surveillance and Society* 1, no. 3 (2003): 247.

14. Gary Shapiro, *Archaelogies of Vision: Foucault and Nietzsche on Seeing and Saying* (Chicago: University of Chicago Press, 2003), 295. On "control society," see Stanley Cohen, *Visions of Social Control: Crime, Punishment and Classification* (Cambridge: Polity, 1985); Gilles Deleuze, "Postscript on the Societies of Control," in *Negotiations, 1972–1990*, trans. Martin Joughin (New York: Columbia University Press, 1995); and W. Bogard, *The Simulation of Surveillance: Hypercontrol in Telematic Societies* (Cambridge: Cambridge University Press, 1996). See also Z. Bauman, "On Postmodern Uses of Sex," in *Love and Eroticism*, ed. Mike Featherstone (London: Sage, 1999); and R. Boyne, "Post-Panopticism," *Economy and Society* 29, no. 2 (May 2000): 286.

15. Hille Koskela, "Cam Era—the Contemporary Urban Panopticon," *Surveillance and Society* 1, no. 3 (2003): 295. See also D. Lyon, *Surveillance Society: Monitoring Everyday Life* (Philadelphia: Open University Press, 2001). For an astute critique of the hasty characterization of CCTV as "panoptic," see Majid Yar, "Panoptic Power and the Pathologisation of Vision: Critical Reflections on the Foucauldian Thesis," *Surveillance and Society* 1, no. 3 (2003): 254–71. On nanopanopticism, see Michael Mehta, "Privacy vs. Surveillance: How to Avoid a Nano-Panoptic Future," *Canadian Chemical News*, November/December 2002, 31–33.

16. See, respectively, Lalvani, *Photography*; Susan Bordo, *Unbearable Weight: Feminism, Western Culture and the Body* (Berkeley and Los Angeles: University of California Press, 1993), 27; John Campbell and Matt Carlson, "Panopticon.com: Online Surveillance and the Commodification of Privacy," *Journal of Broadcasting and Electronic Media* 46, no. 4 (2002): 586–606; Clive Norris and Gary Armstrong, *The Maximum Surveillance Society: The Rise of CCTV* (New York: Berg, 1999); Jeremy Crampton, "Cartographic Rationality and the Politics of Geosurveillance and Security," *Cartography and Geographic Information Science* 30, no. 2 (2003): 135–48; Holly Blackford, "Playground Panopticism: Ring-around-the-Children, a Pocketful of Women," *Childhood: International Journal of Childhood Studies* 11, no. 2 (2004): 227–49; Tony Fabijancic, "The Prison in the Arcade: A Carceral Diagram of Consumer Space," *Mosaic* 34 (2001): 141–58; Garry Crawford, *Consuming Sport: Fans, Sport and Culture* (New York: Routledge, 2004); Vikki Bell, *Interrogating Incest: Feminism, Foucault and the Law* (London: Routledge, 1993); Marilyn Strathern, ed., *Audit Cultures: Anthropological Studies in Accountability, Ethics, and the Academy* (New York: Routledge, 2000); Chris Rojek and John Urry, *Touring Cultures: Transformations of Travel and Theory* (New York: Routledge, 1997); Peter Stokes, "Bentham, Dickens, and the Uses of the Workhouse," *Studies in English Literature, 1500–1900* 41, no. 4 (2001): 711–27; and Gareth Cordery, "Foucault, Dickens, and David Copperfield," *Victorian Literature and Culture*

26, no. 1 (1998): 71–85. This list is far from exhaustive or authoritative. It includes some splendid and astute scholarship.

17. For an encyclopedic review of the concept of malign vision in French thought, see Martin Jay, *Downcast Eyes: The Denigration of Vision in Twentieth-Century French Thought* (London: University of California Press, 1994). See also Shapiro, *Archaeologies of Vision*, 294.

18. Lauren Goodlad, *Victorian Literature and the Victorian State: Character and Governance in a Liberal Society* (Baltimore: Johns Hopkins University Press, 2003), x. For other critiques of the panoptic paradigm, see Jerrold Seigel, *The Idea of the Self: Thought and Experience in Western Europe since the Seventeenth Century* (New York: Cambridge University Press, 2005), 621; and Oz Frankel, *States of Inquiry: Social Investigations and Print Culture in Nineteenth-Century Britain and the United States* (Baltimore: Johns Hopkins University Press, 2006), 140.

19. Markus, *Buildings and Power*, 121–25.

20. "The Panopticon of Science and Art," in *Year-Book of Facts in Science and Art*, ed. John Timbs (London: David Bogue, 1855), 9–11.

21. See Thomas Holt, *The Problem of Freedom: Race, Labour and Politics in Jamaica and Britain, 1832–1938* (Baltimore: Johns Hopkins University Press, 1992), 105–12; and Philippa Levine, *Prostitution, Race and Politics: Policing Venereal Disease in the British Empire* (New York: Routledge, 2003), 87. In neither of these otherwise compelling analyses is the use of the term *panopticon* or *panoptic* substantiated or justified. Alain Corbin, in his magisterial *Women for Hire: Prostitution and Sexuality in France after 1850* (trans. Alan Sheridan [Cambridge, MA: Harvard University Press, 1990]), also finds the government of prostitution to be characterized by the "desire for panopticism" (9) but quickly notes that such desires were "frustrated" and that wise heads like Parent-Duchâtelet's appreciated that such strategies were counterproductive (10). On the colonial origins of the panopticon, see Timothy Mitchell, *Colonising Egypt* (Berkeley and Los Angeles: University of California Press, 1991), 35.

22. Boyne, "Post-Panopticism," 285, 303.

23. Bentham, *Panopticon*, 38.

24. Marcel Gauchet and Gladys Swain, *Madness and Democracy: The Modern Psychiatric Universe*, trans. Catherine Porter (Princeton, NJ: Princeton University Press, 1999), 84.

25. See Elden, "Plague, Panopticon, Police," 247. Foucault's nonpanoptic stances are unsurprising, given his basically Nietzschean position. See Shapiro, *Archaeologies of Vision*.

26. On the *synopticon*, which basically refers to televisual perception, see T. Matheisen, "The Viewer Society," *Theoretical Criminology* 1, no. 2 (1997): 215–34; and Boyne, "Post-Panopticism," 301. On the *polyopticon*, which refers to perception along multiple axes, see M. Allen, "'See You in the City!': Perth's Citiplace and the Space of Surveillance," in *Metropolis Now: Planning*

and the Urban in Contemporary Australia, ed. K. Gibson and S. Watson (Leichhart: Pluto Australia, 1994), 137–47. The *omnicon*, a rather unlikely state of affairs in which everyone watches everyone, is briefly invoked in Nic Groombridge, "Crime Control or Crime Culture TV?" *Surveillance and Society* 1, no. 1 (2002): 43. For the *oligopticon* (or, more properly, "oligoptica"), see Bruno Latour and Emilie Hermant, *Paris ville invisible* (Paris: Institut Synthélabo pour le progrès de la connaissance, 1998); and Bruno Latour, *Reassembling the Social: An Introduction to Actor-Network Theory* (Oxford: Oxford University Press, 2005); as well as chapter 2 below.

27. Yar, "Panoptic Power," 260–61.

28. Charles Baudelaire, "The Painter of Modern Life," in *The Painter of Modern Life and Other Essays*, trans. and ed. Jonathan Mayne (New York: Da Capo, 1986), 9 (cited also in Dana Brand, *The Spectator and the City in Nineteenth-Century American Literature* [Cambridge: Cambridge University Press, 1991], 5).

29. Benjamin, *Arcades Project*, 442. On intoxication, see ibid., 417. On the gendered nature of the flâneur, see Janet Wolff, "The Invisible *Flâneuse*," *Theory, Culture and Society* 2, no. 3 (1985): 37–47.

30. Benjamin, *Arcades Project*, 430, 446.

31. Susan Buck-Morss, "Dream World of Mass Consumption: Walter Benjamin's Theory of Modernity and the Dialectics of Seeing," in *Modernity and the Hegemony of Vision*, ed. David Levin (Berkeley and Los Angeles: University of California Press, 1993), 310. See also Williams, *Dream Worlds*; and Schwartz, *Spectacular Realities*.

32. Alan Blum, *The Imaginative Structure of the City* (Montreal: McGill-Queens University Press, 2003), 162. On the seductive commodity, see Jean Baudrillard, *Seduction*, trans. Brian Singer (New York: St. Martin's, 1990).

33. Brand, *The Spectator and the City*, 2. See also Georg Simmel, "The Metropolis and Mental Life," in *The Sociology of Georg Simmel*, ed. and trans. Kurt H. Wolff (New York: Free Press, 1950); and Keith Tester, introduction to *The Flâneur*, ed. Keith Tester (London: Routledge, 1994), 7. The bombardment of the senses is dealt with in numerous contexts, e.g., Robert Jütte, *A History of the Senses from Antiquity to Cyberspace*, trans. James Lynn (Cambridge: Polity, 2005), 182–86. The classic text here is, of course, Guy Debord, *Society of the Spectacle*, trans. Donald Nicholson-Smith (New York: Zone, 1994).

34. Brand, *The Spectator and the City*, 16.

35. George Augustus Sala, "The Secrets of the Gas," *Household Words* 9 (March 4, 1854): 159, cited in Rick Allen, "Observing London Street-Life: G. A. Sala and A. J. Munby," in *The Streets of London: From the Great Fire to the Great Stink*, ed. Tim Hitchcock and Heather Shore (London: Rivers Oram, 2003), 210.

36. Brand, *The Spectator and the City*, 43.

37. Susan Buck-Morss, *The Dialectics of Seeing: Walter Benjamin and the Arcades Project* (Cambridge, MA: MIT Press, 1989), 344–45.

38. For further critique of the flâneur, see Joachim Schlör, *Nights in the Big City: Paris, Berlin, London, 1840–1930*, trans. Pierre Gottfried Imhof and Dafydd Rees Roberts (London: Reaktion, 1998), 244, 263.

39. There are others, notably economic, that will also be touched on throughout this book, albeit rather cursorily. For the economics of utility management, see R. Millward, "The Political Economy of Urban Utilities," in *The Cambridge Urban History of Britain*, vol. 3, *1840–1950*, ed. Martin Daunton (Cambridge: Cambridge University Press, 2000), 315–49; J. Wilson, *Lighting the Town: A Study of Management in the North West Gas Industry, 1805–1880* (London: P. Chapman, 1991); Derek Matthews, "The Technical Transformation of the Late Nineteenth-Century Gas Industry," *Journal of Economic History* 47, no. 4 (December 1987): 967–80; and John Poulter, *An Early History of Electricity Supply: The Story of Electric Light in Victorian Leeds* (London: P. Peregrinus, 1986).

40. *British Architect and Northern Engineer* 18 (December 22, 1882): 607.

41. W. H. Chaloner, *The Social and Economic Development of Crewe, 1780–1923* (Manchester: Manchester University Press, 1950), 208.

42. *Engineer* 77 (August 2, 1895): 112.

43. Edward Nettleship, *Diseases of the Eye*, 5th ed. (Philadelphia: Lea Bros., 1890), 276.

44. William Hosgood Young Webber, *Town Gas and Its Uses for the Production of Light, Heat and Motive Power* (New York: D. Van Nostrand, 1907), 122.

45. Dibdin, *Public Lighting*, 398–99.

46. On Morris, see Asa Briggs, *Victorian Things* (London: Penguin, 1988), 229. For Stephenson, see Robert Louis Stephenson, "A Plea for Gas Lamps," in *Virginibus Puerisque* (London: C. K. Paul, 1881).

47. *Electrician* 4 (May 7, 1880): 325.

48. Schivelbusch cites Dondey-Dupré's scheme to illuminate the whole of Paris via a series of lighthouses as well as more realized schemes for "tower lighting" in American cities (see *Disenchanted Night*, 121–34). Impressive and interesting as such schemes are, they were totally unrepresentative of contemporary British engineering practice.

49. Isaiah Berlin, "Two Concepts of Liberty," in *Liberty, Incorporating Four Essays on Liberty*, ed. Henry Hardy (Oxford: Oxford University Press, 2002), 121.

50. For useful introductory collections of essays, see Peter Mandler, ed., *Liberty and Authority in Victorian Britain* (Oxford: Oxford University Press, 2006); and Richard Bellamy, ed., *Victorian Liberalism: Nineteenth-Century Political Thought and Practice* (London: Routledge, 1990). See also Goodlad, *Victorian Literature and the Victorian State*; T. A. Jenkins, *The Liberal Ascendancy, 1830–1886* (Basingstoke: Macmillan, 1994); and Jonathan Parry, *The Rise and Fall of Liberal Government in Victorian Britain* (New Haven, CT: Yale University Press, 1993).

51. Key primary texts here include Edmund Burke, "Thoughts and Details on Scarcity," in *The Portable Edmund Burke*, ed. Isaac Kramnick (New York: Penguin, 1999), 194–212; J. S. Mill, "On Liberty," in *Essential Works of John Stuart Mill*, ed. Max Lerner (London: Bantam, 1961); Herbert Spencer, *The Man versus the State, with Six Essays on Government, Society and Freedom* (Indianapolis: Liberty Classics, 1981); and L. T. Hobhouse, *Liberalism* (New York: H. Holt, 1911).

52. John Stuart Mill, *Considerations on Representative Government* (New York: Harper & Bros., 1862), 286.

53. The importance of negative liberty to Victorian government is emphasized in Philip Harling, "The Powers of the Victorian State," in Mandler, ed., *Liberty and Authority*, 25–50.

54. On negative and positive liberty, see Berlin, "Two Concepts," 169–81. See also Stephan Collini, *Liberalism and Sociology: L. T. Hobhouse and Political Argument in England, 1880–1914* (Cambridge: Cambridge University Press, 1979), 46–49.

55. See Richard Bellamy, "J. S. Mill, T. H. Green and Isaiah Berlin on the Nature of Liberty and Liberalism," in *Rethinking Liberalism* (London: Pinter, 2000), 22–46.

56. Peter Mandler, "Introduction: State and Society in Victorian Britain," in Mandler, ed., *Liberty and Authority*, 16.

57. Mill, *Considerations on Representative Government*, 178.

58. The exclusivity of liberal subjectivity is obviously most marked in the areas of race and gender. For an exemplary discussion of liberalism in its imperial context, see Uday Singh Mehta, *Liberalism and Empire: A Study in Nineteenth-Century British Liberal Thought* (Chicago: University of Chicago Press, 1999). For a succinct discussion of the "despotism" at the heart of liberalism, see Mariana Valverde, "Despotism and Ethical Liberal Governance," *Economy and Society* 25, no. 3 (1996): 357–72. On liberalism and gender, see Susan Kingsley Kent, *Gender and Power in Britain, 1640–1900* (London: Routledge, 1999); Carole Pateman, *The Sexual Contract* (Stanford, CA: Stanford University Press, 1988); and Elizabeth Maddock Dillon, *The Gender of Freedom: Fictions of Liberalism and the Literary Public Sphere* (Stanford, CA: Stanford University Press, 2004).

59. See Stefan Collini, "The Idea of 'Character' in Victorian Political Thought," *Transactions of the Royal Historical Society* 35 (1985): 29–50.

60. Samuel Smiles, *Self-Help: With Illustrations of Character, Conduct and Perseverance*, ed. Peter Sinnema (Oxford: Oxford University Press, 2002), 315. On the following page, Smiles declares that "character is power," in pointed contradistinction to the Baconian-Hobbesian formation that "knowledge is power."

61. This is a central premise of Patrick Joyce's *The Rule of Freedom: Liberalism and the Modern City* (London: Verso, 2003), 2.

62. Adam Smith, *Theory of Moral Sentiments* (Amherst, NY: Prometheus, 2000), 342–43 (pt. 6, sec. 2).

63. Alan Sykes, *The Rise and Fall of British Liberalism, 1776–1988* (New York: Longman, 1997), 1. Here, liberalism is described as an eighteenth-century "doctrine of opposition."

64. This definition of *discourse* owes much to Foucault. See Michel Foucault, *The Order of Things: An Archaeology of the Human Sciences* (London: Routledge, 1970), and *The Archaeology of Knowledge*, trans. Alan Sheridan (New York: Pantheon, 1972).

65. Attempts to equate liberalism with the Liberal Party are notoriously difficult. See Sykes, *Rise and Fall of British Liberalism*; and Parry, *Rise and Fall of Liberal Government*. Both Sykes and Parry admit the difficulty of their project, and both arguably fail. See also E. Biagini, *Liberty, Retrenchment and Reform: Popular Liberalism in the Age of Gladstone, 1860–1880* (Cambridge: Cambridge University Press, 1992).

66. This is a vital point, made in Joyce, *Rule of Freedom*, 102–3, 113.

67. See Oliver MacDonagh, "The Nineteenth-Century Revolution in Government: A Reappraisal," *Historical Journal* 1, no. 1 (1958): 59–73, and *Early Victorian Government, 1830–1870* (New York: Holmes & Meier, 1977).

68. Adam Smith, *An Enquiry into the Nature and Causes of the Wealth of Nations*, ed. Edwin Cannan, 2 vols. in 1 (Chicago: University of Chicago Press, 1976), 2:244–45, 253 (quote).

69. Mill, *Considerations on Representative Government*, 292–93. See also Christine Bellamy, *Administering Central-Local Relations, 1871–1919: The Local Government Board in Its Fiscal and Cultural Context* (Manchester: Manchester University Press, 1988), 9.

70. Winston Churchill, "Liberalism and Socialism" (speech delivered at St. Andrew's Hall, Glasgow, October 11, 1906), reprinted in Robert Rhodes James, ed., *Winston S. Churchill: His Complete Speeches, 1897–1963*, 8 vols. (London: Chelsea House, 1974), 1:675–76. See also Winston Churchill, *Liberalism and the Social Problem* (London: Hodder & Stoughton, 1909).

71. On population and "biopower," see Michel Foucault, *The History of Sexuality*, vol. 1, *An Introduction*, trans. Robert Hurley (New York: Vintage, 1990); Theodore Porter, *The Rise of Statistical Thinking* (Princeton, NJ: Princeton University Press, 1986); and Mary Poovey, *Making a Social Body: British Cultural Formation, 1830–1864* (Chicago: University of Chicago Press, 1995). On the earlier concept of territoriality, see Chandra Mukerji, *Territorial Ambitions and the Gardens of Versailles* (Cambridge: Cambridge University Press, 1997). On territory and population, see Dreyfus and Rabinow, *Michel Foucault*, 133–39.

72. Manu Goswami, *Producing India: From Colonial Economy to National Space* (Chicago: University of Chicago Press, 2004), esp. 31–131. See also Patrick Carroll, *Science, Culture and Modern State Formation* (Berkeley and Los Angeles: University of California Press, 2006); Christopher Hamlin, *Public Health*

and Social Justice in the Age of Chadwick: Britain, 1800–1854 (Cambridge: Cambridge University Press, 1998), 264–65; and Ben Marsden and Crosbie Smith, *Engineering Empires: A Cultural History of Technology in Nineteenth-Century Britain* (New York: Palgrave Macmillan, 2005). For hydraulic engineering, see Matthew Gandy, "Rethinking Urban Metabolism: Water, Space and the Modern City," *City* 8, no. 3 (December 2004): 371–87. For a thoughtful and expanded discussion of technology and politics, see Andrew Barry, *Political Machines: Governing a Technological Society* (London: Athlone, 2001).

73. The *Oxford English Dictionary* defines *infrastructure* as "a collective term for the subordinate parts of an undertaking; substructure, foundation; *spec.* the permanent installations forming a basis for military operations, as airfields, naval bases, training establishments, etc." The term's military origins should remind us of the close connections between states, engineers, and armies. The word was first used in English in 1927, but I am using it, a touch anachronistically, to refer to any large-scale system of collective technological provision (communication, water, energy, information).

74. Samuel Smiles, *Lives of the Engineers: Metcalfe-Telford: History of Roads* (London: John Murray, 1904), 2.

75. This "growth of government" is covered in greater detail in chapter 3 below. It is an old debate but still significant in light of recent scholarship on expertise, governmentality, and the social. The key works here are Albert Venn Dicey, *Lectures on the Relation between Law and Public Opinion in England during the Nineteenth Century* (London: Macmillan, 1905); MacDonagh, "Revolution in Government"; and Henry Parris, "The Nineteenth Century Revolution in Government: A Reappraisal Reappraised," *Historical Journal* 2 (1960): 17–37. See also Karl Polanyi, *The Great Transformation: The Political and Economic Origins of Our Time*, new ed. (Boston: Beacon, 2001).

76. For an excellent introduction to the development of "expertise" in nineteenth-century government, see Roy MacLeod, ed., *Government and Expertise: Specialists, Administrators, and Professionals, 1860–1919* (Cambridge: Cambridge University Press, 1988).

77. Edwin Chadwick, *Report on the Sanitary Condition of the Labouring Population of Great Britain*, ed. M. W. Flinn (Edinburgh: Edinburgh University Press, 1965), 380.

78. See R. A. Buchanan, "Institutional Proliferation in the British Engineering Profession, 1847–1914," *Economic History Review*, n.s., 38, no. 1 (1985): 42–60, and "Engineers and Government in Nineteenth-Century Britain," in MacLeod, ed., *Government and Expertise.*

79. Samuel Smiles, *Lives of the Engineers: Vermuyden-Myddleton-Perry-James Brindley* (London: John Murray, 1904), xx.

80. Henry Armstrong, "The Reign of the Engineer," *Quarterly Review* 198, no. 396 (October 1903): 462.

81. Chadwick, *Sanitary Report*, 164 (emphasis added).

82. On governmentality, see Michel Foucault, "Governmentality," in *The Foucault Effect: Studies in Governmentality*, ed. Graham Burchell, Colin Gordon, and Peter Miller (Chicago: University of Chicago Press, 1991). See also Graham Burchell, "Governmental Rationality: An Introduction," in ibid.; Nikolas Rose, *Powers of Freedom: Reframing Political Thought* (Cambridge: Cambridge University Press, 1999); Mitchell Dean, *Governmentality: Power and Rule in Modern Society* (London: Sage, 1999); and Simon Gunn, "From Hegemony to Governmentality: Changing Conceptions of Power in Social History," *Journal of Social History* 39, no. 3 (2006): 705–20.

83. A notable and fascinating exception is Joyce's *Rule of Freedom*, which treats urban systems (sewers, markets) and rationalities (maps, statistics) as technologies essential to liberal rule but also the source of instability within it. Much of the rest of this work gestures to the material before ultimately reducing it to an expression of prior, pure rationalities. See, e.g., Dean, *Governmentality*, 30. See also Mitchell Dean, "Putting the Technological into Government," *History of the Human Sciences* 9, no. 3 (1996): 47–68; and Rose, *Powers of Freedom*, 55.

84. Foucault, *Discipline and Punish*, 23.

85. Michel Foucault, "Space, Knowledge and Power," in *The Foucault Reader*, ed. Paul Rabinow (New York: Pantheon, 1984), 240.

86. This was Engels's interpretation of Saint-Simon in Karl Marx and Friedrich Engels, *Werke*, 43 vols. (Berlin: Dietz, 1956–90), 19:195. The English citation is found in Isaiah Berlin, "Political Ideas in the 20th Century," in *Liberty*, 85.

87. See, e.g., Bruno Latour, *We Have Never Been Modern*, trans. Catherine Porter (London: Harvester, 1993), *Pandora's Hope: Essays on the Reality of Science Studies* (London: Harvard University Press, 1999), and *Reassembling the Social*, esp. 63–86. The phrase *material agency* is Andrew Pickering's. See *The Mangle of Practice: Time, Agency and Science* (Chicago: University of Chicago Press, 1995). This maneuver is not uncontroversial or without metaphysical implications. It is important to note that, in adopting a kind of performative symmetry between humans and nonhumans, such writers are not completely, irresponsibly, and foolishly doing away with the distinction: humans feel pain and have intentionality and consciousness, among innumerable other traits.

88. Bentham, *Panopticon*, 37.

89. Henry Austin, "An Instance of Faulty Arrangements of Dwellings, and Plan for Its Improvement," *Health of Towns Commission Report* (1845), 2:356, cited in Charles Girdlestone, *Letters on the Unhealthy Condition of the Lower Class of Dwellings, Especially in Large Towns* (London: Longman, Brown, Green & Longmans, 1845), 77.

90. William Preece, "The Sanitary Aspects of Electric Lighting," *Electrician* 25 (August 29, 1890): 464.

91. L. T. Hobhouse, *Democracy and Reaction*, ed. P. F. Clarke (London: Harvester, 1972), 120 (cited also in Goodlad, *Victorian Literature and the Victorian State*, 209).

92. On technocracy in Britain, see Jon Agar, *The Government Machine: A Revolutionary History of the Computer* (Cambridge, MA: MIT Press, 2003), 3, 10, 427. See also Langdon Winner, *Autonomous Technology: Technics-out-of-Control as a Theme in Political Thought* (Cambridge, MA: MIT Press, 1978), 135–72.

93. D. G. Ritchie, *The Moral Function of the State* (1889), cited in Michael Taylor, ed., *Herbert Spencer and the Limits of the State: The Late Nineteenth-Century Debate between Individualism and Collectivism* (Bristol: Thoemmes, 1996), 174.

94. Michel Foucault, "The Subject and Power," in Dreyfus and Rabinow, *Michel Foucault*, 221. See also Dean, *Governmentality*, 29.

95. The canonical text remains Georges Canguilhem, *The Normal and the Pathological*, trans. Carolyn R. Fawcett (New York: Zone, 1991). The tension between norm and law is explored in Jan Goldstein, *Console and Classify: The French Psychiatric Profession in the Nineteenth Century*, new ed. (Chicago: University of Chicago Press, 2001). See also François Ewald, "Norms, Discipline and the Law," *Representations* 30 (Spring 1990): 138–61.

96. Bentham cited in Frederick Rosen, "The Origins of Liberal Utilitarianism: Jeremy Bentham and Liberty," in Bellamy, ed., *Victorian Liberalism*, 59.

97. On liberal governmentality, see Rose, *Powers of Freedom*; Dean, *Governmentality*; and my "Making Liberalism Durable: Vision and Civility in the Late-Victorian City," *Social History* 27, no. 1 (2002): 1–15.

98. Albert Borgmann, *Technology and the Character of Contemporary Life: A Philosophical Enquiry* (Chicago: University of Chicago Press, 1984), 34.

CHAPTER 1

1. W. Whalley, *A Popular Description of the Human Eye, with Remarks on the Eyes of Inferior Animals* (London: J. & A. Churchill, 1874), 1.

2. This salience of the visual was not essentially new. Several Western traditions of thought (Platonic, Cartesian) have emphasized vision's salience. Descartes stated: "All the management of our lives depends on the senses, and since that of sight is the most comprehensive and noblest of these, there is no doubt that the inventions which serve to augment its power are among the most useful that there can be" (René Descartes, *Discourse on Method, Optics, Geometry and Meteorology*, trans. Paul J. Olscamp [Indianapolis: Hackett, 2001], 65). Hegel equated eyes, human bodily form, and the organic development of civilization a century before Freud made the same, and more famous, connection: "In the formation of the animal head the predominant thing is the mouth, as the tool for chewing, with the upper and lower jaw, the teeth and masticatory muscles. The other organs are added to this principal organ only as servants and helpers: the nose especially as sniffing out

food, the eye, less important, for spying it" (G. W. F. Hegel, *Aesthetics: Lectures on Fine Art*, trans. T. M. Knox, 2 vols. [Oxford: Clarendon, 1975], 2:728).

3. Charles Darwin, *The Origin of Species by Means of Natural Selection; or, The Preservation of Favoured Races in the Struggle for Life* (London: Penguin, 1985), 217.

4. I take the phrase from Levin, ed., *Modernity and the Hegemony of Vision*. See also David Levin, ed., *Sites of Vision: The Discursive Construction of Sight in the History of Philosophy* (Cambridge, MA: MIT Press, 1997); and Jay, *Downcast Eyes*. The hegemony of vision is reinforced by titles like Jozef Cohen's *Sensation and Perception*, vol. 2, *Audition and the Minor Senses* (Chicago: Rand-McNally, 1969).

5. Georg Simmel, "Sociology of the Senses: Visual Interaction," in *Introduction to the Science of Sociology*, ed. Robert E. Park and Ernest W. Burgess (Chicago: University of Chicago Press, 1921), 360–61. See also Benjamin, *Arcades Project*, 433.

6. The literature here is vast. See, e.g., Foucault, *Birth of the Clinic*; Jean-Paul Sartre, *Being and Nothingness: An Essay on Phenomenological Ontology*, trans. Hazel E. Barnes (New York: Philosophical Library, 1956); Jacqueline Rose, *Sexuality and the Field of Vision* (London: Verso, 1986); Griselda Pollock, *Vision and Difference: Feminism, Femininity and Histories of Art*, new ed. (London: Routledge, 2003); Frank Graziano, *The Lust of Seeing: Themes of the Gaze and Sexual Rituals in the Fiction of Felisberto Hernández* (London: Associated University Presses, 1997); Frantz Fanon, *Black Skin, White Masks*, trans. Charles Markmann (New York: Grove, 1967); and Mary-Louise Pratt, *Imperial Eyes: Travel Writing and Transculturation* (London: Routledge, 1992).

7. Henri Lefebvre, *The Production of Space* (Oxford: Blackwell, 1991), 139.

8. Christoph Asendorf, *Batteries of Life: On the History of Things and Their Perception in Modernity*, trans. Don Reneau (London: University of Berkeley Press, 1993), 94. See also Buck-Morss, *Dialectics of Seeing*, 287.

9. The very division into sight, hearing, touch, taste, and smell is a Western historical product and should never be uncritically depicted as timeless or natural. The number of discrete ways in which an organism physically interacts with its environment can be divided very differently. See, e.g., Robert Rivlin and Karen Gravelle, "The Seventeen Senses," in *Deciphering the Senses: The Expanding World of Human Perception* (New York: Simon & Schuster, 1984), 9–28; and L. Watson, *Jacobson's Organ, and the Remarkable Nature of Smell* (New York: Norton, 2000). Anthropologists have often remarked on the cultural specificity of the West's five senses. The Hausa, e.g., recognize only two senses, while the Javanese describe talking as a sense. See David Howes, ed., *The Varieties of Sensory Experience: A Sourcebook in the Anthropology of the Senses* (Toronto: University of Toronto Press, 1991); and Kathryn Geurts, *Culture and the Senses: Bodily Ways of Knowing in an African Community* (Berkeley and Los Angeles: University of California Press, 2002). Additionally, the Western

sensorium privileges the outward dimensions of sensation: this dimension of experience is often referred to as *exteroperception* (for Westerners, the five senses connecting with an extrinsic world). But other axes of feeling exist, notably *interoperception* (inner-body sensations, murmurs, twitches, and gurglings) and *proprioception* (the sense of balance and equipoise, muscular tension, etc.). The former is easily conflated with the sense of touch, which may, perhaps, explain its relative lack of salience in sensory discourse. These terms were devised by Charles Sherrington. See his *The Integrative Action of the Nervous System* (Oxford: Oxford University Press, 1906), 316–24. For phenomenological analysis of them, see Drew Leder, *The Absent Body* (Chicago: University of Chicago Press, 1990), 36–68.

10. Jonathan Crary, *Suspensions of Perception: Attention, Spectacle and Modern Culture* (Cambridge, MA: MIT Press, 1999), 11–12. See also Jonathan Crary, *Techniques of the Observer: On Vision and Modernity in the Nineteenth Century* (London: MIT Press, 1990).

11. Crary draws on Foucault's *The Order of Things*. Foucault argues there that, with the modern episteme, all knowledge is mediated by the body: "[The study of] perception, sensorial mechanisms, neuro-motor diagrams, and the articulation common to things and the organism . . . led to the discovery that knowledge has anatomo-physiological conditions, that it is formed gradually within the structures of the body, that it may have a privileged place within it, but that its forms cannot be dissociated from its peculiar functioning" (319). The classical episteme is also taken from *The Order of Things*: it refers temporally to the period before the nineteenth century and conceptually to a system of thought that exists outside and apart from man, who is, hence, not subject to its (linguistic, physical) laws and, thus, does not exist (itself a problematic and, perhaps, deliberately counterintuitive provocation). While this may brilliantly capture an important strain in Western thinking, it remains a totalizing, undialectical concept, and, for this reason, it is rejected here. Crary and Foucault illustrate a prevalent trend that is far more compromised and gradual than the conception of epistemes allows. It is worth noting that, if the physiological dimensions of vision existed within the classical model, then aspects of the classical model also reappear in the later, modern episteme. See John Tyndall, *Light and Electricity: Notes on Two Courses of Lectures before the Royal Institution of Great Britain* (New York: D. Appleton, 1871), in which Tyndall is not epistemologically prevented from asserting that "the eye is a camera obscura" (48), albeit one that is temporally mediated (51) and "by no means a perfect optical instrument" (52). Helmholtz made similar observations. See Laura Otis, *Networking: Communicating with Bodies and Machines in the Nineteenth Century* (Ann Arbor: University of Michigan Press, 2001), 43. To be fair, Crary acknowledges the "partial" persistence of the classical "well into the nineteenth century" (*Suspensions of Perception*, 12 n. 1), which somewhat undermines the epistemic model on which his argument is based.

12. John Locke, *An Essay concerning Human Understanding*, ed. Peter H. Nidditch (Oxford: Clarendon, 1975), 171. On early modern perception, see Robert L. Martensen, *The Brain Takes Shape: An Early History* (Oxford: Oxford University Press, 2004).

13. For an insightful discussion of Locke's views of personhood, see Seigel, *Idea of the Self*, esp. 99.

14. Locke, *Essay concerning Human Understanding*, 43.

15. These thinkers included Leonardo da Vinci and Vesalius. See Nicholas Wade, *A Natural History of Vision* (Cambridge, MA: MIT Press, 1998), 26–37.

16. Arthur Zajonc, *Catching the Light: The Entwined History of Light and Mind* (New York: Oxford University Press, 1993), 30.

17. Locke, *Essay concerning Human Understanding*, 163.

18. For Hartley, see Roy Porter, *Flesh in the Age of Reason: The Modern Foundations of Body and Soul* (New York: Norton, 2004), 347–73. See also Christopher Lawrence, "The Nervous System and Society in the Scottish Enlightenment," in *Natural Order: Historical Studies of Scientific Culture*, ed. Barry Barnes and Steven Shapin (London: Sage, 1979), 19–40.

19. The politics of early-nineteenth-century physiology are explored in L. S. Jacyna, "Immanence or Transcendence: Theories of Life and Organization in Britain, 1790–1835," *Isis* 74, no. 243 (September 1983): 311–29.

20. Thomas Reid, *An Inquiry into the Human Mind*, ed., with an introduction by Timothy Duggan (Chicago: University of Chicago Press, 1970), 145, 157.

21. E. Clarke and L. S. Jacyna, *Nineteenth-Century Origins of Neurological Concepts* (Berkeley and Los Angeles: University of California Press, 1987), 215–16.

22. Robert Young, *Mind, Brain and Adaptation in the Nineteenth Century: Cerebral Localization from Gall to Ferrier* (Oxford: Clarendon, 1970), 64–65, 74–80.

23. Crary, *Techniques of the Observer*, 81.

24. Joseph Le Conte, *Sight: An Exposition of the Principles of Monocular and Binocular Vision*, 2nd ed. (New York: D. Appleton, 1897), 1–2.

25. The *Oxford English Dictionary* defines *sensation* as "an operation of any of the senses; a psychical affection or state of consciousness consequent on and related to a particular condition of some portion of the bodily organism, or a particular impression received by one of the organs of sense. Now commonly in more precise use, restricted to the subjective element in any operation of one of the senses, *a physical 'feeling' considered apart from the resulting 'perception' of an object*" (emphasis added). *Perception* is defined in numerous related ways, including "the taking cognizance or being aware of a sensible or quasi-sensible object" and "in strict philosophical language (first brought into prominence by Reid): The action of the mind by which it refers its sensations to an external object as their cause. Distinguished from *sensation, conception* or imagination, and *judgement* or inference." See also Alison Winter, *Mesmerized: Powers of Mind in Victorian Britain* (Chicago: University of Chicago Press, 1998), 39; and R. Olson, *Scottish Philosophy and British Physics, 1750–*

1880: A Study in the Foundations of the Victorian Scientific Style (Princeton, NJ: Princeton University Press, 1975).

26. This critique existed within the immanent model and can be considered apart from the transcendentalist critique of the physiological paradigm (see Jacyna, "Immanence or Transcendence"). Critics here would include antimaterialists, antiatheists, vitalists, etc.

27. Pierre Cabanis, *Rapports du physique et du moral de l'homme* (1802), cited in Alain Corbin, *The Foul and the Fragrant: Odor and the French Social Imagination*, trans. Miriam L. Kochan (Cambridge, MA: Harvard University Press, 1986), 139.

28. Maurice Merleau-Ponty, *Phenomenology of Perception*, trans. Colin Smith (London: Routledge & Kegan Paul, 1962), 235.

29. Of course, the eye had functioned as a critical object in earlier forms of discourse, e.g., as a sevenfold set of concentric spheres in Aristotelian cosmology, making it the homologous microcosm of the celestial world. It was also apprehended in clinical detail by Galen, who perused the eyes of freshly slaughtered oxen and saw them as orbs, not spheroids. The modern physiological version was not "truer" but, rather, a historically novel truth system, established via empirical scientific techniques. See David Park, *The Fire within the Eye: A Historical Essay on the Nature and Meaning of Light* (Princeton, NJ: Princeton University Press, 1997).

30. Thomas Young, *A Course of Lectures on Natural Philosophy and the Mechanical Arts*, 2 vols. (London: Johnson, 1807), 1:447, cited in Wade, *Natural History of Vision*, 85.

31. Human "double vision" is today described by the terms *photopic* and *scotopic*. Photopic perception is operative by daylight, when the entire chromatic spectrum is physiologically activated by the cones grouped around the fovea centralis, while scotopic perception is activated at dusk when chromatic perception dims and is replaced by a wider, greyer, more peppery visual field. See H. Ripps and R. A. Weale, "The Visual Photoreceptors," in *The Eye*, vol. 2A, *Visual Function in Man*, ed. Hugh Davson (New York: Academic, 1969), 5–8.

32. W. H. R. Rivers, "Vision," in *Text-Book of Physiology*, ed. E. A. Schäfer, 2 vols. (London: Young J. Pentland, 1900), 2:1083–88.

33. Jan Purkinje's text is *Beobachtungen und Versuche zur Physiologie der Sinne: Neue Beiträge zur Kenntniss des Sehens in subjectiver Hinsicht* (Berlin: Reimer, 1825). The term *Purkinje shift* (or *Purkinje phenomenon*) was coined in 1882. See Wade, *Natural History of Vision*, 173–75.

34. F. C. Donders, *On the Accommodation and Refraction of the Eye, with a Preliminary Essay on Physiological Dioptrics*, trans. William Daniel Moore (London: New Sydenham Society, 1864), 39.

35. Wade, *Natural History of Vision*, 183.

36. Donders, *Accommodation and Refraction*, 197; Tyndall, *Light and Electricity*, 52. See also Crary, *Suspensions of Perception*, 215–18, 220.

37. David Brewster, "The Sight and How to See," *North British Review* 26 (1856): 170; Robert Carter, *A Practical Treatise on Diseases of the Eye* (Philadelphia: Henry C. Lea, 1876), 101. See also Elizabeth Green Musselman, *Nervous Conditions: Science and the Body Politic in Early Industrial Britain* (Albany: State University of New York Press, 2006), 125–26.

38. The degree of pupillary contraction was proportionate to the logarithm of intensity. Rivers, "Vision," 1042–43.

39. André Blondel, "Street Lighting by Arc Lamps," *Electrician* 36 (November 15, 1895): 90.

40. William Preece commenting on J. Swinburne, "Electrical Measuring Instruments," *Minutes of Proceedings* (Institute of Civil Engineers) 110 (April 26, 1892): 55–56. See also Webber, *Town Gas*, 125–28.

41. Donders, *Accommodation and Refraction*, 162.

42. Charles Wheatstone, "Contributions to the Physiology of Vision: Part the First. On Some Remarkable, and Hitherto Unobserved, Phenomena of Binocular Vision," *Philosophical Transactions of the Royal Society of London* 128 (1838): 380.

43. The zonule of Zinn is a ligament holding the lens in place and facilitating ocular accommodation. See R. E. Dudgeon, *The Human Eye; Its Optical Construction Popularly Explained* (London: Hardwicke & Bogue, 1878), 63.

44. Henry Maudsley, *Body and Mind—an Enquiry into Their Connection and Mutual Influence, Specially in Reference to Mental Disorders* (New York: D. Appleton, 1871), 32.

45. Crary, *Suspensions of Perception*, 291.

46. Otis, *Networking*, 25–28; Anson Rabinbach, *The Human Motor: Energy, Fatigue and the Origins of Modernity* (Berkeley and Los Angeles: University of California Press, 1992), 93. The temporal implications of vision for optical technologies are explored in Lynda Nead, "Velocities of the Image, c. 1900," *Art History* 27, no. 5 (November 2004): esp. 760–62.

47. William Steavenson, "Electricity, and Its Manner of Working in the Treatment of Disease," *Electrician* 14 (March 28, 1886): 408.

48. Brewster, "The Sight and How to See," 161.

49. Donders, *Accommodation and Refraction*, 84, 210, 81. The term *emmetropia* was first used in English in Donders's text (1864).

50. Ibid., 175 (emphasis added).

51. W. F. Southard, *The Modern Eye: With an Analysis of 1300 Errors of Refraction* (San Francisco: W. A. Woodward, 1893), 4.

52. Otto Haab, *Atlas and Epitome of Ophthalmoscopy and Ophthalmoscopic Diagnosis*, ed. G. E. Schweintz, 2nd American ed. (Philadelphia: W. B. Saunders, 1910), 67.

53. William Porterfield, "An Essay concerning the Motions of Our Eyes: Part II. Of Their Internal Motions," *Edinburgh Medical Essays and Observations* 4 (1738), discussed in Wade, *Natural History of Vision*, 43.

54. Donders, *Accommodation and Refraction*, 10.

55. William MacKenzie, *A Practical Treatise on the Diseases of the Eye*, 4th ed., rev. and enlarged (London: Blanchard & Lea, 1855), 850.

56. Donders, *Accommodation and Refraction*, 245.

57. Lindsay Granshaw, "'Fame and Fortune by Means of Bricks and Mortar': The Medical Profession and Specialist Hospitals in Britain, 1800–1948," in *The Hospital in History*, ed. Lindsay Granshaw and Roy Porter (New York: Routledge, 1989), 204.

58. Lawrence cited in W. F. Bynum, *Science and the Practice of Medicine in the Nineteenth Century* (Cambridge: Cambridge University Press, 1994), 193. See also Foucault, *Birth of the Clinic*.

59. Alexander Wynter Blyth, *A Dictionary of Hygiène and Public Health, Comprising Sanitary Chemistry, Engineering, and Legislation, the Dietetic Values of Foods, and the Detection of Adulterations* (London: C. Griffin, 1876), 93.

60. MacKenzie, *Practical Treatise*, 366–68. *Ophthalmia* was a generic term for any inflammatory and usually contagious condition of any part of the eye. More specific terms were used for particular areas affected, e.g., *iritis, corneitis,* and *conjunctivitis.*

61. Carter, *Practical Treatise*, 139.

62. MacKenzie, *Practical Treatise*, 732–33 (on cataracts), 370 (on strabismus), 799 (on eyelid defects and artificial pupils).

63. Nettleship, *Diseases of the Eye*, 127–28. An example of a failed transplant of a rabbit's eye (involving Charles H. May, a physician at New York Polyclinic) was documented in *Archives of Ophthalmology* 16 (1887): 47, 182.

64. Nettleship, *Diseases of the Eye*, 206.

65. H. Haynes Walton, *A Treatise on Operative Ophthalmic Surgery*, 1st American ed., ed. S. Little (Philadelphia: Lindsay & Blakiston, 1853), 489.

66. MacKenzie, *Practical Treatise*, 654.

67. Carter, *Practical Treatise*, 412–13.

68. William Cumming, "On a Luminous Appearance of the Human Eye," *Transactions of the Medico-Chirurgical Society* 29 (1846): 283; Carter, *Practical Treatise*, 76–77. See also T. Wharton Jones, "Report on the Ophthalmoscope," *British and Foreign Medico-Chirurgical Review* 14 (October 1854): 549–57.

69. For the optical geometry, see Carter, *Practical Treatise*, 79. See also Edgar Browne, *How to Use the Ophthalmoscope: Being Elementary Instructions in Ophthalmoscopy* (Philadelphia: Henry O. Lea, 1877); and John Hulke, *A Practical Treatise on the Use of the Ophthalmoscope* (London, 1861).

70. Hulke, *Use of the Ophthalmoscope*, 7–20.

71. E. Williams, "The Ophthalmoscope," *Medical Times and Gazette*, July 1, 1854, 7.

72. Haab, *Atlas of Ophthalmoscopy*, 22.

73. See http://www.steeles.com/welchallyn/WA_PanOpticOphth.html.

74. Carter, *Practical Treatise*, 57 (on Professor Jaeger), 59 (quote and on Snellen).

75. For the early-nineteenth-century discovery of astigmatism, see Wade, *Natural History of Vision*, 62–64.

76. Donders, *Accommodation and Refraction*, 170.
77. Brewster, "The Sight and How to See," 176.
78. Donders, *Accommodation and Refraction*, 389.
79. Brewster, "The Sight and How to See," 183.
80. Margaret Mitchell, *History of the British Optical Association, 1895–1978* (Chorley: British Optical Association, 1982), 56.
81. John Browning, *How to Use Our Eyes and How to Preserve Them by the Aid of Spectacles* (London: Chatto & Windus, 1883), 66.
82. Charles A. Long, *Spectacles: When to Wear and How to Use Them* (London: Bland & Long, 1855), 22.
83. Many investigators of light, including Brewster, Plateau, and Fechner, seriously damaged their eyes as a consequence of their activities. See Crary, *Techniques of the Observer*, 141.
84. *Archives of Ophthalmology* 19 (1890): 350.
85. Southard, *Modern Eye*, 2.
86. George Harlan, *Eyesight, and How to Care for It* (New York: Julius King Optical Co., 1887), 78, 105–6.
87. On neurasthenia, visual bombardment, and attention, see Rabinbach, *Human Motor*, 153–78.
88. The eyes of animals were also objects of concern. The veterinarian John Gamgee noted that certain eye diseases in animals, e.g., periodic ophthalmia in horses, were treatable with ointments and lotions. He recommended that such beasts be placed in darkness for the duration of their (sometimes incurable) affliction. See John Gamgee, *Our Domestic Animals in Health and Disease*, 4 vols. (Edinburgh: MacLachlan & Stewart, 1867), 4:372–78. Elsewhere, some zoophiles equipped their animals with spectacles (Briggs, *Victorian Things*, 415), while equine eyestrain might be considered when painting stables (Francis Vacher, *A Healthy Home* [London: The Sanitary Record, (1894?)], 164), and porcine emotions were occasionally regarded in abattoir construction (e.g., "New Public Works in Carlisle," *Builder* 55 [September 8, 1888], 180).
89. Browning, *How to Use Our Eyes*, 22, 25.
90. Ibid., 32. The nature of the "artificial light" was unspecified, but, given the date of publication (1883), we can assume that Browning meant gas, oil, or candles. He gives no indication of the candlepower, so this was an extremely rough, subjective test.
91. Robert Brudenell Carter, *Eyesight: Good and Bad: A Treatise on the Exercise and Preservation of Vision*, 2nd ed. (London: Macmillan, 1880), 264.
92. H. Haynes Walton, "Larvae of Insects under the Eyelids," in *Operative Ophthalmic Surgery*, 79–80; William MacKenzie, "Entozoa in the Organ of Vision," in *Practical Treatise*, 1006–13.
93. Browning, *How to Use Our Eyes*, 27.
94. Robert MacNish, *The Anatomy of Drunkenness*, new ed. (Glasgow: W. R. M'Phun, 1847), 175.

95. Nettleship, *Diseases of the Eye*, 269. The smoking-eye defect connection was a common ophthalmologic motif. See "Optic Nerve Atrophy in Smokers," *British Medical Journal*, 1890, no. 1 (May 10): 1072. See also Jordanna Bailkin, "Colour Problems: Work, Pathology and Perception in Modern Britain," *International Labour and Working-Class History* 68 (2005): 99.

96. Carter, *Eyesight*, 168–69.

97. MacKenzie, *Practical Treatise*, 364.

98. Edward W. Hope and Edgar A. Browne, *A Manual of School Hygiene: Written for the Guidance of Teachers in Day-Schools* (Cambridge: Cambridge University Press, 1901), 107, 89.

99. Ibid., 117.

100. Arthur Newsholme, *School Hygiene: The Laws of Health in Relation to School Life* (London: Swan Sonnenschein, Lowrey, 1887), 110.

101. D. F. Lincoln, *School and Industrial Hygiene* (Philadelphia: P. Blakiston, 1896), 56.

102. Hope and Browne, *Manual of School Hygiene*, 100.

103. See Richard Liebreich, *School Life in Its Influence on Sight and Figure: Two Lectures* (London: J. & A. Churchill, 1878).

104. Hope and Browne, *Manual of School Hygiene*, 121.

105. Newsholme, *School Hygiene*, 113.

106. Southard, *Modern Eye*, 11.

107. See, e.g., "Defects of Vision in Candidates for the Public Services," *Lancet*, 1886, no. 2 (August 28): 409.

108. Brewster, "The Sight and How to See," 181–82.

109. Nystagmus is a form of involuntary lateral eye movement, often caused by working for prolonged periods of time in the dark. Asthenopia is the inability to focus on nearby objects or, more generally, eyestrain or weakening of sight.

110. J. T. Arlidge, *The Hygienic Diseases and Mortality of Occupations* (London: Perceval, 1892), 193.

111. See R. I. McKibbin, "Social Class and Social Observation in Edwardian England," *Transactions of the Royal Historical Society* 28 (1978): 191.

112. On the chiasm, see Maurice Merleau-Ponty, *The Visible and the Invisible*, ed. Claude Leforte, trans. Alphonso Lingis (Evanston, IL: Northwestern University Press, 1968), 130–55.

113. Leder, *Absent Body*, 150.

114. Brewster, "The Sight and How to See," 148.

115. This topic is discussed in Winter, *Mesmerized*.

116. Hans Jonas, *The Phenomenon of Life: Toward a Philosophical Biology* (New York: Harper & Row, 1966), 149–50.

117. Thomas Huxley, *Lessons in Elementary Physiology*, enlarged and rev. ed. (New York: Macmillan, 1917), 438.

118. Le Conte, *Sight*, 6.

119. Alexis De Tocqueville, *Democracy in America*, ed. J. P. Mayer, trans. George Lawrence (New York: Harper & Row, 1966), 459.
120. Richard Rorty notes that Cartesianism involves "a single inner space in which bodily and perceptual sensations ('confused ideas of sense and imagination,' in Descartes's phrase), mathematical truths, moral rules, the idea of God, moods of depression, and all the rest of what we now call 'mental' were objects of quasi-observation" (*Philosophy and the Mirror of Nature* [Princeton, NJ: Princeton University Press, 1979], 50).
121. Jean-Paul Sartre, "The Look," in *Being and Nothingness*, 252–302. See also Jay, *Downcast Eyes*, 263–328.
122. Harlan, *Eyesight*, 104.
123. Smith, *Theory of Moral Sentiments*, 164–65.
124. Benjamin Constant, *The Spirit of Conquest and Usurpation (and Their Relation to European Civilisation)* (1815), in *Benjamin Constant: Political Writings*, ed. and trans. Biancamaria Fontana (Cambridge: Cambridge University Press, 1988), 104–5, cited in Pierre Manent, *An Intellectual History of Liberalism*, trans. Rebecca Balinski (Princeton, NJ: Princeton University Press, 1995), 90. Constant's views on spectatorship were significantly more critical than Smith's. See Seigel, *Idea of the Self*, 268–91.
125. Smiles, *Self-Help*, 326.
126. Anthony, Earl of Shaftesbury, *Characteristics of Men, Manners, Opinions, Times* (1711), ed. John M. Robertson (1900; reprint, Indianapolis: Bobbs-Merrill, 1964), 2:137, cited in G. J. Barker-Benfield, *The Culture of Sensibility: Sex and Society in Eighteenth-Century Britain* (Chicago: University of Chicago Press, 1992), 105.
127. Norbert Elias, *The Civilizing Process: Sociogenetic and Psychogenetic Investigations*, trans. Edmund Jephcott, rev. ed. (Oxford: Blackwell, 2000), 67–68.
128. Crary, *Suspensions of Perception*, 17.
129. Jürgen Habermas, *The Structural Transformation of the Public Sphere: An Inquiry into a Category of Bourgeois Society*, trans. Thomas Berger with the assistance of Frederick Lawrence (Cambridge, MA: MIT Press, 1989), 49.
130. Gabriel Tarde, "Catégories logiques et institutions sociales," *Revue Philosophique* 28 (August 1889): 303, cited in Crary, *Suspensions of Perception*, 243.
131. Rose, *Powers of Freedom*, 73; Elias, *Civilizing Process*, 415.
132. Smiles, *Self-Help*, 5–6 (emphasis added). Emulation was also discussed by Mill. See Goodlad, *Victorian Literature and the Victorian State*, 43.
133. Richard Sennett, *The Fall of Public Man* (New York: Norton, 1976), 25.
134. For fashion and reserved conduct, see Sennett, *Fall of Public Man*, esp. 161–94. See also T. J. Clark, *The Painting of Modern Life: Paris in the Art of Manet and His Followers* (Princeton, NJ: Princeton University Press, 1984), 253.
135. Simon Gunn, *The Public Culture of the Victorian Middle Class: Ritual and Authority and the English Industrial City, 1840–1914* (Manchester: Manchester University Press, 2000), 30.

136. Johann Caspar Lavater, *Essays on Physiognomy, for the Promotion of the Knowledge and the Love of Mankind* (Boston: William Spotswood & David West, 1794), 80.

137. Lucy Hartley, *Physiognomy and the Meaning of Expression in Nineteenth-Century Culture* (Cambridge: Cambridge University Press, 2001), 41. Similarly, phrenology was a technique associated with liberal or dissenting thinkers who generally shared a detranscendentalizing impulse toward unmediated knowledge of the natural world and the human body. See Roger Cooter, *The Cultural Meaning of Popular Science: Phrenology and the Organisation of Consent in Nineteenth-Century Britain* (Cambridge: Cambridge University Press, 1984).

138. Sennett, *Fall of Public Man*, 146.

139. Joseph Simms, *Physiognomy Illustrated* (1872), 37, cited in Daniel Pick, *Faces of Degeneration: A European Disorder, c. 1848–c. 1918* (Cambridge: Cambridge University Press, 1989), 52.

140. Charles Dickens, "The Demeanour of Murderers," *Household Words* 13, no. 325 (June 14, 1856): 505.

141. Smiles, *Self-Help*, 227.

142. This is the subject of Steven Shapin and Simon Schaffer's *Leviathan and the Air-Pump: Hobbes, Boyle, and the Experimental Life* (Princeton, NJ: Princeton University Press, 1985). See also Steven Shapin, *A Social History of Truth: Civility and Science in Seventeenth-Century England* (Chicago: University of Chicago Press, 1994); and Foucault, *The Order of Things*, 132–38.

143. For the bodily nature of scientific perception, see Graeme Gooday, "Spot-Watching, Bodily Postures and the 'Practiced Eye': The Material Practice of Instrument Reading in Late Victorian Electrical Life," in *Bodies/Machines*, ed. Iwan Rhys Morus (Oxford: Berg, 2002), 165–96. On medical perception, see the essays in W. F. Bynum and Roy Porter, eds., *Medicine and the Five Senses* (Cambridge: Cambridge University Press, 1993). On scopes, see Carroll, *Science, Culture and Modern State Formation*, esp. 47–51.

144. Virchow cited in Bynum, *Science and the Practice of Medicine*, 100. For scientific perception, see Ian Hacking, *Representing and Intervening: Introductory Topics in the Philosophy of Natural Science* (Cambridge: Cambridge University Press, 1983), 167–209. See also Crary, *Techniques of the Observer*; Donna Haraway, *Simians, Cyborgs and Women: The Reinvention of Nature* (New York: Free Association, 1991); and Musselman, *Nervous Conditions*, 17–19.

145. Shapin and Schaffer, *Leviathan and the Air-Pump*, 60–65.

146. On photography and perception, see Jennifer Tucker, *Nature Exposed: Photography as Eyewitness in Victorian Science* (Baltimore: Johns Hopkins University Press, 2006); and Lynda Nead, "Animating the Everyday: London on Camera circa 1900," *Journal of British Studies* 43 (2004): 65–90.

147. "Photography in Medical Science," *Lancet*, 1859, no. 1 (January 22): 89.

148. Yaron Ezrahi, "Technology and the Civil Epistemology of Democracy," in *Technology and the Politics of Knowledge*, ed. Andrew Feenberg and Alastair Hannay (Bloomington: Indiana University Press, 1995), 162. See also Yaron

Ezrahi, *The Descent of Icarus: Science and the Transformation of Contemporary Democracy* (Cambridge, MA: Harvard University Press, 1990).

149. Michael Warner, *The Letters of the Republic: Publication and the Public Sphere in Eighteenth-Century America* (Cambridge, MA: Harvard University Press, 1990), 41. See also Charles Taylor, *Modern Social Imaginaries* (Durham, NC: Duke University Press, 2004), 89; and Frankel, *States of Inquiry*, 42.

150. See Winter, *Mesmerized*, 335.

151. On mapping and liberalism, see Joyce, *Rule of Freedom*, 35–56.

152. Edward Mogg, *Mogg's New Picture of London and Guide to Its Sights* (1847), cited in David Smith, *Victorian Maps of the British Isles* (London: Batsford, 1985), 74.

153. Corbin, *The Foul and the Fragrant*, 5. Corbin's conception of the social imaginary is more antagonistic and political than Charles Taylor's. See Taylor, *Modern Social Imaginaries*.

154. Alain Corbin, "Towards a History and Anthropology of the Senses," in *Time, Desire and Horror: Towards a History of the Senses* (Cambridge: Polity, 1995), and *The Foul and the Fragrant*.

155. Peter Stallybrass and Allon White, *The Politics and Poetics of Transgression* (Ithaca, NY: Cornell University Press, 1986), 21.

156. George Sims, *Horrible London*, in *How the Poor Live; and, Horrible London* (London: Garland, 1986), 120 (emphasis added).

157. Andrew Mearns et al., *The Bitter Cry of Outcast London* (Leicester: Leicester University Press, 1970), 58.

158. James Burn Russell, "Uninhabitable Houses: Who Inhabit Them? Who Own Them? What Is to Be Done with Them?" (paper read at the Congress of the Sanitary Association of Scotland, August 23–24, 1894), in *Public Health Administration in Glasgow: A Memorial Volume of the Writings of James Burn Russell*, ed. A. K. Chalmers (Glasgow: James Maclehose, 1905), 232.

159. Friedrich Engels, *The Condition of the Working Class in England*, ed. David MacLellan (Oxford: Oxford University Press, 1993), 58.

160. Liddle cited in Chadwick, *Sanitary Report*, 136.

161. This point is made in greater depth by Georges Vigarello, *Concepts of Cleanliness: Changing Attitudes in France since the Middle Ages*, trans. Jean Birrell (Cambridge: Cambridge University Press, 1988), 194.

162. Charles Booth, *Life and Labour of the People of London: First Series: Poverty: Streets and Population Classified* (London: Macmillan, 1902), 75.

163. "Prosecution of a Dog Fancier," *Sanitary Record* 5 (September 23, 1876): 204.

164. Robert Roberts, *The Classic Slum: Salford Life in the First Quarter of the Century* (Harmondsworth: Penguin, 1971), 38.

165. "London Noise," *Sanitary Record* 5 (October 28, 1876): 278.

166. Jon Agar, "Bodies, Machines and Noise," in Morus, ed., *Bodies/Machines*, 198.

167. John M. Picker, "The Soundproof Study: Victorian Professionals, Work Space and Urban Noise," *Victorian Studies* 42, no. 3 (Spring 1999/2000): 427–48.

See also Michael Toyka-Seid, "Noise Abatement and the Search for Quiet Space in the Modern City," in *Resources of the City: Contributions to an Environmental History of Modern Europe*, ed. Dieter Schott, Bill Luckin, and Geneviève Massard-Guilbaud (Aldershot: Ashgate, 2005), 215–29.

168. Sims, *How the Poor Live*, in *How the Poor Live; and, Horrible London*, 21.

169. Richard Howard, *An Enquiry into the Morbid Effects of the Deficiency of Food, Chiefly with Respect to Their Occurrence among the Labouring Poor* (London: Simpkin, Marshall, 1839), 5.

170. *Illustrated London News*, May 31, 1851, 569, cited in Thomas Richards, *The Commodity Culture of Victorian Britain: Advertising and Spectacle, 1851–1914* (Stanford, CA: Stanford University Press, 1990), 37.

171. Chadwick, *Sanitary Report*, 297.

172. Girdlestone, *Unhealthy Condition of Dwellings*, 58.

173. Mearns et al., *Bitter Cry*, 61.

174. Alexander Patterson, *Across the Bridges; or, Life by the South London Riverside* (London: Edward Arnold, 1911), 59.

175. H. C. Bartlett, "The Chemistry of Dirt," *British Architect and Northern Engineer* 10 (October 18, 1878): 152.

176. Cornelius Fox, *Sanitary Examinations of Water, Air and Food*, 3rd ed. (London: J. & A. Churchill, 1878), 236.

177. Shaw cited in Chadwick, *Sanitary Report*, 164–65 (emphasis added).

178. Maudsley, *Body and Mind*, 59.

179. Herbert Spencer, *The Principles of Sociology*, 3 vols. (New York: Appleton, 1896), 1:75, 77.

180. This opens up the Western sensory tradition to the same kind of critique as is seen in Dipesh Chakrabarty, *Provincializing Europe: Postcolonial Thought and Historical Difference* (Princeton, NJ: Princeton University Press, 2000). See also n. 2 above.

181. For the desensitization of the prostitute, see Corbin, *Women for Hire*, 302.

182. Barker-Benfield, *Culture of Sensibility*, 23.

183. Joseph Turnley, *The Language of the Eye: The Importance and Dignity of the Eye as Indicative of General Character, Female Beauty, and Manly Genius* (London: Partridge, 1856), cited in Kate Flint, *The Victorians and the Visual Imagination* (Cambridge: Cambridge University Press, 2000), 26.

184. Herbert Spencer, *The Principles of Psychology* (1870), cited in George W. Stocking, *Victorian Anthropology* (New York: Free Press, 1987), 229.

185. Havelock Ellis, *Man and Woman: A Study of Human Secondary Sexual Characteristics*, 7th ed. (Boston: Houghton Mifflin, 1929), 265, cited in Cynthia Eagle Russett, *Sexual Science: The Victorian Construction of Womanhood* (Cambridge, MA: Harvard University Press, 1989), 46.

186. J. Milner Fothergill, *The Town Dweller: His Needs and His Wants* (New York: D. Appleton, 1889), 97.

187. Nikolas Rose, *Governing the Soul: The Shaping of the Private Self*, 2nd ed. (London: Free Association, 1999), 141–42.

188. Crary, *Techniques of the Observer*, 141–49. See also Michael Heidelberger, *Nature from Within: Gustav Theodor Fechner and His Psychophysical Worldview*, trans. Cynthia Klohr (Pittsburgh: University of Pittsburgh Press, 2004).

189. William I. Miller, *Anatomy of Disgust* (London: Harvard University Press, 1997), 33 (quote), 43–45.

190. Dominique Laporte, *History of Shit*, trans. Rodolphe el-Khoury and Nadia Benabid (Cambridge, MA: MIT Press, 2000), 32.

191. It is the discourse that has definitively changed, rather than physical reality (which may itself have also changed), as Corbin demonstrates. See *The Foul and the Fragrant*, 155–56.

192. Judge cited in Joel Franklin Brenner, "Nuisance Law and the Industrial Revolution," *Journal of Legal Studies* 3 (1973): 414.

193. Alain Corbin, *Village Bells: Sound and Meaning in the Nineteenth-Century French Countryside*, trans. Martin Thom (New York: Columbia University Press, 1998), 298.

194. Henry Jephson, *The Sanitary Evolution of London* (London: T. Fisher Unwin, 1907), 21–22.

195. The evolutionary point was sometimes made very explicitly. Benjamin Kidd, e.g., noted how "increased sensitiveness [*sic*] to stimulus" was a sign of progress inseparable from the collective development of the social body. Benjamin Kidd, *Social Evolution*, 2nd American ed. (New York: Macmillan, 1894), 181.

196. Proving that a certain trade (gluemaking, slaughtering, etc.) was actually offensive was another question entirely, usually solved by the protean, unanalytic conception of the "nuisance," something discussed in chapter 3 below. What matters here is the spatialization of sensory irritation and its concomitant calculation (however arbitrary the actual figures), enshrined in building law. The world *should ideally* assume a certain form, and, thus, the world forever becomes a failed translation of abstract into concrete space, a failure fueling liberal interventionary-regulatory techniques.

197. City of Birmingham, *Byelaws for Good Rule and Government, 1897*, cited in Martin Daunton, *House and Home in the Victorian City: Working-Class Housing, 1850–1914* (London: Edward Arnold, 1983), 267.

198. Lefebvre, *Production of Space*, 56.

199. Robert Fishman, *Bourgeois Utopias: The Rise and Fall of Suburbia* (New York: Basic, 1987), esp. 39–72. See also Anthony Wohl, *The Eternal Slum: Housing and Social Policy in Victorian London* (New Brunswick, NJ: Transaction, 2002), 285–316; and H. J. Dyos and D. A. Reeder, "Slums and Suburbs," in *The Victorian City: Images and Realities*, ed. H. J. Dyos and D. A. Reeder, 2 vols. (London: Routledge & Kegan Paul, 1973), 1:359–86.

200. Manchester City Council, *Proceedings*, 1868–69, 272 (emphasis added). See also R. Rumney, "On the Ashpit System of Manchester," *Engineer* 30 (September 30, 1870): 223.

CHAPTER 2

1. See James H. Cassedy, "Hygeia: A Mid-Victorian Dream of a City of Health," *Journal of the History of Medicine and Allied Sciences* 17 (1962): 217–28.
2. Benjamin Ward Richardson, *Hygeia: A City of Health* (London: Macmillan, 1876), 20, 30, 24, 42 (see also 27–28).
3. Engels, *Condition of the Working Class*, 39 (emphasis added).
4. Ibid., 59.
5. Chadwick, *Sanitary Report*, 306.
6. Lord Aberdare, "Social Science in Brighton," *Builder* 33 (October 9, 1875): 901. On Jack the Ripper, see Judith Walkowitz, *City of Dreadful Delight: Narratives of Sexual Danger in Late-Victorian London* (Chicago: University of Chicago Press, 1992), 191–228.
7. Smiles, *Self-Help*, 297.
8. Edward Watkin cited in *Minutes of Evidence and Appendix as to England and Wales*, vol. 2 of *Royal Commission on the Housing of the Working Classes* (London: Eyre & Spottiswoode, 1885), 359.
9. R. A. Slaney, *Report on the State of Birmingham and Other Large Towns* (London: W. Clowes, 1845), 7.
10. Sims, *Horrible London*, in *How the Poor Live; and, Horrible London*, 115. On shame, see n. 131 of chapter 1 above.
11. Florence Nightingale, *Notes on Nursing: What It Is, and What It Is Not* (London: Churchill Livingstone, 1980), 71.
12. John Hassan, *The Seaside, Health and Environment in England and Wales since 1800* (Aldershot: Ashgate, 2003), 83.
13. Leigh cited in E. D. Simon and J. Inman, *The Rebuilding of Manchester* (London: Longman, 1935), 7. On sunlight treatment in Manchester, see Arthur Redford, *The History of Local Government in Manchester*, 3 vols. (London: Longmans, Green, 1939), 3:318.
14. Harry G. Critchley, *Hygiene in School: A Manual for Teachers* (London: Allman, 1906), 91.
15. Forbes Winslow, *Light: Its Influence on Life and Health* (London: Longmans, Green, Reader & Dyer, 1867), 5.
16. A. Emrys-Jones, "Smoke and Impure Air," *Exhibition Review* 4 (1882), cited in Stephen Mosley, *The Chimney of the World: A History of Smoke Pollution in Victorian and Edwardian Manchester* (Cambridge: White Horse, 2001), 66.
17. Charles Aikman, *Milk: Its Nature and Composition* (London: Adam & Charles Black, 1895), 80. Sunlight's healthful impact on humans was not, obviously, unknown before this period. See, e.g., Ekirch, *At Day's Close*, 14.
18. "A Plea for Sunshine," *Plumber and Sanitary Engineer* 2 (May 1879): 162. An identical formula is found in John Tyndall, *Heat Considered as a Mode of Motion* (London: Longman, 1863), 431–32. For earlier experiments on the relation between light and plant motion, see François Delaporte, *Nature's Second*

Kingdom: Explorations of Vegetality in the Eighteenth Century, trans. Arthur Goldhammer (Cambridge, MA: MIT Press, 1982), 164–69.

19. D. F. Lincoln, "Sanitary School Construction," *Plumber and Sanitary Engineer* 2 (November 1, 1879): 391.

20. James Kay, *The Moral and Physical Condition of the Working Classes in Manchester, 1832* (London: Frank Cass, 1970), 27.

21. Alexis De Tocqueville, *Journeys to England and Ireland*, trans. George Lawrence and K. P. Mayer (London: Faber & Faber, 1958), 105. See also Harold Platt, *Shock Cities: The Environmental Transformation and Reform of Manchester and Chicago* (Chicago: University of Chicago Press, 2005), 7–8.

22. J. Firth, *Municipal London; or, London Government as It Is, and London under a Municipal Council* (London: Longmans, Green, 1876), 296, 297.

23. E. P. Hennock, *Fit and Proper Persons: Ideal and Reality in Nineteenth-Century Urban Government* (Montreal: McGill-Queen's University Press, 1973), 186. See also Harling, "Powers of the Victorian State," 45.

24. For details of these forms of legislation, see John Prest, *Liberty and Locality: Parliament, Permissive Legislation and Ratepayers' Democracies in the Nineteenth Century* (Oxford: Clarendon, 1990), 6–12.

25. Léon Faucher, *Manchester in 1844: Its Present Condition and Future Prospects*, trans. "a Member of the Manchester Athenaeum" (Manchester: Abel Heywood, 1844), xii.

26. For municipal politics, see Hennock, *Fit and Proper Persons*; and David Owen, *The Government of Victorian London, 1855–1889: The Metropolitan Board of Works, the Vestries and the City Corporation* (London: Belknap, 1982).

27. "Fifth Annual Report of the Registrar-General," *Parliamentary Papers*, 1843, vol. 31, cited in Anthony Wohl, *Endangered Lives: Public Health in Victorian Britain* (London: Dent, 1983), 144.

28. See James Winter, *London's Teeming Streets* (London: Routledge, 1993), chap. 11. For an excellent study of the "socialization" of food, see James Vernon, "The Ethics of Hunger and the Assembly of Society: The Techno-Politics of the School Meal in Modern Britain," *American Historical Review* 110, no. 3 (June 2005): 693–725.

29. Hamlin, *Public Health and Social Justice*, 258.

30. The 1890 Housing of the Working Classes Act, e.g., allowed the condemnation of houses on the grounds of lack of light. See F. H. Millington, *The Housing of the Poor* (London: Cassell, 1891), 33.

31. Joseph Boule, *British Architect and Northern Engineer* 11 (April 11, 1879): 155.

32. Manchester City Council, *Proceedings*, 1869–70, 100. Manchester's many inhabited cellars were a result of cellars being originally constructed to house looms (humidity helped the weaving process). See Platt, *Shock Cities*, 31.

33. Henry Lemmoin-Cannon, *The Sanitary Inspector's Guide: A Practical Treatise on the Public Health Act, 1875, and the Public Health Act, 1890, So Far as They Affect the Inspector of Nuisances* (London: P. S. King, 1902), 62–65.

34. Jacinta Prunty, *Dublin Slums, 1800–1925: A Study in Urban Geography* (Dublin: Irish Academic Press, 1997), 118.

35. W. M. Frazer, *Duncan of Liverpool: Being an Account of the Work of Dr. W. H. Duncan, Medical Officer of Health of Liverpool, 1847–63* (London: Hamish Hamilton Medical, 1947), 43.

36. Fleeming Jenkin, "On Sanitary Inspection," *Sanitary Record* 8 (April 19, 1878): 241.

37. 2 & 3 Will. 4, c. 71, sec. 3 (1832), cited in Banister Fletcher, *Light and Air: A Text-Book for Architects and Surveyors* (London: B. T. Batsford, 1879), 6.

38. Cranworth cited in John McLaren, "Nuisance Law and the Industrial Revolution—Some Lessons from Social History," *Oxford Journal of Legal Studies* 3, no. 2 (1983): 185.

39. Fletcher, *Light and Air*, 102.

40. F. H. Hummel, "Ancient Lights," *Architect*, July 28, 1877, 40.

41. James Silk Buckingham, *National Evils and Practical Remedies, with a Plan of a Model Town* (London: Peter Jackson, 1849), 107, 193.

42. Benjamin Ward Richardson, "A Model City of Health and Comfort," *Builder* 33 (October 16, 1875): 924 (emphasis added).

43. Paul Rabinow, *French Modern: Norms and Forms of the Social Environment* (Cambridge, MA: MIT Press, 1989), 12, 212.

44. *Sanitary Record* 5 (October 7, 1876): 238.

45. The term is most fully developed in Latour, *Reassembling the Social*, esp. 181. See also Bruno Latour, "On Recalling ANT," in *Actor-Network Theory and After*, ed. John Law and John Hassard (Oxford: Blackwell, 1999), 18; and Latour and Hernant, *Paris ville invisible*.

46. Foucault, *Discipline and Punish*, 201–2. Foucault's own analysis sometimes suggests, however, a kind of reciprocal panopticism, which is spatially incomprehensible. For example: "[a] panopticism in which the vigilance of intersecting gazes was soon to render useless both the eagle and the sun" (ibid., 217).

47. Ibid., 205. On the active, creative dimension of such perception, see Yar, "Panoptic Power," 261.

48. Tony Bennett, "The Exhibitionary Complex," *New Formations* 4 (Spring 1988): 81.

49. T. Roger Smith, *Acoustics in Relation to Architecture and Building: The Laws of Sound as Applied to the Arrangement of Buildings*, new ed. (London: Virtue, 1878), 115.

50. Martin Daunton, "Public Place and Private Space: The Victorian City and the Working-Class Household," in *The Pursuit of Urban History*, ed. Derek Fraser and Anthony Sutcliffe (London: Edward Arnold, 1983), 214–15.

51. Manchester City Council, *Proceedings*, February 6, 1884, cited in Redford, *History of Local Government in Manchester*, 3:152–53 (emphasis added). On gardens and supervision, see Peter Thorsheim, "Green Space and Class in

Imperial London," in *The Nature of Cities*, ed. Andrew C. Isenberg (Rochester, NY: University of Rochester Press, 2006), 28–29.

52. For a fascinating discussion of one such supervisory creature, the *portière* in Parisian apartments, see Marcus, *Apartment Stories*, 42–50.
53. See n. 149 of chapter 1 above.
54. "The Bethnal Green Museum," *Builder* 30 (May 18, 1872): 380.
55. Richard Biernacki, *The Fabrication of Labor: Germany and Britain, 1640–1914* (Berkeley and Los Angeles: University of California Press, 1995), 133–41.
56. Frank J. Burgoyne, *Library Construction, Architecture, Fittings, and Furniture*, new ed. (London: George Allen, 1905).
57. James Dugald Stewart, "Planning: Reference Departments," in *Open Access Libraries: Their Planning, Equipment and Organisation*, by James Dugald Stewart et al. (London: Grafton, 1915), 39.
58. E. C. Robins, "Buildings for Secondary Schools," *British Architect and Northern Engineer* 13 (April 23, 1880): 199.
59. Markus gives the example of Samuel Wilderspin, for whom the playground functioned as a laboratory and the masters acted as scientist-gods. See *Buildings and Power*, 71–74.
60. Adrian Forty, "The Modern Hospital in France and England: The Social and Medical Uses of Architecture," in *Buildings and Society: Essays on the Social Development of the Built Environment*, ed. Anthony D. King (London: Routledge & Kegan Paul, 1980), 70, 75; John Sutherland and Douglas Galton, "Principles of Hospital Construction: IV. The Ward Plan," *Lancet*, 1874, no. 1 (April 18): 552–53. On the scout system, see Florence Nightingale, *Notes on Hospitals*, 3rd ed. (London: Longman, Green, Longman, Roberts, & Green, 1863), 50.
61. H. R. Aldridge, *The Case for Town Planning: A Practical Manual* (London: National Housing and Town Planning Council, 1915), 133, cited in W. Ashworth, *The Genesis of Modern British Town Planning: A Study in Economic and Social History of the Nineteenth and Twentieth Centuries* (London: Routledge & Kegan Paul, 1954), 170.
62. Percy Edwards, *History of London Street Improvements, 1855–1897* (London: P. S. King, 1898), 10 (quote [discussing the 1838 select committee report]), 163.
63. Nash cited in H. J. Dyos, *Exploring the Urban Past: Essays in Urban History*, ed. David Cannadine and David Reeder (Cambridge: Cambridge University Press, 1982), 82.
64. "Sanitary Consolidation," *Quarterly Review* 88, no. 176 (March 1851): 455.
65. For eighteenth-century street improvement, see Peter Borsay, *The English Urban Renaissance: Culture and Society in the Provincial Town, 1660–1770* (Oxford: Oxford University Press, 1989), 60–79.
66. George Joachim Goschen, *Laissez-Faire and Government Interference* (1883), cited in Taylor, ed., *Herbert Spencer*, 79.

67. William Atkinson, "The Orientation of Buildings and of Streets in Relation to Sunlight," *Technology Quarterly and Proceedings of the Society of Arts* 18 (September 1, 1905): 214.

68. Firth, *Municipal London*, 324.

69. "On Some Nuisances and Absurdities in London," *Builder* 33 (September 26, 1874): 803.

70. *Sphinx*, January 16, 1869, 202, 203.

71. J. Caminada, *Twenty-five Years of Detective Life* (Manchester: John Heywood, 1895), 16.

72. John Leigh and N. Gardiner, *A History of the Cholera in Manchester, in 1849* (Manchester: Sims & Dinham, 1850).

73. Manchester City Council, *Proceedings*, 1867–68, 408–9.

74. W. E. A. Axon, ed., *Annals of Manchester: A Chronological Record from the Earliest Times to the End of 1885* (Manchester: John Heywood, 1886), 319.

75. *Manchester City News*, March 1, 1873.

76. Manchester City Council Improvement and Buildings Committee, *Minutes* 13 (January 26, 1880): 245.

77. *Manchester Guardian*, June 28, 1890.

78. Jewsbury Hulse cited in Manchester City Council Improvement and Buildings Committee, *Minutes* 11 (July 25, 1870): 132.

79. *Manchester City News*, March 26, 1881.

80. *Manchester Critic*, October 19, 1877, 151–52.

81. Leigh cited in Manchester City Council, *Proceedings*, 1880–81, 417.

82. Concerns about smoke go back at least to Edward I's reign; perhaps the first important treatise on the subject was John Evelyn's *Fumifigium* (1661). See Carolyn Merchant, *The Death of Nature: Women, Ecology, and the Scientific Revolution* (New York: HarperCollins, 1990). For earlier attitudes toward smoke, see William H. Te Brake, "Air Pollution and Fuel Crises in Preindustrial London, 1250–1650," *Technology and Culture* 16, no. 3 (1975): 337–59; and Peter Brindlecombe, "Attitudes and Responses towards Air Pollution in Medieval England," *Journal of the Air Pollution Control Association* 26, no. 10 (1976): 941–45. On modern smoke, see esp. Mosley, *Chimney of the World*; and Peter Thorsheim, *Inventing Pollution: Coal, Smoke, and Culture in Britain since 1800* (Athens: Ohio University Press, 2006).

83. *The Times*, April 24, 1845, cited in E. Ashby and M. Anderson, *The Politics of Clean Air* (Oxford: Clarendon, 1981), 11.

84. Alfred Carpenter, "London Fogs," *Westminster Review*, n.s., 61 (January 1882): 145.

85. This list comes from ibid., 144; and Edmund A. Parkes, *A Manual of Practical Hygiene*, 7th ed. (Philadelphia: P. Blakiston, 1887), 136.

86. *Lancet*, 1874, no. 2 (July 18): 94.

87. Rollo Russell, *London Fogs* (London: Edward Stanford, 1880), 22. Russell's eyesight was, apparently, itself very poor. See Bill Luckin, "The Shaping of

a Public Environmental Sphere in Late Nineteenth-Century London," in *Medicine, Health and the Public Sphere in Britain, 1600–2000*, ed. Steve Sturdy (London: Routledge, 2002), 228.

88. Carpenter, "London Fogs," 143–44.
89. See Wohl, *Endangered Lives*, 205–32, 273–79; and Mosley, *Chimney of the World*, 58–67, 96–107.
90. John Leigh, *Coal-Smoke: Report to the Health and Nuisance Committees of the Corporation of Manchester* (Manchester: John Heywood, 1883), 5.
91. Russell, *London Fogs*, 33.
92. *Manchester Guardian*, May 28, 1842, cited in Mosley, *Chimney of the World*, 20.
93. Thomas Bateman, *Reports on the Diseases of London, and the State of the Weather, from 1804 to 1816* (London: Longman, 1819), 185, cited in Bill Luckin, "'The Heart and Home of Horror': The Great London Fogs of the Late Nineteenth Century," *Social History* 28, no. 1 (January 2003): 34.
94. Russell, *London Fogs*, 22, 25.
95. William Marcet, "On Fogs," *Quarterly Journal of the Royal Meteorological Society* 15, no. 70 (April 1889): 64.
96. "The Reign of Darkness," *Spectator*, January 19, 1889, 85.
97. Marcet, "On Fogs," 63–64.
98. For a physiological explanation of this, see Huxley, *Lessons in Elementary Physiology*, 447.
99. Marcet, "On Fogs," 69. Marcet refers here to experiments in John Tyndall, *Sound*, 4th ed. (London: Longmans, Green, 1883), 331–37.
100. See R. M. MacLeod, "The Alkali Acts Administration, 1863–84: The Emergence of the Civil Scientist," *Victorian Studies* 9 (December 1965): 85–112; and David Vogel, *National Styles of Regulation: Environmental Policy in Great Britain and the United States* (Ithaca, NY: Cornell University Press, 1986), 32, 33, 79.
101. Carpenter, "London Fogs," 150.
102. A summary of these contrivances is provided in Carlos Flick, "The Movement for Smoke Abatement in Nineteenth-Century Britain," *Technology and Culture* 21, no. 1 (1980): 39–46.
103. "Smoke Prevention," *Engineer* 78 (July 24, 1896): 90.
104. For the British tradition of "best practicable means," see Vogel, *National Styles of Regulation*, 79–81.
105. Shaw-Lefebvre cited in "Smoke Abatement," *Nature* 23 (January 13, 1881): 246.
106. Ronald Douglas and Susan Frank, *A History of Glassmaking* (Henley-on-Thames: Foulis, 1972).
107. "On the Technology of Glass," *Builder* 31 (January 25, 1873): 70.
108. Walter Rosenhain, *Glass Manufacture* (London: Archibald Constable, 1908), 32, 34.
109. Harry J. Powell, *The Principles of Glass-Making* (London: George Bell, 1883).

110. For example, "Toughened Glass," *Sanitary Record* 3 (December 11, 1875): 428.

111. James Loudon, *Encyclopaedia of Gardening* (London, 1822), quoted in J. Hix, *The Glass House* (Cambridge, MA: MIT Press, 1981), 29. See also William M. Taylor, *The Vital Landscape: Nature and the Built Environment in Nineteenth-Century Britain* (Aldershot: Ashgate, 2004), 51–53, 67–71.

112. David Pike, *Subterranean Cities: The World beneath Paris and London, 1800–1945* (Ithaca, NY: Cornell University Press, 2005), 89–90.

113. "Plate Glass," *Engineering* 1 (January 12, 1866): 21.

114. *British Architect and Northern Engineer* 10 (December 6, 1878): 221.

115. *The Times* cited in Mitchell, *Colonising Egypt*, 20.

116. W. Hughes, *On the Principles and Management of the Marine Aquarium* (London: John Van Voorst, 1875); J. Taylor, *The Aquarium: Its Inhabitants, Structure and Management* (London: W. H. Allen, 1884); Philip Gosse, *A Handbook to the Marine Aquarium: Containing Instructions for Constructing, Stocking, and Maintaining a Tank, and for Collecting Plants and Animals* (London: John Van Voorst, 1855); Ann Kelly, "Ponds in the Parlour: The Victorian Aquarium," *Things* 2 (1995): 55–68.

117. "The Manchester Aquarium," *Builder* 32 (February 28, 1874): 174.

118. John Haywood, "On Health and Comfort in House-Building," *Builder* 31 (August 23, 1873): 669.

119. John Henry Walsh, *A Manual of Domestic Economy: Suited to Families Spending from £100 to £1000 a Year*, 2nd ed. (London: G. Routledge, 1857), 81, 121.

120. See W. Lascelles, "Glass Roofs for London," *Builder* 34 (December 16, 1876): 1226.

121. Douglas Galton, *Healthy Hospitals: Observations on Some Points Connected with Hospital Construction* (Oxford: Clarendon, 1893), 195.

122. Douglas Galton, "Some of the Sanitary Aspects of House Construction," *Builder* 32 (February 21, 1874): 157 (quote), and *Healthy Hospitals*, 200. See also Aston Webb, "Relation between Size of Windows and Rooms," *Plumber and Sanitary Engineer* 2 (March 1879): 93.

123. Edward Bowmaker, *The Housing of the Working Classes* (London: Methuen, 1895), 90.

124. John W. Papworth and Wyatt Papworth, *Museums, Libraries, and Picture Galleries, Public and Private; Their Establishment, Formation, Arrangement and Architectural Content* (London: Chapman & Hall, 1853), 73, 74.

125. Carl Pfeiffer, "Light: Its Sanitary Influence and Importance in Building," *Builder* 35 (July 21, 1877): 731.

126. "Lighting in School-Rooms," *Builder* 30 (September 7, 1872): 705.

127. Harlan, *Eyesight*, 128.

128. Resolution cited in J. George Hodgins, *Hints and Suggestions on School Architecture and Hygiene: With Plans and Illustrations* (Toronto: Printed for the Education Department, 1886), 50.

129. "Rules to Be Observed in Planning and Fitting Up Schools," *Builder* 30 (May 4, 1872): 349.

130. Richard Sennett, *The Conscience of the Eye: The Design and Social Life of Cities* (London: Faber & Faber, 1991), 108.

131. Henry Greenway, "A New Mode of Hospital Construction," *Builder* 30 (June 29, 1872): 505.

132. General building rules for London schools included the provision for glazing the upper halves of doors. "Rules to Be Observed," 349. For a good example of workhouse glazing, see "Prestwich Union Workhouse," *Builder* 30 (August 17, 1872): 645.

133. Walsh, *Manual of Domestic Economy*, 122. For Parisian shutters, see Marcus, *Apartment Stories*, 164.

134. *Sanitary Record* 10 (March 21, 1879): 204.

135. Rosenhain, *Glass Manufacture*, 149.

136. "London Noise," 277. Not everyone, obviously, hated street noise. Dickens apparently found the tranquil environment of Lausanne unconducive to the writing of *Dombey and Son* and longed for London's clamorous streets. Benjamin, *Arcades Project*, 426.

137. "Needless Noise," *Lancet*, 1876, no. 2 (September 23): 440.

138. "The Strand: And Some Items Noteworthy in It," *Builder* 35 (May 19, 1877): 502.

139. "Asphalte and Other Pavements, II," *Engineer* 33 (April 26, 1872): 287.

140. "Noiseless Pavements," *Sanitary Record* 4 (February 6, 1875): 98–99.

141. H. Royle, "Street Pavements as Adopted in the City of Manchester," *Proceedings* (Association of Municipal and Sanitary Engineers and Surveyors) 3 (July 6 and 7, 1876): 54–58. See also D. K. Clark, "Stone Pavements of Manchester," in *The Construction of Roads and Streets*, by Henry Law and D. Clark (London: Lockwood, 1890), 198–201.

142. E. B. Ellice-Clark, "On the Construction and Maintenance of Public Highways," *Proceedings* (Association of Municipal and Sanitary Engineers and Surveyors) 3 (July 6, 1876): 68. Noiseless tire experimentation thrived: in 1888, e.g., the Shrewsbury and Talbot Cab and Noiseless Tyre Co. was formed. See Henry Charles Moore, *Omnibuses and Cabs: Their Origin and History* (London: Chapman & Hall, 1902), 267–68.

143. "Terra Metallic Pavings," *Builder* 35 (February 3, 1877): 100; "Tar Pavements and Tar Varnish," *Engineering News* 26 (December 19, 1891): 598–99. For cork pavements, see *Engineering News* 27 (January 16, 1892): 62. Metallic pavings were sometimes used for coach houses, stables, etc.

144. "Queer Pavements," *Engineering Record* 32 (September 19, 1896): 288. Clinkers are bricks or blocks made from the ash left following the incineration of refuse.

145. "Noiseless Pavements," *Lancet*, 1873, no. 2 (December 13): 849. See also J. Earle, *Black Top: A History of the British Flexible Roads Industry* (Oxford: Blackwell, 1974).

146. "Asphalte and Other Pavements, II," 287.
147. Manchester City Council Paving, Sewering and Highways Committee, *Minutes*, August 16, 1871, 46.
148. "Carriage-Way Pavements," *Engineering* 54 (August 12, 1892): 189.
149. H. Allnutt, *Wood Pavement as Carried Out on Kensington High Road, Chelsea &c.* (London: E. & F. Spon, 1880), 15–16.
150. For the Manchester rinking craze, see *City Lantern* 2 (January 7, 1876): 113. The first such rink appears to have been constructed in Brighton in 1874. See Bruce Haley, *The Healthy Body and Victorian Culture* (Cambridge, MA: Harvard University Press, 1978), 135.
151. Richardson, *Hygeia*, 20; Chadwick, *Sanitary Report*, 131.
152. "Wood and Asphalt Paving," *Engineer* 74 (May 20, 1892): 435.
153. "Silent Paving," *Lancet*, 1875, no. 2 (August 21): 297. The velocity of sound through numerous kinds of wood was also calculated. See Tyndall, *Sound*, 40.
154. Manchester City Council Paving, Sewering and Highways Committee, *Minutes*, October 3, 1877, 73, and May 24, 1882, 380.
155. "Wood Pavement," *Lancet*, 1882, no. 2 (September 30): 549.
156. "Street Paving and Hygiene," *Engineer* 78 (December 28, 1894): 573.
157. "Carriage-Way Pavements," *Engineering* 54 (July 29, 1892): 144.
158. "Microbes in Wood Pavement," *Engineering Record* 32 (September 5, 1896): 261.
159. George Turnbull, "Pavements," in *London in the Nineteenth Century*, by Walter Besant (London: Adam & Charles Black, 1909), 343; Ian McNeil, "Roads, Bridges and Vehicles," in *An Encyclopedia of the History of Technology*, ed. Ian McNeil (London: Routledge, 1990), 436. American cities were still laying wood pavements in the early twentieth century. See A. Blanchard and H. Drowne, *Text-Book on Highway Engineering* (New York: John Wiley, 1913), 503–29.
160. "Asphalte and Other Pavements, II," 287.
161. E. B. Ellice-Clark, "Asphalt and Its Application to Street Paving," *Proceedings* (Association of Municipal and Sanitary Engineers and Surveyors) 5 (July 31, 1879): 51.
162. C. T. Kingzett, "Some Analyses of Asphalte Pavings," *Analyst* 8 (January 1883): 4–10.
163. For a revisionist view of the "failure" of local government to tackle sanitation, see Christopher Hamlin, "Muddling in Bumbledom: On the Enormity of Large Sanitary Improvements in Four British Towns, 1855–1886," *Victorian Studies* 32 (Autumn 1988): 55–83.
164. Simon and Inman, *Rebuilding of Manchester*, 81.
165. See, e.g., E. D. Simon and Marion Fitzgerald, *The Smokeless City* (London: Longmans, Green, 1922).
166. Owen, *Government of Victorian London*, 92–93.
167. "The Atmosphere of Great Towns," *Engineer* 78 (May 15, 1896): 496.

168. Alain Corbin, "Public Opinion, Policy and Industrial Pollution in the Pre-Haussmann Town," in *Time, Desire and Horror*.

169. Sennett (*Fall of Public Man*, esp. 259–68) and Gunn (*Public Culture of the Victorian Middle Class*, esp. 187–97) both emphasize the historical transience of specific forms of respectable nineteenth-century collective practice.

CHAPTER 3

1. B. Kirkman Gray, *Philanthropy and the State; or, Social Politics* (London: P. S. King, 1908), 29, 31.

2. Josef Redlich, *Local Government in England*, ed. with additions by Francis W. Hirst, 2 vols. (London: Macmillan, 1903), 1:96, 2:249.

3. Earlier forms of inspection included surveys stretching back to the Domesday Book, house-to-house inspections during times of plague, the activities of private groups like the late-seventeenth-century Society for the Reformation of Manners, and the extraordinary toils of eighteenth-century excisemen. See, respectively, Philip Corrigan and Derek Sayer, *The Great Arch: English State Formation as Cultural Revolution* (Oxford: Blackwell, 1985), 21; Paul Slack, *The Impact of Plague in Tudor and Stuart England* (London: Routledge & Kegan Paul, 1985), 226; Alan Hunt, *Governing Morals: A Social History of Moral Regulation* (Cambridge: Cambridge University Press, 1999), 28–56; and John Brewer, *The Sinews of Power: War, Money and the English State, 1688–1783* (Cambridge, MA: Harvard University Press, 1988), 101–14.

4. W. L. Burn, *The Age of Equipoise: A Study of the Mid-Victorian Generation* (London: George Allen & Unwin, 1964), 17.

5. See Dicey, *Lectures*; MacDonagh, "Revolution in Government," and *Early Victorian Government*; William C. Lubenow, *The Politics of Government Growth: Early Victorian Attitudes toward State Intervention, 1833–1848* (Newton Abbot: David & Charles Archon, 1971), 15–16; and David Roberts, *Victorian Origins of the British Welfare State* (New Haven, CT: Yale University Press, 1960), 327–33.

6. MacDonagh, *Early Victorian Government*, 49; Corrigan and Sayer, *Great Arch*, 125.

7. Herman Finer, *Theory and Practice of Modern Government*, rev. ed. (New York: Henry Holt, 1949), 47.

8. A list of works devoting significant numbers of pages to inspection would be enormous. See, e.g., Peter Bartrip, "British Government Inspection, 1832–1875: Some Observations," *Historical Journal* 25, no. 3 (1982): 605–26; Corrigan and Sayer, *Great Arch*; Hennock, *Fit and Proper Persons*; John Harris, *British Government Inspection as a Dynamic Process: The Local Services and the Central Departments* (New York: Frederick A. Praeger, 1955); Roberts, *Victorian Origins*; H. E. Boothroyd, *A History of the Inspectorate: Being a Short Account of the Inspecting Service of the Board of Education* (London: Board of

Education Inspectors, 1923); R. K. Webb, "A Whig Inspector," *Journal of Modern History* 27 (1955): 352–64; and Herbert Preston-Thomas, *The Work and Play of a Government Inspector* (London: William Blackwood, 1909).

9.   See Seth Koven, *Slumming: Sexual and Social Politics in Victorian London* (Princeton, NJ: Princeton University Press, 2004); F. K. Prochaska, *Women and Philanthropy in Nineteenth-Century England* (Oxford: Clarendon, 1980); and Dorice Elliott, *The Angel out of the House: Philanthropy and Gender in Nineteenth-Century England* (Charlottesville: University of Virginia Press, 2002).

10.   Even in texts dealing with the minutiae of urban administration, like Owen's *Government of Victorian London*, the urban inspection is seldom dealt with as a discrete phenomenon, worthy of substantial discussion. Inspectors appear regularly in Owen's text, but little analytic energy is exercised on explicating their activities. Some important articles exist on urban inspection, however. See Gerry Kearns, "Cholera, Nuisances and Environmental Management in Islington, 1830–55," in *Living and Dying in London*, ed. W. F. Bynum and Roy Porter (London: Welcome Institute for the History of Medicine, 1991), 94–125; Christopher Hamlin, "Public Sphere to Public Health: The Transformation of 'Nuisance,'" in Sturdy, ed., *Medicine, Health and the Public Sphere*, 189–204, and "Sanitary Policing and the Local State, 1873–1874: A Statistical Study of English and Welsh Towns," *Social History of Medicine* 28, no. 1 (2005): 39–61; and Tom Crook, "Sanitary Inspection and the Public Sphere in Late Victorian and Edwardian Britain: A Case Study in Liberal Governance," *Social History* 32, no. 4 (2007): 369–93.

11.   For this convoluted history, see Hamlin, "Public Sphere to Public Health," 189–204; Brenner, "Nuisance Law"; and McLaren, "Nuisance Law."

12.   Blyth, *Dictionary of Hygiène*, 404.

13.   Hamlin, "Public Sphere to Public Health," 195.

14.   Albert Taylor, *The Sanitary Inspector's Handbook*, 3rd ed. (London: H. K. Lewis, 1901), 49.

15.   Frazer, *Duncan of Liverpool*, 40–41; John Butt, "Working-Class Housing in Glasgow, 1851–1914," in *The History of Working-Class Housing*, ed. Stanley D. Chapman (Totowa, NJ: Rowman & Littlefield, 1971), 58.

16.   Firth, *Municipal London*, 307.

17.   Frazer, *Duncan of Liverpool*, 109.

18.   George Reid, "The Position of Sanitary Inspectors," *Lancet*, 1896, no. 2 (September 19): 803.

19.   Prunty, *Dublin Slums*, 159. Women were also being appointed as inspectors at the state level. The first female factory inspectors were appointed in 1893, while women were employed by the education inspectorate to inspect cookery and needlework classes. See, e.g., Mary Drake McFeely, *Lady Inspectors: The Campaign for a Better Workplace, 1893–1921* (Athens: University of Georgia Press, 1991); and Boothroyd, *History of the Inspectorate*, 73–75.

20. Taylor, *Sanitary Inspector's Handbook*, 44.
21. François Delaporte, *Disease and Civilization: The Cholera in Paris, 1832*, trans. Arthur Goldhammer (Cambridge, MA: MIT Press, 1986), 38. See also Corbin, *The Foul and the Fragrant*, 159.
22. Kay, *Moral and Physical Condition*, 14, 67.
23. On inspections during epidemics, see Kearns, "Cholera, Nuisances and Environmental Management," 111. See also Peter Baldwin, *Contagion and the State in Europe, 1830–1930* (Cambridge: Cambridge University Press, 1999), esp. 136–39.
24. Peter Baldwin, "The Victorian State in Comparative Perspective," in Mandler, ed., *Liberty and Authority*, 58.
25. Blyth, *Dictionary of Hygiène*, 290. For his Gloucestershire statistics, Blyth cited the *Annual Report of Medical Officer of Health to Gloucester Combined Sanitary Authorities, 1874*.
26. Taylor, *Sanitary Inspector's Handbook*, 57.
27. John William Tripe cited in *Minutes of Evidence and Appendix as to England and Wales*, 318.
28. Duncan cited in Frazer, *Duncan of Liverpool*, 74.
29. Taylor, *Sanitary Inspector's Handbook*, 45.
30. Gerard Jensen, *Modern Drainage Inspection and Sanitary Surveys* (London: The Sanitary Publishing Co., 1899), 28–29.
31. Parkes, *Practical Hygiene*, 210.
32. Jensen, *Modern Drainage Inspection*, 78.
33. Fox, *Sanitary Examinations*, 222. On arsenic poisoning, see Peter Bartrip, "How Green Was My Valance? Environmental Arsenic Poisoning and the Victorian Domestic Ideal," *English Historical Review* 111 (1994): 891–913.
34. H. C. Bartlett, "Poisonous and Non-Poisonous Paints and Wall Papers," *Sanitary Record* 6 (January 6, 1877): 12.
35. Lemmoin-Cannon, *Sanitary Inspector's Guide*, 86.
36. Chadwick, *Sanitary Report*, 411.
37. John Burnett, *A Social History of Housing, 1815–1970* (London: Methuen, 1978), 63; Koven, *Slumming*, 55.
38. A tenement was a house divided into separate dwellings for families or individuals. A lodging house involved a similar breaking up, but the dwelling spaces were shared by various individuals who were not related. This distinction was, however, porous, never entirely clear, and, moreover, subject to change over time according to legal prescription. See Burnett, *Social History of Housing*, 64. On later legislation (in 1885 and 1890), see Wohl, *Eternal Slum*, 248, 252.
39. Wohl, *Eternal Slum*, 74.
40. Frank Stockman, *A Practical Guide for Sanitary Inspectors*, 3rd ed. (London: Butterworth, 1915), 183.
41. J. Treble, "Liverpool Working-Class Housing, 1801–1851," in Chapman, ed., *History of Working-Class Housing*, 183; Frazer, *Duncan of Liverpool*, 119.

42. James Burn Russell, "On the 'Ticketed Houses' of Glasgow, with an Interrogation of the Facts for Guidance towards the Amelioration of the Lives of Their Occupants" (presidential address read before the Philosophical Society of Glasgow, November 7, 1888), in *Public Health Administration in Glasgow*, 216.

43. Local government board guidelines recommended three hundred cubic feet as the minimum. See Taylor, *Sanitary Inspector's Handbook*, 80.

44. Joseph Robinson, *The Sanitary Inspector's Practical Guide, with Inspection of Lodging-Houses (under the Sanitary Acts), and the Sale of Food and Drugs Amendment Act 1879*, 2nd ed. (London: Shaw, 1884), 56.

45. *First Report of Her Majesty's Commissioners for Inquiring into the Housing of the Working Classes* (London: Eyre & Spottiswoode, 1885), 8.

46. Bates and Powell cited in *Minutes of Evidence and Appendix as to England and Wales*, 142.

47. Bylaws cited in Stockman, *Practical Guide for Sanitary Inspectors*, 194.

48. Cited in Carroll, *Science, Culture and Modern State Formation*, 119.

49. Chadwick, *Sanitary Report*, 349.

50. For more details, see Henry Letheby, *On Noxious and Offensive Trades and Manufactures, with Special Reference to the Best Practicable Means of Abating the Several Nuisances Therefrom* (London: Statham, 1875).

51. [Edward?] Ballard divided olfactory nuisances into those relating to animal keeping, slaughter, and industries working with animal, vegetable, or mineral matter or some combination thereof.

52. Brenner, "Nuisance Law," 408.

53. For example, *R v. Cross* (1826). See McLaren, "Nuisance Law," 171.

54. Letheby, *Noxious and Offensive Trades*, 2.

55. Manchester Borough Council, *Proceedings*, June 2, 1847, cited in Redford, *History of Local Government in Manchester*, 2:84.

56. Manchester City Council, *Proceedings*, 1867–68, 375.

57. Stockman, *Practical Guide for Sanitary Inspectors*, 39.

58. Taylor, *Sanitary Inspector's Handbook*, 70. See also Mosley, *Chimney of the World*, 168.

59. Carpenter, "London Fogs," 155.

60. Thorsheim, *Inventing Pollution*, 118.

61. See, e.g., J. D. Blaisdell, "To the Pillory for Putrid Poultry: Meat Hygiene and the Medieval London Butchers, Poulterers and Fishmongers' Companies," *Veterinary History* 9 (1997): 114–24.

62. Engels, *Condition of the Working Class*, 81.

63. William J. Ashworth, *Customs and Excise: Trade, Production and Consumption in England, 1640–1845* (Oxford: Oxford University Press, 2003).

64. John Burnett, *Plenty and Want: A Social History of Diet in England from 1815 to the Present Day*, rev. ed. (London: Scolar, 1979), 94–122.

65. Arthur Hill Hassall, *Food: Its Adulterations, and the Methods for Their Detection* (London: Longmans, Green, 1876), 854 (quote), 163.

66. Thomas Walley, *A Practical Guide to Meat Inspection* (Edinburgh: Young J. Pentland, 1890), 3.

67. Francis Vacher, *The Food Inspector's Handbook*, 4th ed. (London: The Sanitary Publishing Co., 1905), 2.

68. Blyth, *Dictionary of Hygiène*, 541.

69. Frazer, *Duncan of Liverpool*, 129.

70. Walley, *Meat Inspection*, 172. On the increasing importance of municipal veterinary practice, see Ann Hardy, "Pioneers in the Victorian Provinces: Veterinarians, Public Health and the Urban Animal Economy," *Urban History* 29, no. 3 (2002): 372–87.

71. Letheby cited in Hassall, *Food*, 475.

72. Robinson, *Sanitary Inspector's Practical Guide*, 77.

73. Walley, *Meat Inspection*, 22–24. See also Kier Waddington, "Unfit for Human Consumption: Tuberculosis and the Problem of Infected Meat in Late-Victorian Britain," *Bulletin of the History of Medicine* 77, no. 3 (2003): 636–61.

74. Walley, *Meat Inspection*, 59.

75. Cited in Taylor, *Sanitary Inspector's Handbook*, 251.

76. For the impact of legislation and analysis, see Michael French and Jim Phillips, *Cheated Not Poisoned? Food Regulation in the United Kingdom, 1875–1938* (Manchester: University of Manchester Press, 2000), 33–66.

77. James Burn Russell, "The Detective System v. the Clearing House System" (address to the annual congress of the Sanitary Association of Scotland, Dumfries, September 1896), in *Public Health Administration in Glasgow*, 606.

78. A. M. Trotter, "Meat and Milk Inspection," in *Municipal Glasgow: Its Evolution and Enterprises* (Glasgow: Glasgow Corporation, 1914), 242.

79. A. M. Trotter, "The Inspection of Meat and Milk in Glasgow," *Journal of Comparative Anatomy and Therapeutics* 14 (1901): 86.

80. On tuberculin and dairy farm regulation, see Kier Waddington, "To Stamp Out 'So Terrible a Malady': Bovine Tuberculosis and Tuberculin Testing in Britain, 1890–1939," *Medical History* 48 (2004): 29–48.

81. William Moyle, "Inspecting London," in *Living London: Its Work and Its Play, Its Humour and Its Pathos, Its Sights and Its Scenes*, ed. George Sims, 3 vols. (London: Cassell, 1902), 3:235.

82. See, e.g., Foucault, *Birth of the Clinic*.

83. R. Stephen Ayling, *Public Abattoirs: Their Planning, Design and Equipment* (London: E. & F. Spon, 1908), 4.

84. Russell, "Detective System," 594.

85. Nevertheless, clinical detachment was obviously necessary. For an anthropological investigation of the abattoir, see Noellie Vialles, *Animal to Edible*, trans. J. A. Underwood (Cambridge: Cambridge University Press, 1994).

86. Benjamin Ward Richardson, "Public Slaughter-Houses: A Suggestion for Farmers," *New Review* 8 (January-June 1893): 632.

87. Ayling, *Public Abattoirs*, 34.

88. Algernon Henry Grosvenor cited in *Minutes of Evidence and Appendix as to England and Wales*, 381.

89. Jenkin, "On Sanitary Inspection," 241.

90. William Eassie, *Healthy Houses: A Handbook to the History, Defects and Remedies of Drainage, Ventilation, Warming, and Kindred Subjects: With Estimates for the Best Systems in Use, and Upward of Three Hundred Illustrations* (New York: D. Appleton, 1872), 12.

91. Ibid., 62, 75, 79.

92. Jensen, *Modern Drainage Inspection*, 67.

93. Taylor, *Sanitary Inspector's Handbook*, 127.

94. Local government board regulations (1878) cited in Stockman, *Practical Guide for Sanitary Inspectors*, 218.

95. Hassall, *Food*, 832.

96. Pure Beer Act (1887) cited in *Analyst* 12 (June 1887): 100.

97. Sale of Horseflesh Act (1889) cited in Walley, *Meat Inspection*, 184.

98. Jensen, *Modern Drainage Inspection*, 34.

99. Taylor, *Sanitary Inspector's Handbook*, 68.

100. Hence, Latour's notion of the "immutable mobile" is not strictly appropriate here. See Bruno Latour, *Science in Action: How to Follow Scientists and Engineers through Society* (Milton Keynes: Open University Press, 1987), 227.

101. Joyce, *Rule of Freedom*, 232.

102. Stockman, *Practical Guide for Sanitary Inspectors*, 10.

103. Carter, *Eyesight*, 235–36.

104. On actor-network theory, see Latour, *Reassembling the Social*; and Michel Callon, "Society in the Making: The Study of Technology as a Tool for Sociological Analysis," in *The Social Construction of Technological Systems: New Directions in the Sociology and History of Technology*, ed. W. Bijker et al. (Cambridge, MA: MIT Press, 1987).

105. Pitt cited in Enid Gauldie, *Cruel Habitations: A History of Working-Class Housing, 1780–1914* (London: George Allen & Unwin, 1974), 118.

106. Newcastle Commission cited in Goodlad, *Victorian Literature and the Victorian State*, 6.

107. *Hansard Parliamentary Debates*, 3rd ser., vol. 93 (1847), cols. 1,103–4, cited in Lubenow, *Politics of Government Growth*, 86.

108. Baldwin, "Victorian State in Comparative Perspective," 54.

109. On prostitution, see Judith Walkowitz, *Prostitution and Victorian Society: Women, Class and the State* (Cambridge: Cambridge University Press, 1980).

110. Cited in Nadja Durbach, *Bodily Matters: The Anti-Vaccination Movement in England, 1853–1907* (Durham, NC: Duke University Press, 2005), 74.

111. Cromer cited in Mitchell, *Colonising Egypt*, 95.

112. Dorothy Porter, "'Enemies of the Race': Biologism, Environmentalism, and Public Health in Edwardian England," *Victorian Studies* 34, no. 2 (Winter 1991): 171.

113. Bentham, *Panopticon*, 42.

114. Foucault, *Discipline and Punish*, 214.
115. William Rendle, *London Vestries, and Their Sanitary Work: Are They Willing and Able to Do It? And May They Be Trusted in the Face of a Severe Epidemic?* (London: John Churchill, 1865), 18.
116. R. A. Selby cited in *Minutes of Evidence and Appendix as to England and Wales*, 539.
117. Reid, "Position of Sanitary Inspectors," 802–3.
118. Alexander Winter Blyth, *A Manual of Public Health* (London: Macmillan, 1890), 603.
119. Gloria Clifton, *Professionalism, Patronage and Public Service in Victorian London: The Staff of the Metropolitan Board of Works, 1856–1889* (London: Athlone, 1992), 136.
120. *Mining Journal*, January 12, 1867, cited in Bartrip, "British Government Inspection," 615.
121. William Compton cited in *Minutes of Evidence and Appendix as to England and Wales*, 35.
122. Reid, "Position of Sanitary Inspectors," 802.
123. Francis Sheppard, "St. Leonard, Shoreditch," in Owen, *Government of Victorian London*, 330.
124. Prest, *Liberty and Locality*, 76.
125. John C. Thresh, *An Inquiry into the Causes of Excess Mortality in No. 1 District, Ancoats* (Manchester: Heywood, 1889), 40–41, cited in Platt, *Shock Cities*, 325.
126. Platt, *Shock Cities*, 452.
127. Bowmaker noted that volunteer effort (by charities, clergymen, and parochial officers) should complement official inspection. If the local authority be found wanting, however, then "independent action" was justified. See *Housing of the Working Classes*, 53.
128. Robert Angus Smith, "Annual Report for 1864," cited in Ashby and Anderson, *Politics of Clean Air*, 26.
129. "RCNV. II. Minutes of Evidence," *Parliamentary Papers*, 1878, vol. 44, cited in Wohl, *Endangered Lives*, 219.
130. "Noxious Vapours," *Sanitary Record* 6 (June 8, 1877): 384.
131. "Report of the Royal Commission on the Working of the Factory and Workshop Acts," *Parliamentary Papers*, 1876, vol. 29, cited in P. W. J. Bartrip, "State Intervention in Mid-Nineteenth Century Britain: Fact or Fiction?" *Journal of British Studies* 23, no. 1 (1983): 73.
132. "Assaulting a Sanitary Inspector," *Sanitary Record* 8 (June 14, 1878): 380.
133. Vacher, *Food Inspector's Handbook*, 15.
134. Hamlin, "Sanitary Policing and the Local State," 41, 55, and *Public Health and Social Justice*, 280.
135. Taylor, *Sanitary Inspector's Handbook*, 84.
136. A. T. Rook, letter to the town clerk of Manchester, cited in *Minutes of Evidence and Appendix as to England and Wales*, 711.

137. For a much more thorough analysis of the "ethics" of inspection, see Tom Crook, "Norms, Forms and Bodies: Public Health, Liberalism and the Victorian City, 1830–1900" (Ph.D. diss., University of Manchester, 2004).

138. Lemmoin-Cannon, *Sanitary Inspector's Guide*, 130.

139. Firth, *Municipal London*, 312.

140. The 1891 Public Health (London) Act, e.g., stated that, after January 1, 1895, a sanitary inspector should hold a certificate approved by the local government board or have been an inspector for at least three years. Stockman, *Practical Guide for Sanitary Inspectors*, 3–4.

141. Stockman, *Practical Guide for Sanitary Inspectors*, 21.

142. Taylor, *Sanitary Inspector's Handbook*, 33. Tactless conduct by inspectors was publicly ridiculed. See, e.g., "Want of Tact on the Part of Sanitary Inspectors," *Lancet*, 1893, no. 2 (December 30): 1643.

143. Robinson, *Sanitary Inspector's Practical Guide*, 131.

144. Hennock, *Fit and Proper Persons*, 252.

145. Vogel, *National Styles of Regulation*, 26.

146. Trotter, "Inspection of Meat and Milk," 86.

147. See, e.g., Taylor, *Sanitary Inspector's Handbook*, 262.

148. Stockman, *Practical Guide for Sanitary Inspectors*, 23 (emphasis added).

149. Wohl, *Eternal Slum*, 189

150. Russell, "Uninhabitable Houses," 235.

151. Stockman, *Practical Guide for Sanitary Inspectors*, 126–27.

152. Ibid., 20. A justice could give a written order allowing the inspector to enter forcibly in the case of a proven nuisance. See also "Important Question as to Entry by an Inspector of Nuisances," *Sanitary Record* 3 (July 10, 1875): 30. A butcher locked the doors to his premises on a Sunday and refused to open them. Despite Vacher's claim that Sunday was "not necessarily an reasonable time" for inspection (*Food Inspector's Handbook*, 16), it was noted that only in cases of felony could doors be legitimately broken on Sundays.

153. J. E. Salway cited in *Minutes of Evidence and Appendix as to England and Wales*, 315.

154. Butt, "Working-Class Housing in Glasgow," 68–69.

155. Thomas Carnelly, J. S. Haldane, and A. M. Anderson, "The Carbonic Acid, Organic Matter, and Micro-Organisms in Air, More Especially of Dwellings and Schools," *Philosophical Transactions of the Royal Society of London* 178 (1887): 69–70.

156. E. P. Thompson, *The Making of the English Working Class* (New York: Vintage, 1966), 662.

157. Parliamentary commission cited in ibid., 82.

158. Oastler cited in Lubenow, *Politics of Government Growth*, 171.

159. MacDonagh, *Early Victorian Government*, 55–57. See also Webb, "Whig Inspector."

160. "The Public, and 'Public Analysts,'" *Analyst* 1 (November 30, 1876): 155.

161. Burnett, *Plenty and Want*, 117.

162. Hassall, *Food*, 868, 870.

163. Lubenow, *Politics of Government Growth*, 91.

164. Taylor, *Sanitary Inspector's Handbook*, 32.

165. For the full rainbow, see Walley, *Meat Inspection*, 27–31. *Rinderpest* is another term for cattle plague.

166. Vacher, *Food Inspector's Handbook*, 5.

167. Walley, *Meat Inspection*, 36.

168. Vacher, *Food Inspector's Handbook*, 6.

169. Joyce, *Rule of Freedom*, 21.

170. The best study of this is Thomas Laqueur, *Solitary Sex: A Cultural History of Masturbation* (New York: Zone, 2003).

171. Corbin, *The Foul and the Fragrant*, 175; Marcus, *Apartment Stories*, 94; Daunton, *House and Home*, 54; Stefan Muthesius, *The English Terraced House* (New Haven, CT: Yale University Press, 1982), 99.

172. Stevenson cited in Burnett, *Social History of Housing*, 191.

173. Waterlow cited in *Minutes of Evidence and Appendix as to England and Wales*, 426.

174. Vigers, representing the Peabody Trust, replying to Charles Gatliff, "On Improved Dwellings and Their Beneficial Effect on Health and Morals, with Suggestions for Their Extension," *Journal of the Statistical Society of London* 38, no. 1 (March 1875): 61.

175. On regional variety in terraced housing form, see Daunton, *House and Home*, 38–59.

176. Vigarello, *Concepts of Cleanliness*, 215–25. Of the private bathroom, Vigarello notes: "The history of cleanliness [was] never associated to such a degree with that of a space" (ibid., 216). See also Tom Crook, "Power, Privacy and Pleasure: Liberalism and the Modern Cubicle," *Cultural Studies* 21, nos. 4–5 (2007): 549–69; Joyce, *Rule of Freedom*, 75; Lawrence Wright, *Clean and Decent: The Fascinating History of the Bathroom and the Water-Closet*, new ed. (London: Penguin, 2000); and Laporte, *History of Shit*.

177. For a typology, see Daunton, *House and Home*, 246–61.

178. Corbin, *The Foul and the Fragrant*, 174.

179. Metropolitan Building Act cited in Stockman, *Practical Guide for Sanitary Inspectors*, 249.

180. "Small Room, Big Issue," *Manchester Guardian*, November 21, 2006. Most early workplace toilets were built without handbasins. See Anne Hardy, "Food, Hygiene, and the Laboratory: A Short History of Food Poisoning in Britain, *circa* 1850–1950," *Social History of Medicine* 12, no. 2 (1999): 307.

181. Fletcher, *Light and Air*, 12.

182. Herbert Arnold, *Popular Guide to House Painting, Decoration, Varnishing, Whitewashing, Colour Mixing, Floor Finishing, Paperhanging, Etc.* (Manchester: John Heywood, 1905).

183. Crook, "Power, Privacy and Pleasure," 552, 559.

184. For some elements of this apparatus in France, see Alain Corbin, "Backstage," in *A History of Private Life*, ed. Philippe Ariès and Georges Duby, vol. 4, *From the Fires of Revolution to the Great War*, ed. Michelle Perrot (Cambridge, MA: Belknap, 1990).
185. Some schools, recalled Robert Roberts, maintained faith in the doorless privy as an antimasturbatory apparatus. See Roberts, *Classic Slum*, 135. See also Crook, "Power, Privacy and Pleasure," 565–66.
186. Crook, "Power, Privacy and Pleasure," 554. See also Deborah Brunton, "Evil Necessaries and Abominable Erections: Public Conveniences and Private Interests in the Scottish City, 1830–1870," *Social History of Medicine* 18, no. 2 (2005): 187–202; and "Lavatories at Railway Stations," *Lancet*, 1892, no. 2 (August 20): 436.
187. "Public Latrines in London," *Engineering News* 31 (May 31, 1894): 445. For an example of the occlusion of urination, see "Dray's Patent Concealed Urinal," *Sanitary Record* 7 (November 2, 1877): 292. For gender aspects, see Barbara Penner, "A World of Unmentionable Suffering: Women's Public Conveniences in Victorian London," *Journal of Design History* 14, no. 1 (2001): 35–52.
188. Ellice Hopkins, *Work in Brighton* (London: Hatchards, 1877), 86, cited in Frank Mort, *Dangerous Sexualities: Medico-Moral Politics in England since 1830* (London: Routledge & Kegan Paul, 1987), 124.
189. Henry Roberts, *The Dwellings of the Labouring Classes, Their Arrangement and Construction; Illustrated by a Reference to the Model Houses of the Society for Improving the Condition of the Labouring Classes*, 3rd ed. (London: Society for Improving the Condition of the Labouring Classes, 1853), 9.
190. Lemmoin-Cannon, *Sanitary Inspector's Guide*, 74.
191. Lisa Forman Cody, "Living and Dying in Georgian London's Lying-in Hospitals," *Bulletin of the History of Medicine* 78 (2004): 322. In the later nineteenth century, Frederic Mouat claimed to have seen four children in a single hospital bed. See Frederick Mouat, "The Principles of Hospital Construction and Management in Relation to the Successful Treatment of Disease," in *Hospital Construction and Management*, by Frederic Mouat and H. Saxon Snell (London: J. & A. Churchill, 1883), 37.
192. See, e.g., Nightingale, *Notes on Hospitals*. See also G. C. Cook, "Henry Currey FRIBA (1820–1900): Leading Victorian Hospital Architect, and Early Exponent of the 'Pavilion Principle,'" *Postgraduate Medical Journal* 78, no. 920 (June 2002): 352–59. On hospital design, see Forty, "Modern Hospital"; and Markus, *Buildings and Power*, 106–18. See also Bynum, *Science and the Practice of Medicine*, 132–33.
193. Roy Porter, "The Rise of Physical Examination," in Bynum and Porter, eds., *Medicine and the Five Senses*, 180.
194. Nightingale, *Notes on Hospitals*, 79. See also Forty, "Modern Hospital," 77. On French curtained beds, see Armand Husson, *Étude sur les hopitaux, considérés sous le rapport de leur construction, de la distribution de leurs batiments,*

de l'ameublement, de l'hygiène & du service des salles de malades (Paris: Paul Dupont, 1862), 87–90; and Corbin, *The Foul and the Fragrant*, 101.

195. W. J. Gordon, *How London Lives: The Feeding, Cleansing, Lighting and Police of London; with Chapters on the Post Office and Other Institutions*, new ed. (London: Religious Tract Society, 1897), 160.

196. Cited in H. Saxon Snell, "Typical Examples of the General Hospitals in Various Countries," in Mouat and Snell, *Hospital Construction*, 76.

197. Philippe Ariès, *The Hour of Our Death*, trans. Helen Weaver (Oxford: Oxford University Press, 1991), 568.

198. Ruth Richardson, *Death, Dissection, and the Destitute*, 2nd ed. (Chicago: University of Chicago Press, 2000); Ian Burney, *Bodies of Evidence: Medicine and the Politics of the English Inquest, 1830–1926* (Baltimore: Johns Hopkins University Press, 2000). For an American comparison, see Ellen Stroud, "Dead Bodies in Harlem: Environmental History and the Geography of Death," in Isenberg, ed., *Nature of Cities*, 70.

199. Burney, *Bodies of Evidence*, 86–91.

200. Sims, *How the Poor Live*, in *How the Poor Live; and, Horrible London*, 60.

201. Foucault, *Discipline and Punish*, 9. See also Randall McGowen, "Civilizing Punishment: The End of the Public Execution in England," *Journal of British Studies* 33, no. 3 (1994): 257–82.

202. "London Private Slaughter-Houses," *Sanitary Record* 1 (December 19, 1874): 436.

203. A. Darbyshire, "On Public Abattoirs," *Builder* 33 (February 6, 1875): 113.

204. "Slaughter-Houses in the Metropolis," *Builder* 30 (August 17, 1872): 643.

205. "Metropolitan Board of Works Regulations for Slaughter-Houses," *Sanitary Record* 1 (November 14, 1874): 346.

206. J. S. Mill, "Civilization," in *The Collected Works of John Stuart Mill*, vol. 18, *Essays on Politics and Society*, ed. J. M. Robson (Toronto: University of Toronto Press, 1977), 131.

207. Corbin, *Women for Hire*, 61 (quote), 124.

208. Cited in Harriet Ritvo, *The Animal Estate: The English and Other Creatures in the Victorian Age* (Cambridge, MA: Harvard University Press, 1987), 215.

209. Girdlestone, *Unhealthy Condition of Dwellings*, 32.

210. Mosley, *Chimney of the World*, 44. For consumptive monkeys, see also Parkes, *Practical Hygiene*, 162.

211. Preston-Thomas, *The Work and Play of a Government Inspector*, 363–64 (emphasis added).

CHAPTER 4

1. Frederick Accum, *Description of the Process of Manufacturing Coal Gas, for the Lighting of Streets, Houses and Public Buildings*, 2nd ed. (London: Thomas Boys, 1820), 6. For more on Accum, see Christopher Hamlin, *A Science of*

Impurity: Water Analysis in Nineteenth-Century Britain (Berkeley and Los Angeles: University of California Press, 1990), 52.

2. Accum examined by Mr. Warren, in "Abstract of the Minutes of Evidence Given to a Committee of the House of Commons on the First Application of the Chartered Gas-Light and Coke Company, for an Act of Parliament to Incorporate Them, in 1809," cited in William Matthews, *An Historical Sketch of the Origin, Progress and Present State of Gas-Lighting*, 2nd ed. (London: Simpkin & Marshall, 1832), 286. On the 1809 bill, see Frederick Clifford, *A History of Private Bill Legislation*, 2 vols. (London: Butterworths, 1885), 1:204–8.

3. Accum cross-examined by Mr. Brougham, in "Abstract . . . 1809," cited in Matthews, *Gas-Lighting*, 289.

4. For Trotter's cycling exploits, see Paul Nahin, *Oliver Heaviside: The Life, Work, and Times of an Electrical Genius of the Victorian Age* (Baltimore: Johns Hopkins University Press, 2002), 165.

5. Schivelbusch, *Disenchanted Night*.

6. Accum, *Coal Gas*, 38.

7. For the history of control systems, see James Beniger, *The Control Revolution: Technological and Economic Origins of the Information Society* (Cambridge, MA: Harvard University Press, 1986).

8. Dillon, *Artificial Sunshine*, 127.

9. Jean Tardin, *Histoire naturelle de la fontaine qui brûl près de Grenoble* (1618), discussed in Fernand Braudel, *The Structures of Everyday Life: The Limits of the Possible*, trans. Siân Reynolds (New York: Harper & Row, 1981), 434.

10. Thomas Shirley, "A Description of a Well and Earth in Lancashire Taking Fire by a Candle Approached to It," *Philosophical Transactions of the Royal Society of London* (1667), cited in Matthews, *Gas-Lighting*, 4; "Extract from a Letter by the Rev. Dr. John Clayton," *Philosophical Transactions* (1739), cited in ibid., 15.

11. William Murdoch, "An Account of the Application of the Gas from Coal to Economical Purposes," *Philosophical Transactions of the Royal Society of London* 98 (1808): 131.

12. Owen Merriman, *Gas-Burners: Old and New* (London: Walter King, 1884), 13–14.

13. Charles Hunt, *A History of the Introduction of Gas Lighting* (London: Walter King, 1907), 101.

14. Accum, *Coal Gas*, 59; W. C. Holmes, *Instructions for the Management of Gas Works* (London: E. & F. Spon, 1874), 20.

15. Dibdin, *Public Lighting*, 104–5.

16. Accum, *Coal Gas*, 165.

17. Wilson, *Lighting the Town*, 24.

18. Richard Trench and Ellis Hillman, *London under London: A Subterranean Guide*, new ed. (London: J. Murray, 1993), 82; Samuel Clegg Jr., *A Practical Treatise on the Manufacture and Distribution of Coal-Gas; Its Introduction and*

Progressive Improvement, Illustrated by Engravings from Working Drawings (London: John Weale, 1841), 17.

19. T. Newbigging and W. T. Fewtrell, eds., *King's Treatise on the Science and Practice of the Manufacture and Distribution of Coal Gas*, 3 vols. (London: W. B. King, 1878–82), 2:334–36.
20. Accum, *Coal Gas*, 246.
21. Walter Grafton, *A Handbook of Practical Gas-Fitting: A Treatise on the Distribution of Gas in Service Pipes, the Use of Coal Gas, and the Best Means of Economizing Gas from Main to Burner*, 2nd ed. (London: Batsford, 1907), 36.
22. Trench and Hillman, *London under London*, 96.
23. William T. Sugg, *The Domestic Uses of Coal Gas, as Applied to Lighting, Cooking and Heating, Ventilation* (London: Walter King, 1884), 40.
24. Grafton, *Practical Gas-Fitting*, 8.
25. Accum, *Coal Gas*, 225.
26. Sugg, *Domestic Uses of Coal Gas*, 8. For the calculus relating to gas's motion through the system, see Clegg, *Practical Treatise*, 169–82.
27. For early feedback technologies, see Beniger, *Control Revolution*, 174–77.
28. Andrew Ure, *The Philosophy of Manufactures; or, An Exposition of the Scientific, Moral, and Commercial Economy of the Factory System of Great Britain* (London: Frank Cass, 1967), 1.
29. Stirling Everard, *The History of the Gas Light and Coke Company, 1812–1949* (London: Ernest Benn, 1949), 71.
30. Holmes, *Management of Gas Works*, 37. See also Clegg, *Practical Treatise*, 145–48.
31. Newbigging and Fewtrell, eds., *King's Treatise*, 2:423.
32. Sugg, *Domestic Uses of Coal Gas*, 90.
33. Dibdin, *Public Lighting*, 207 (emphasis added).
34. Dean Chandler, *Outline of the History of Lighting by Gas* (London: Chantry Lane Printing Works, 1936), 65.
35. Newbigging and Fewtrell, eds., *King's Treatise*, 3:1.
36. Dibdin, *Public Lighting*, 150.
37. J. Rutter, *Gas-Lighting: Its Progress and Its Prospects; with Remarks on the Rating of Gas-Mains, and a Note on the Electric-Light* (London: John Parker, 1849), 15.
38. Thomas Peckston, *A Practical Treatise on Gas-Lighting*, 3rd ed. (London: Hebert, 1841), 378.
39. James Brown, *Instructions to Gas Consumers on the Principle and Use of the Meter and to Non-Consumers, on the Economy of Coal-Gas over All Other Modes of Artificial Illumination* (London: W. Strange, 1850), 14.
40. A. Wood, *A Guide to Gas-Lighting* (Lewes: George P. Bacon, 1872), 30.
41. See Simon Schaffer, "Babbage's Dancer and the Impresarios of Mechanism," in *Cultural Babbage: Technology, Time and Invention*, ed. Francis Spufford and Jenny Uglow (London: Faber & Faber, 1996), 53–80.
42. *Journal of Gas Lighting* 36 (August 10, 1880): 221.

43. G. Glover, *On National Standards of Gas Measurement and Gas Meters* (London: W. Trounce, 1866), 3. The 1859 act defined the cubic foot as 62.321 pounds avoirdupois weight of distilled rainwater at 62 degrees F with the barometer measuring 30 inches. Models of this capacity were made in bottle form so that meters could be tested. Chandler, *History of Lighting by Gas*, 78. The 1859 act was permissive, but, by the early twentieth century, it was operative across Britain. See Jacques Abady, *Gas Analyst's Manual: Incorporating F. W. Hartley's "Gas Analyst's Manual" and "Gas Measurement"* (London: E. & F. Spon, 1902), 385.
44. Everard, *Gas Light and Coke Company*, 105; Stanley Harris, *The Development of Gas Supply on North Merseyside, 1815–1949: A Historical Survey of the Former Gas Undertakings of Liverpool, Southport, Prescot, Ormskirk and Skelmersdale* (Liverpool: North Western Gas Board, 1956), 71–72.
45. Abady, *Gas Analyst's Manual*, 383–443; Grafton, *Practical Gas-Fitting*, 306–7.
46. Dibdin, *Public Lighting*, 184.
47. *Manchester Municipal Code: Being a Digest of the Local Acts of Parliament, Charters, Commissions, Orders, Bye-Laws, Regulations, and Public Instructions and Forms in Force within the City of Manchester*, ed. Thomas Hudson, 11 vols. (Manchester: Solicitors' Law Stationery Society, 1894–1928), 3:424.
48. Sugg, *Domestic Uses of Coal Gas*, 35, 38.
49. Dibdin, *Public Lighting*, 278.
50. Francis Sheppard, "St. Pancras," in Owen, *Government of Victorian London*, 293.
51. See, e.g., "Preventing the Surreptitious Opening of Padlocks," *Journal of Gas Lighting* 74 (October 17, 1899): 943.
52. Milo Roy Maltbie, "Gas Lighting in Great Britain," *Municipal Affairs* 4 (1900): 560.
53. *Report from the Select Committee on Metropolitan Gas Companies* (1899), cited in Daunton, *House and Home*, 238.
54. George Turnbull, "Lighting," in Besant, *London*, 322.
55. "Gas Consumption through Prepayment Meters," *Engineering Record* 36 (November 20, 1897): 530. See also "A Dearth of Pence," *Journal of Gas Lighting* 74 (December 19, 1899): 1485.
56. G. H. W. Gerhardi, *Electricity Meters: Their Construction and Management* (London: Benn Bros., 1917), 231.
57. A. D. Webster, *Town Planting: And the Trees, Shrubs, Herbaceous and Other Plants That Are Best Adapted for Resisting Smoke* (London: George Routledge, 1910), 8.
58. Accum, *Coal Gas*, 234–35.
59. On the argand lamp, see Schivelbusch, *Disenchanted Night*, 9–14.
60. Sugg, *Domestic Uses of Coal Gas*, 84.
61. Merriman, *Gas-Burners*, 25, 38.
62. Wilson, *Lighting the Town*, 25, 156.

63. Samuel Hughes, *Gas Works: Their Construction and Arrangement; and the Manufacture and Distribution of Coal Gas*, revised, rewritten, and much enlarged by William Richards (London: Crosby, Lockwood, 1885), 378–79.

64. M. E. Falkus, "The British Gas Industry before 1850," *Economic History Review*, n.s., 20, no. 3 (1967): 506; Wilson, *Lighting the Town*, 12, 16.

65. Wilson, *Lighting the Town*, 156.

66. Everard, *Gas Light and Coke Company*, 130.

67. Schivelbusch, *Disenchanted Night*; Everard, *Gas Light and Coke Company*; Wilson, *Lighting the Town*.

68. Matthews, *Gas-Lighting*, 131. Matthews's figure was probably an exaggeration; the correct figure is, perhaps, closer to 120–30. See Falkus, "British Gas Industry," 498.

69. Wilson, *Lighting the Town*, 130.

70. Newbigging and Fewtrell, eds., *King's Treatise*, 2:282.

71. John Hollingshead, *Underground London* (London: Groombridge, 1862), 197.

72. Charles Hunt, *The Construction of Gas-Works* (London: Institution of Civil Engineers, 1894), 25, 34; Lynda Nead, *Victorian Babylon: People, Streets and Images in Nineteenth-Century London* (London: Yale University Press, 2000), 92–94.

73. "Light in London (III)," *Builder* 33 (November 13, 1875): 1009–10.

74. Figures cited in *Builder* 50 (June 19, 1886): 875.

75. "The Beckton Works of the Chartered Gas Company (I)," *Engineer* 29 (February 4, 1870): 60.

76. "The Opening of the Beckton Gasworks," *Engineer* 31 (February 10, 1871): 95.

77. "Progress of the Gas Supply," *Engineer* 33 (February 28, 1872): 135.

78. Beckton is featured on Paul Talling's wonderful Web site: http://www.derelictlondon.com/misc_pics.htm.

79. Maltbie, "Gas Lighting," 542; Wilson, *Lighting the Town*, 101.

80. S. B. Langlands, "Lighting of Streets," in *Municipal Glasgow*, 135, 134–35.

81. Wilson, *Lighting the Town*, 177; Maltbie, "Gas Lighting," 546–49.

82. For the early history of metropolitan gaslight companies, see Everard, *Gas Light and Coke Company*; and Firth, *Municipal London*, 341–45. See also Turnbull, "Lighting," 318–23.

83. Harry Chubb, "The Supply of Gas to the Metropolis," *Journal of the Statistical Society of London* 39, no. 2 (June 1876): 354.

84. Clifford, *Private Bill Legislation*, 1:229.

85. Firth, *Municipal London*, 671.

86. Graeme Gooday, *The Morals of Measurement: Accuracy, Irony, and Trust in Late Victorian Electrical Practice* (Cambridge: Cambridge University Press, 2004), 3.

87. Gooday terms the belief that numbers and standards alone create trust and constitute consecrated practice the *metrological fallacy*. *Measurement*, he argues, embraces a wider set of ethical, practical, and social questions than *metrology*. See *Morals of Measurement*, 11.

88. *Architect*, September 8, 1877, 130.
89. For nineteenth-century debates on the nature of light, see Park, *Fire within the Eye*; G. N. Cantor and M. J. S. Hodge, eds., *Conceptions of Ether: Studies in the History of Ether Theories, 1740–1900* (Cambridge: Cambridge University Press, 1981); and Jed Buchwald, *The Rise of the Wave Theory of Light: Optical Theory and Experiment in the Early Nineteenth Century* (London: University of Chicago Press, 1989).
90. In the otherwise exemplary *Nineteenth-Century Scientific Instruments* (Berkeley and Los Angeles: University of California Press, 1983), Gerard Turner ignores the photometer altogether. Photometers are included in Robert Bud and Deborah Warner, eds., *Instruments of Science: An Historical Encyclopedia* (London: Science Museum, 1998), 456–58.
91. Nineteenth-century experimenters developed two basic forms of radiometer, the thermopile (1829) and the bolometer (1880). The thermopile, devised by Leopoldo Nobili, utilized the thermoelectric effect to detect infrared radiation. Bolometry operated on the principle that light was analogous to heat and, thus, that the magnitude of the latter was an index of the former. The first practical bolometer was devised by Samuel Langley in 1880. See Silvanus Thompson, "Experiments in Bolometry," *Electrician* 11 (October 6, 1883): 491.
92. Johann Heinrich Lambert, *Photométrie; ou, De la mesure et de la gradation de la lumière, des couleurs, et de l'ombre* (Paris: L'Harmattan, 1997). The text was originally published in Latin in 1760.
93. For explication of the inverse-square law, see Dibdin, *Public Lighting*, 20–21; and Park, *Fire within the Eye*, 161–62.
94. On the jet photometer, see Dibdin, *Public Lighting*, 92. Dibdin also correctly noted that, since it did not measure light, the jet photometer was, technically, not a photometer at all.
95. W. J. Dibdin, *Practical Photometry: A Guide to the Study of the Measurement of Light* (London: Walter King, 1889), 2.
96. Dibdin, *Practical Photometry*, 3.
97. A list of early-twentieth-century photometers is provided in J. A. Fleming, *A Handbook for the Electrical Laboratory*, 2 vols. (London: "The Electrician" Printing and Publishing Co., 1901), 264–301.
98. Dibdin, *Practical Photometry*, 141.
99. Alexander Pelham Trotter, *Illumination: Its Distribution and Measurement* (London: Macmillan, 1911), 138.
100. Dibdin, *Practical Photometry*, 121.
101. "Photometric Units and Dimensions," *Electrician* 33 (September 28, 1894): 632.
102. Louis Bell, *The Art of Illumination* (New York: McGraw, 1902), 313.
103. Walsh, *Manual of Domestic Economy*, 124.
104. A grain is the smallest unit of weight in Britain, measuring one-seven thousandth of a pound avoirdupois. France and Germany utilized different

standards, the carcel oil lamp and the Hefner paraffin candle, respectively. Bell, *Art of Illumination*, 316.

105. Cited in Dibdin, *Practical Photometry*, 88.

106. The term *candle* persisted as the basic unit of light, despite efforts to replace it with *lumen*, which, with *lux* (illumination), was mooted as an elegant epithet for absolute light. The terms were decisively revived following the 1896 Geneva Congress.

107. "Photometric Standards," *Sanitary Engineer* 5 (December 1, 1881): 10.

108. James Barr and Charles Phillips, "The Brightness of Light: Its Nature and Measurement," *Electrician* 32 (March 9, 1894): 524. An inch was defined as three barleycorns in 1324.

109. Dibdin, *Practical Photometry*, 90.

110. "Tests of Glow Lamps," *Electrician* 37 (October 2, 1896): 731.

111. O. Lummer and F. Kurlbaum, "The Search for a Unit of Light," *Electrician* 34 (November 9, 1894): 37.

112. J. E. Petavel, "An Experimental Research on Some Standards of Light," *Proceedings* (Royal Society of London) 65 (1899): 469–503. For a discussion of the unsatisfactory nature of the platinum standard at that time, see esp. ibid., 478–80.

113. Dibdin, *Practical Photometry*, 115.

114. "Comparison of Harcourt and Methven Photometric Standards," *Electrician* 17 (October 15, 1886): 481.

115. Fleming, *Handbook for the Electrical Laboratory*, 262–63; Edward Pyatt, *The National Physical Laboratory: A History* (Bristol: Adam Hilger, 1983), 44.

116. Fleming, *Handbook for the Electrical Laboratory*, 238.

117. Abady, *Gas Analyst's Manual*, 116.

118. C. H. Stone, *Practical Testing of Gas and Gas Meters* (New York: John Wiley, 1909), 36.

119. Matthews, *Gas-Lighting*, 217–18.

120. The Gas Light and Coke Company was bound to supply gas at sixteen candles, while the Imperial and South Metropolitan level was fourteen. Firth, *Municipal London*, 372.

121. Firth, *Municipal London*, 355.

122. Harold Royle, *The Chemistry of Gas Manufacture: A Practical Manual for the Use of Gas Engineers, Gas Managers, and Students* (London: Crosby, Lockwood, 1907), 242.

123. Webber, *Town Gas*, 16; Maltbie, "Gas Lighting," 551. The gas mantle meant that luminous flames no longer represented the kind of light many people used: hence agreements to *lower* legal maxima, on the grounds that the mantle would provide illumination well above that provided by the test burner. Webber, *Town Gas*, 9–10.

124. *Engineer* 72 (August 28, 1891): 174.

125. Everard, *Gas Light and Coke Company*, 193.

126. Dibdin, *Practical Photometry*, 12.

127. Firth, *Municipal London*, 261.
128. Sean F. Johnston, *A History of Light and Colour Measurement: Science in the Shadows* (London: Institute of Physics Publishing, 2001), 43.
129. A list was provided in Royle, *Chemistry of Gas Manufacture*, 240.
130. This was not a straightforward application of expertise, however. The very first examiners, Evans, Patterson, and Pierce, appear to have been rather less than "experts," according to a scathing report accusing them of being "absolutely ignorant of all matters connected with the purification, testing and supply of Gas in any shape." *Journal of Gas Lighting*, September 24, 1872, 801, cited in Firth, *Municipal London*, 368.
131. Johnston, *History of Light and Colour Measurement*, 48.
132. W. Foster, "Notes on Photometry," *Engineer* 32 (October 6, 1871): 229–30.
133. Gasworks Clauses Act (1871), schedule A, pt. 2, cited in Dibdin, *Public Lighting*, 32.
134. The routine was described, with an example, in Thomas Newbigging, *The Gas Manager's Handbook; Consisting of Tables, Rules and Useful Information for Gas Engineers, Managers, and Others Engaged in the Manufacture and Distribution of Coal Gas*, 3rd ed. (London: Walter King, 1883), 247.
135. Clifford, *Private Bill Legislation*, 1:222.
136. See Graeme Gooday, "The Premisses of Premises: Spatial Issues in the Historical Construction of Laboratory Credibility," in *Making Space for Science*, ed. Crosbie Smith and Jon Agar (Basingstoke: Macmillan, 1998).
137. Manchester City Council, *Proceedings*, 1867, 312.
138. "The Gas Supply of the City," *Chemical News and Journal of Physical Science* 4 (March 1869): 160.
139. *British Architect and Northern Engineer* 17 (February 10, 1882): 63. See also W. Wallace, "On the Heating Power of Coal Gas of Different Qualities," *Proceedings of the Royal Philosophical Society of Glasgow* 12 (1879): 208–11.
140. "London Gas," *Lancet*, 1877, no. 2 (October 20): 590.
141. Here, I echo Gooday's measured and sympathetic critique of Latour. See *Morals of Measurement*, 17–34, 37.
142. Dibdin, *Practical Photometry*, 49.
143. William Preece, "Report on Electric Lighting in the City," *Electrician* 15 (April 25, 1885): 497. For the vagaries of the law relating to city light levels, see n. 120 above.
144. "Photometric Standards," 10.
145. Trotter, *Illumination*, 183.
146. See Otto Sibum, "Reworking the Mechanical Value of Heat: Instruments of Precision and Gestures of Accuracy in Early Victorian England," *Studies in History and Philosophy of Science* 26 (1995): 73–106; and Christopher Lawrence and Steven Shapin, *Science Incarnate: Historical Embodiments of Natural Knowledge* (Chicago: University of Chicago Press, 1998).
147. Gooday, "Spot-Watching," 174.
148. Trotter, *Illumination*, 67.

149. H. Tranin, "Mesures photométriqes dans les differentes régions du spectre," *Journal de physique* 5 (1876): 304, cited (in English) in Johnston, *History of Light and Colour Measurement*, 55.

150. Trotter, *Illumination*, 192.

151. T. Seymour Hawker, "Hawker's Sine Photometer," *Electrician* 13 (July 26, 1884): 253.

152. Bell, *Art of Illumination*, 331 (emphasis added).

153. Fleming, *Handbook for the Electrical Laboratory*, 287.

154. Trotter, *Illumination*, 196.

155. On binocular photometry, see *Engineer* 78 (June 19, 1896): 610.

156. Alexander Pelham Trotter, "The Distribution and Measurement of Illumination," *Minutes of Proceedings* (Institute of Civil Engineers) 110 (May 10, 1892): 99.

157. William Abney, *Colour Measurement and Mixture* (London: Society for Promoting Christian Knowledge, 1891), 79, cited in Johnston, *History of Light and Colour Measurement*, 53.

158. A. E. Kennelly and S. E. Whiting, *Some Observations on Photometric Precision* (New York: National Electric Light Association, 1908), esp. 9–10. These results were also discussed in Trotter, *Illumination*, 191.

159. William Edward Barrows, *Electrical Illuminating Engineering* (New York: McGraw, 1908), 31. For the Purkinje effect, see n. 33 of chapter 1.

160. Blondel, "Street Lighting" (November 15), 90.

161. F. K. Richtmeyer and E. C. Crittenden, "The Precision of Photometric Measurements," *Journal of the Optical Society of America and Review of Scientific Instruments* 4 (1920), cited in Johnston, *History of Light and Colour Measurement*, 63.

162. Johnston, *History of Light and Colour Measurement*, 129.

163. W. H. Preece, "Report on Electric Lighting in the City," *Electrician* 15 (April 18, 1885): 480.

164. Dibdin, *Practical Photometry*, 62.

165. William Preece, "On a New Standard of Illumination and the Measurement of Light," *Proceedings* (Royal Society of London) 36 (1884): 271.

166. William Preece, "The Measure of Illumination," *Electrician* 23 (September 13, 1889): 478.

167. William Preece, Report to the Streets Committee of the Commissioners of Sewers of the City of London (August 1884), cited in Trotter, *Illumination*, 200.

168. Trotter, *Illumination*, 199 (emphasis added). See also J. S. Dow and V. H. Mackinney, "Surface Brightness and a New Instrument for Its Measurement," *Transactions of the Optical Society* 12 (1910): 66–104.

169. Trotter, *Illumination*, 15.

170. Trotter, "Distribution and Measurement of Illumination," 71.

171. Barrows, *Electrical Illuminating Engineering*, 182.

172. Cited in Dibdin, *Public Lighting*, 404.

173. Dibdin, *Public Lighting*, 404.

174. Trotter, *Illumination*, 247.

175. See n. 75 of the introduction.

CHAPTER 5

1. Dibdin, *Public Lighting*, 19.

2. Mordey cited in Rollo Appleyard, *The History of the Institution of Electrical Engineers (1871–1931)* (London: Institution of Electrical Engineers, 1939), 197. For the opposite view, see Hans Jonas, "The Scientific and Techno-logical Revolutions: Their History and Meaning," *Philosophy Today* 2, no. 4 (1971): 98; and Lewis Mumford, *Technics and Civilisation*, new ed. (New York: Harbinger, 1963), 255.

3. A point made cogently by Iwan Rhys Morus, *Frankenstein's Children: Electricity, Exhibition, and Experiment in Early-Nineteenth-Century London* (Princeton, NJ: Princeton University Press, 1998), xiii.

4. Oil gas was a competitor to coal gas from the early nineteenth century. See Accum, *Coal Gas*, 289. For albocarbon illumination, see Dibdin, *Public Lighting*, 218. For acetylene, see Jakle, *City Lights*, 28–29; William E. Gibbs, *Lighting by Acetylene: Generators, Burners and Electric Furnaces* (New York: Nostrand, 1898); and Frederick Dye, *Lighting by Acetylene: A Treatise for the Practical Engineer Etc.* (London: E. & F. N. Spon, 1902). On metal gas, see *British Architect and Northern Engineer* 21 (February 22, 1884): 89.

5. Dibdin, *Public Lighting*, 490–91.

6. *Encyclopaedia Britannica* (1911), http://www.1911encyclopedia.org/Lighting (accessed October 7, 2007).

7. Briggs, *Victorian Things*, 292.

8. Merriman, *Gas-Burners*, 63.

9. Hartley cited in Merriman, *Gas-Burners*, 70.

10. "Illumination and Ventilation," *Builder* 54 (April 14, 1888): 263.

11. "Gas-Lighting by Incandescence," *Builder* 51 (September 18, 1886): 436.

12. Dibdin, *Public Lighting*, 225.

13. Langlands, "Lighting of Streets," 136.

14. Dibdin, *Public Lighting*, 230–31.

15. Archibald Williams, *How It Works: Dealing in Simple Language with Steam, Electricity, Light, Heat, Sound, Hydraulics, Optics, Etc.* (London: Thomas Nelson, 1906), 409.

16. Schivelbusch, *Disenchanted Night*, 48.

17. "At the Inventions Exhibition," *British Architect and Northern Engineer* 23 (June 26, 1885): 306.

18. J. Slater, "Artificial Illumination," *Builder* 22 (February 9, 1889): 106. On improvements in gas purity, see J. Wanklyn, "Need Gas-Making Be a Public Nuisance?" *British Architect and Northern Engineer* 14 (July 3, 1885): 9; and Hunt, *Construction of Gas-Works*, 19.

19. Schivelbusch, *Disenchanted Night*; Bowers, *Lengthening the Day*; and Brian Bowers, *A History of Electric Light and Power* (New York: Peter Peregrinus, 1982).

20. "The Electric Light from a Medical Standpoint," *Electrician* 7 (October 15, 1881): 338.

21. *Electrician* 8 (March 18, 1882): 281.

22. *British Architect and Northern Engineer* 17 (January 27, 1882): 41.

23. W. Crimp, "An Experiment in Street Lighting," *British Architect and Northern Engineer* 14 (August 28, 1885): 98. Paraffin had become vastly cheaper since the discovery of large reserves in Pennsylvania in 1859.

24. Crimp, "Experiment in Street Lighting"; and C. Cooper, "Progress in Oil Lighting at Wimbledon," *Proceedings* (Association of Municipal and Sanitary Engineers and Surveyors) 15 (September 28, 1889): 21.

25. Gordon, *How London Lives*, 109.

26. See Thomas Hughes, *Networks of Power: Electrification and Western Society, 1880–1930* (London: Johns Hopkins University Press, 1983), 64, 250, 257–58.

27. "Gas Companies and Electric Supply," *Electrician* 25 (August 1, 1890): 338.

28. *Engineer* 76 (March 4, 1894): 376.

29. William Perren Maycock, *Electric Wiring, Fittings, Switches and Lamps: A Practical Book for Electric Light Engineers, &c.* (London: Whittaker, 1899), 411–12.

30. On various forms of vapor lamp (mercury, sodium, etc.), see Jakle, *City Lights*, 79–90.

31. Schivelbusch, *Disenchanted Night*, 52.

32. These buildings included the Sunderland Atheneum and London's National Gallery. See "Staite's Improvements in Lighting, Etc.," *Patent Journal and Inventor's Magazine* 4 (1848): 169–73; G. H. Staite, *Staite's Electric Light, 1846–1849* (Chester, 1882); G. Woodward, "Staite and Petrie: Pioneers of Electric Lighting," *Science, Measurement and Technology: IEE Proceedings A* 136, no. 6 (November 1989): 290–96; Morus, *Frankenstein's Children*, 185; and Bowers, *Lengthening the Day*, 66–67.

33. Paget Higgs, *The Electric Light in Its Practical Applications* (London: E. & F. Spon, 1879), 6.

34. *British Architect and Northern Engineer* 15 (April 8, 1881): 182.

35. Board of Trade Regulations for Securing the Safety of the Public cited in Dibdin, *Public Lighting*, 508.

36. "The Electric 'Candle,'" *Builder* 35 (June 30, 1877): 677. The candle was named after its Russian inventor, Paul Jablochkoff.

37. T. C. Hepworth, *The Electric Light: Its Past History and Present Position* (London: Routledge, 1879), 82.

38. *Plumber and Sanitary Engineer* 1 (February 1878): 44.

39. J. Shoolbred, "Electric Lighting, and Its Application to Public Illumination by Municipal and Other Bodies," *Proceedings* (Association of Municipal and Sanitary Engineers and Surveyors) 6 (July 31, 1879): 15–18.

40. J. W. Bazalgette and T. W. Keates, cited in "The Report on the Electric Light on the Thames Embankment," *Electrician* 3 (June 7, 1879): 31.
41. Lewis Angell responding to J. C. Shoolbred, "Illumination by Electricity," *Proceedings* (Association of Municipal and Sanitary Engineers and Surveyors) 8 (June 29, 1882): 185.
42. *Electrician* 11 (October 13, 1883): 526.
43. Killingworth Hedges, *Useful Information on Electric Lighting*, 4th ed. (London: E. & F. N. Spon, 1882), 42.
44. Percy Scrutton, *Electricity in Town and Country Houses* (London: Archibald Constable, 1898), 137.
45. Thomas Edison, "The Success of the Electric Light," *North American Review* 131 (October 1880): 297 (cited also in "Mr Edison to the Rescue," *British Architect and Northern Engineer* 14 [October 22, 1880]: 189). See also Charles Bazerman, *The Languages of Edison's Light* (Cambridge, MA: MIT Press, 1999). I would like to thank Graeme Gooday for pointing me toward this reference.
46. For the British response, see Bazerman, *Edison's Light*, 186.
47. Paul Israel, *Edison: A Life of Invention* (New York: John Wiley, 1998), 167–90; Hughes, *Networks of Power*, 18–78.
48. Hepworth, *The Electric Light*, 107.
49. Southard, *Modern Eye*, 20.
50. See David Brewster, "Observations on the Lines of the Solar Spectrum, and on Those Produced by the Earth's Atmosphere, and by the Action of Nitrous Acid Gas," *Transactions of the Royal Society of Edinburgh* 12 (1834): 519–30. On spectroscopy's uses, including the identification of new elements and impurities in metals, see Henry Roscoe, *Spectrum Analysis: Six Lectures Delivered in 1868 before the Society of Apothecaries of London*, 4th ed. (London: Macmillan, 1885); and William McGucken, *Nineteenth-Century Spectroscopy* (Baltimore: Johns Hopkins University Press, 1969).
51. Johnston, *History of Light and Colour Measurement*, 27; Dibdin, *Practical Photometry*, 149; Musselman, *Nervous Conditions*, 68.
52. Higgs, *The Electric Light*, 5.
53. W. Pickering, "Concerning the Gas-Flame, Electric and Solar Spectra, and Their Effect on the Eye," *Nature* 26 (February 9, 1882): 341.
54. H. Kohn, "Comparative Determinations of the Acuteness of Vision and the Perception of Colours by Day-Light, Gas-Light and Electric Light," *Archives of Ophthalmology* 9 (1880): 60, 62.
55. S. Chandley, "The Combination of Colours," *Builder* 31 (January 18, 1873): 40.
56. James Dixon, *A Guide to the Practical Study of Diseases of the Eye: With an Outline of Their Medical and Operative Treatment* (Philadelphia: Lindsay & Blakiston, 1860), 76; Harlan, *Eyesight*, 97–98.
57. Fletcher, *Light and Air*, 118. On the law of ancient light, see nn. 40 and 41 of chapter 2 above.

58. *Warehouseman and Drapers' Trade Journal* cited in *Electrician* 8 (May 13, 1882): 418 (emphasis added).
59. Mill owner cited in *Electrician* 8 (January 14, 1882): 136.
60. Hepworth, *The Electric Light*, 82.
61. Bell, *Art of Illumination*, 275.
62. Scrutton, *Electricity*, 138.
63. George Grossmith and Weedon Grossmith, *Diary of a Nobody* (Harmondsworth: Penguin, 1965), 183. See also Dillon, *Artificial Sunshine*, 70.
64. John Gamgee, "Cattle Plague and Diseased Meat, in Their Relations with the Public Health . . . : A Letter to . . . Sir G. Grey" (London, 1857), 39, cited in Richard Perren, *The Meat Trade in Britain, 1840–1914* (London: Routledge & Kegan Paul, 1978), 34. Walley noted that yellowing flesh could never be accurately discerned by "artificial light," which presumably meant gas or oil (*Meat Inspection*, 29).
65. *Engineer* 74 (November 18, 1892): 431.
66. W. Preece responding to Trotter, "Distribution and Measurement of Illumination," 142.
67. Festing cited in Appleyard, *History of the Institution of Electrical Engineers*, 142.
68. Ayrton cited in Appleyard, *History of the Institution of Electrical Engineers*, 142.
69. *Builder* 40 (January 7, 1882): 10.
70. Terry cited in Dillon, *Artificial Sunshine*, 183.
71. Harold Platt, *The Electric City: Energy and the Growth of the Chicago Area, 1880–1930* (Chicago: University of Chicago Press, 1991), 30.
72. George Ashdown Audsley, *Colour in Dress: A Manual for Ladies* (London: Sampson Low, Marston, 1912), 38. For more scientific analyses, see H. E. Houston and W. W. Washburn, "The Effect of Various Kinds of Illumination upon Coloured Surfaces," *American Journal of Psychology* 19, no. 4 (October 1908): 536–40; and Arthur Blok, *Elementary Principles of Illumination and Artificial Lighting* (London: Scott, Greenwood, 1914), 8–9. See also Walsh, *Manual of Domestic Economy*, 123.
73. O'Dea, *Social History of Lighting*, 9; John Bury, "The Use of Candle-Light for Portrait-Painting in Sixteenth-Century Italy," *Burlington Magazine* 119 (June 1977): 434–37.
74. For the relation between illumination and art, see Susan Hollis Clayson, "Outsiders: American Painters and Cosmopolitanism in the City of Light, 1871–1914," in *La France dans le regard des États-Unis/France as Seen by the United States*, ed. Frédéric Monneyron and Martine Xiberras (Perpignan: Presses Universitaires de Perpignan; Montpellier: Publications de l'Université Paul Valéry—Montpellier III, 2006), 57–71.
75. William Sugg, *Gas as an Illuminating Agent Compared with Electricity* (London: Walter King, 1882), 3–4, 6.
76. "Illumination in Warfare," *Engineer* 32 (December 29, 1871): 438.
77. See, e.g., Brigadier-General E. S. May, *An Introduction to Military Geography* (London: Hugh Rees, 1919), 16–18.

78. Mel Gorman, "Electric Illumination in the Franco-Prussian War," *Social Studies of Science* 7 (1977): 525–29.
79. Report cited in *Electrician* 8 (April 29, 1882): 387.
80. For more detail on such contrivances, see F. Nerz, *Searchlights: Their Theory, Construction and Applications*, trans. Charles Rogers (London: Archibald Constable, 1907).
81. "The Electric Search Light in Warfare," *Engineer* 73 (March 25, 1892): 252.
82. J. Maier, *Arc and Glow Lamps: A Practical Handbook on Electric Lighting* (London: Whittaker, 1886), 350.
83. For the growing use of electric light as defense against torpedo craft, see William Baker Brown, *History of Submarine Mining in the British Army* (Chatham: W. & J. Mackay, 1910), 73.
84. Nerz, *Searchlights*, 59.
85. Major R. L. Hippisley, *Lecture on "Electricity and Its Tactical Value for Military Operations"* (Aldershot: Gale & Polden, 1891), 3.
86. Appleyard, *History of the Institution of Electrical Engineers*, 215.
87. Johnston, *History of Light and Colour Measurement*, 114–16.
88. See, e.g., Corbin, *Village Bells*, 101; Otis, *Networking*, 37; Elizabeth Green Musselman, "The Governor and the Telegraph: Mental Management in British Natural Philosophy," in Morus, ed., *Bodies/Machines*, 77–78; and Marsden and Smith, *Engineering Empires*, 185.
89. Colonel F. C. Keyser, *Lecture on "Visual Signalling"* (Aldershot: Gale & Polden, 1891), 5.
90. E. J. Solano, ed., *Signalling, Morse, Semaphore Station Work, Despatch Riding, Telephone Cables, Map Reading* (London: John Murray, 1916), 212.
91. Leonard P. Lewis, *Railway Signal Engineering (Mechanical)* (New York: D. Van Nostrand, 1912), 79.
92. Thomas H. Bickerton, *Colour Blindness and Defective Eyesight in Officers and Sailors of the Mercantile Marine* (Edinburgh: James Thin, 1890), 1–2.
93. For fascinating analyses of color blindness, see Musselman, *Nervous Conditions*, 76–134; and Bailkin, "Colour Problems."
94. *Engineer* 27 (August 16, 1895): 166.
95. T. H. Bickerton, "Sailors and Their Eyesight, Including Colour-Blindness," *British Medical Journal*, 1888, no. 2 (November 10): 1038–41.
96. B. Carter, "Colour Blind Engine Drivers," *Engineer* 69 (January 31, 1890): 95.
97. On the Holmgren test, see Bailkin, "Colour Problems," 96.
98. William de Wiveleslie Abney, *Colour Vision: Being the Tyndall Lectures Delivered in 1894, Etc.* (London: Sampson Law, Marston, 1895); Bickerton, *Colour Blindness*; F. W. Edridge-Green, *Colour-Blindness and Colour-Perception* (London: Kegan Paul, 1891); Karl Grossmann, "Colour-Blindness, with Demonstrations of New Tests," *British Medical Journal*, 1888, no. 2 (November 10): 1041–43.

99. Michael Faraday cited in Roy MacLeod, "Science and Government in Victorian England: Lighthouse Illumination and the Board of Trade, 1866–1886," *Isis* 60 (1969): 5.

100. MacLeod, "Science and Government." Nocturnal work on the Manchester Ship Canal in 1887 was undertaken by 3 electric and 320 oil lamps of two thousand candlepower each. See James Winter, *Secure from Rash Assault: Sustaining the Victorian Environment* (Berkeley and Los Angeles: University of California Press, 1999), 116. On oil lamps in American lighthouses, see Jakle, *City Lights*, 189.

101. R. H. Patterson, "Gas-Burners and the Principles of Gas-Illumination," in Newbigging and Fewtrell, eds., *King's Treatise*, 3:156.

102. J. M. Douglas, "Electric Lighting Applied to Lighthouse Illumination," *British Architect and Northern Engineer* 11 (March 21, 1879): 133.

103. Michael Brian Schiffer, "The Electric Lighthouse in the Nineteenth Century: Aid to Navigation and Political Technology," *Technology and Culture* 46, no. 2 (April 2005): 275–305.

104. *Electrician* 31 (September 1, 1893): 478.

105. E. Edwards, *Our Seamarks: A Plain Account of the Lighthouses, Lightships, Beacons, Buoys, and Fog-Signals Maintained on Our Coasts for the Guidance of Mariners* (London: Longmans, Green, 1884), 56.

106. Archibald Barr and William Stroud, *Telemeters or Range-Finders for Naval and Other Purposes* (London: Institute of Mechanical Engineers, 1896), 56.

107. Marcet, "On Fogs," 63.

108. *Shadow*, January 15, 1870.

109. Manchester City Council, *Proceedings*, 1870–71, 153.

110. Mosley, *Chimney of the World*, 29.

111. *Engineer* 74 (December 23, 1892): 562.

112. For a comparison of British and Dutch imperial lighthouses, see Eric Tagliacozzo, "The Lit Archipelago: Coast Lighting and the Imperial Optic in Insular Southeast Asia, 1860–1910," *Technology and Culture* 46, no. 2 (April 2005): 306–28.

113. *Electrician* 8 (January 7, 1882): 114, and 14 (November 29, 1889): 82; Daniel Headrick, *The Tentacles of Progress: Technology Transfer in an Age of Imperialism* (Oxford: Oxford University Press, 1987), 26

114. J. M. Bryant and H. C. Hake, *Street Lighting*, Engineering Experiment Station Bulletin no. 51 (Urbana: University of Illinois, 1911), 21.

115. A. P. Trotter, *The Elements of Illuminating Engineering* (London: Sir Isaac Pitman, 1921), 93.

116. Moore, *Omnibuses and Cabs*, 241.

117. See Beniger, *Control Revolution*; Wolfgang Schivelbusch, *The Railway Journey: The Industrialization and Perception of Time and Space in the Nineteenth Century* (Leamington Spa: Berg, 1986); and Paul Virilio, *Speed and Politics: An Essay in Dromology* (New York: Semiotext[e], 1977), and *The Aesthetics of Disappearance*, trans. Philip Beichtman (New York: Semiotext[e], 1991).

118. Sennett, *Fall of Public Man*, 25; Foucault, *Discipline and Punish*, 139.

119. Smiles, *Self-Help*, 112.

120. Hermann von Helmholtz, *Populäre wissenschaftliche Vorträge* (1876), 89, cited in Dolf Sternberger, *Panorama of the Nineteenth Century*, trans. Joachim Neugroschel (New York: Urizen, 1977), 173–74.

121. Barr and Phillips, "Brightness of Light," 525.

122. Trotter cited in *Electrician* 29 (May 13, 1892): 32. See also Bell, *Art of Illumination*, 254–57.

123. Blondel, "Street Lighting" (November 15), 90.

124. "Street Lighting," *Electrician* 27 (July 3, 1891): 241.

125. *Builder* 32 (October 24, 1874): 897.

126. Nevertheless, affixing notices was regulated. In London, the fixing of illegal bills and the defacing of legal notice boards could lead to a fine of up to forty shillings under the 1862 Metropolis Management Act.

127. *Electrician* 6 (December 25, 1880): 61.

128. "Raznye izvestiia," *Elektrichestvo*, 1883, nos. 21–22, 239, cited in Jonathan Coopersmith, *The Electrification of Russia, 1880–1926* (Ithaca, NY: Cornell University Press, 1992), 48.

129. Schlör, *Nights in the Big City*, 71–163.

130. "London Roads and the Cost of Them," *Builder* 33 (July 24, 1875): 652. See also E. S. De Beer, "Early History of London Street-Lighting," *History* 25 (1941): 312–28.

131. *General Regulations, Instructions, and Orders, for the Government and Guidance of the Metropolitan Police Force* (London: George E. Eyre & William Spottiswoode, 1862), 77.

132. Newbigging and Fewtrell, eds., *King's Treatise*, 1:69.

133. *Builder* 33 (June 12, 1875): 539.

134. Sugg, *Domestic Uses of Coal Gas*, 103.

135. Caminada, *Detective Life*, 99.

136. For a revealing analysis of this, see Deidre Palk, "Private Crime in Public and Private Places: Pickpockets and Shoplifters in London, 1780–1823," in Hitchcock and Shore, eds., *Streets of London*, 142.

137. *Electrician* 6 (January 1, 1881): 73.

138. *Electrician* 28 (February 5, 1892): 342.

139. Edwin Hurry Fenwick, *A Handbook of Clinical Electric-Light Cystoscopy* (London: J. & A. Churchill, 1904), 1–2. See also Francis Cruise, *The Endoscope as an Aid in the Diagnosis and Treatment of Disease* (Dublin: Fannin, 1865); and Felix Semon, "Electric Illumination of the Various Cavities of the Human Body," *Lancet*, 1885, no. 1 (March 21): 512.

140. Fenwick, *Electric-Light Cystoscopy*, 96, 106.

141. Trotter, *Illumination*, 1.

142. Augustus Noll, *How to Wire Buildings: A Manual of the Art of Interior Wiring*, 4th ed. (New York: D. Van Nostrand, 1906), 120.

143. Higgs, *The Electric Light*, 6 (first emphasis added).

144. Dibdin, *Public Lighting*, 401 (quote), 406–7.
145. W. Sumpner, "The Diffusion of Light," *Electrician* 30 (February 3, 1893): 382.
146. Noll, *How to Wire Buildings*, 113.
147. On American floodlighting, see Jakle, *City Lights*, 180–94.
148. For a brief, but informative, history of floodlighting at British and Continental football grounds, see Simon Inglis, *The Football Grounds of England and Wales* (London: Willow, 1983), 37–42. See also Jakle, *City Lights*, 39–40.
149. On billiard table lighting, see Archibald Boyd, "Implements," in *Billiards*, by Major William Broadfoot (London: Longmans, Green, 1896), 66–67. On railway carriage lighting, see John H. White Jr., "'A Perfect Light Is a Luxury': Pintsch Gas Car Lighting," *Technology and Culture* 18, no. 1 (1977): 64–69.
150. Sugg, *Domestic Uses of Coal Gas*, 89, 104.
151. Grafton, *Practical Gas-Fitting*, 102–4.
152. Edward Titchener, *Lectures on the Elementary Psychology of Feeling and Attention* (New York: Macmillan, 1908), 173. See also Gary Hatfield, "Attention in Early Scientific Psychology," in *Visual Attention*, ed. Richard Wright (Oxford: Oxford University Press, 1998), 18–22; and Crary, *Suspensions of Perception*.
153. Bell, *Art of Illumination*, 16.
154. Walter Pillsbury, *Attention* (London: Swan, 1908), 12, 25, 40.
155. On early modern nocturnal work practices (baking, alemaking, scavenging, and smelting), see Ekirch, *At Day's Close*, 155–84.
156. Matthews, *Gas-Lighting*, 233.
157. *Electrician* 7 (October 1, 1881): 305.
158. David Nye, *Electrifying America: Social Meanings of a New Technology, 1880–1940* (London: MIT Press, 1990), 147.
159. Matthews, *Gas-Lighting*, 2.
160. Maycock, *Electric Wiring*, 376.
161. *Halifax Courier* cited in *Electrician* 4 (March 6, 1880): 183.
162. Benjamin A. Dobson, "On the Artificial Lighting of Workshops," *Proceedings* (Institution of Mechanical Engineers) 3–4 (1893): 407–8.
163. *Report from the Select Committee of the House of Lords on the Electric Lighting Act (1882) Amendment Bill* (London: Hansard, 1886), 275.
164. Bell, *Art of Illumination*, 241.
165. Dobson, "Artificial Lighting of Workshops," 397.
166. Barrows, *Electrical Illuminating Engineering*, 6.
167. Trotter, *Elements of Illuminating Engineering*, 76.
168. Rabinbach, *Human Motor*; Crary, *Suspensions of Perception*.
169. Johnston, *History of Light and Colour Measurement*, 110.
170. Nettleship, *Diseases of the Eye*, 336.
171. "The Electric Light," *Electrician* 5 (November 13, 1880): 306. See also Dillon, *Artificial Sunshine*, 163.
172. See, e.g., a report on the Wenham regenerative lamp in the Nottingham Free Library and Museum: "The Wenham System of Lighting and Ventilating," *Builder* 58 (April 12, 1890): 274.

173. Stewart, "Planning: Reference Departments," 43.

174. G. Aitcheson, "Coloured Decorations," *Builder* 31 (March 29, 1873): 238; E. Roberts, "Hints on House Building," *Builder* 30 (February 24, 1872): 141.

175. Sugg, *Domestic Uses of Coal Gas*, 100.

176. Bell, *Art of Illumination*, 185.

177. Scrutton, *Electricity*, 114.

178. Sugg, *Domestic Uses of Coal Gas*, 106.

179. Scrutton, *Electricity*, 27, 28.

180. For Des Esseintes's quest to find the perfect color by which to pursue his nocturnal, narcissistic habits, see J.-K. Huysmans, *Against Nature*, trans. Robert Baldrick (Harmondsworth: Penguin, 1959), 28–30 (chap. 1). After prolonged experiments, the degenerating protagonist chose orange.

181. Dibdin, *Public Lighting*, 17.

182. Trotter, *Elements of Illuminating Engineering*, 26.

183. Webber, *Town Gas*, 127, 133.

184. Peckston, *Practical Treatise*, 7.

185. Robert Hammond, *The Electric Light in Our Homes: Popularly Explained and Illustrated* (London: Warne, 1884), 87.

186. J. A. Fahie, "Electric Lighting from a Sanitary Point of View," *Electrician* 13 (October 18, 1884): 521.

187. W. H. Preece, "The Electric Light at the Paris Exhibition," *Electrician* 8 (December 24, 1881): 91.

188. Abney, *Colour Vision*, 114.

189. Clarence Clewell, *Factory Lighting* (London: McGraw-Hill, 1913), 103, 154.

190. Nettleship, *Diseases of the Eye*, 276.

191. *Lancet*, 1882, no. 2 (November 11): 815.

192. Walsh, *Manual of Domestic Economy*, 134; Sugg, *Domestic Uses of Coal Gas*, 94.

193. J. Willis, *Hints to Trustees of Chapel Property and Chapel Keepers' Manual* (1884), 17, cited in Bowers, *Lengthening the Day*, 50.

194. J. Slater, "Electric Lighting Applied to Buildings," *Electrician* 6 (April 30, 1881): 305.

195. A. White, "The Silent Electric Arc," *Electrician* 14 (November 29, 1884): 56.

196. Edwin J. Houston and A. E. Kennelly, *Electric Arc Lighting* (New York: W. J. Johnston, 1896), 56–57, 113.

197. André Blondel, "Public Lighting by Arc Lamps," *Electrician* 36 (April 3, 1896): 757.

198. Dobson, "Artificial Lighting of Workshops," 399.

199. Gas Light and Coke Company, reporting in 1878, cited in Trench and Hillman, *London under London*, 166.

200. *Builder* 48 (June 13, 1885): 825.

201. Pickering, "Gas-Flame, Electric and Solar Spectra," 341; Vacher, *Healthy Home*, 104; Carter, *Eyesight*, 217.

202. Long, *Spectacles*, 19.

203. "The Electric Light and the Eye," *Lancet*, 1886, no. 2 (September 18): 544.

204. Hepworth, *The Electric Light*, 41.

205. Dobson, "Artificial Lighting of Workshops," 398.

206. Lee, "Abstract . . . 1809," cited in Matthews, *Gas-Lighting*, 304.

207. Carnelly et al., "Carbonic Acid, Organic Matter and Micro-Organisms," 87.

208. Thomas Ryder, "Electric Lighting of Schools," *Electrician* 22 (December 14, 1888): 173.

209. Fox, *Sanitary Examinations*, 181–82.

210. George Fownes, *Fownes' Manual of Chemistry: Theoretical and Practical*, ed. Robert Bridges, New American ed., from the 12th English ed. (Philadelphia: Henry C. Lea, 1878), 162; Charles Meymott Tidy, *Handbook of Modern Chemistry: Inorganic and Organic for the Use of Students* (London: J. & A. Churchill, 1878), 101, 102.

211. B. Thwaite, "Hygiene Applied to Dwellings," *British Architect and Northern Engineer* 11 (January 10, 1879): 17.

212. Thomas Beames, *The Rookeries of London: Past, Present and Prospective* (London: T. Bosworth, 1852), 85.

213. "Gas in Libraries," *Electrician* 11 (September 23, 1883): 419. See also Burgoyne, *Library Construction*, 13; and William H. Greenhough, "On the Ventilation, Heating and Lighting of Free Public Libraries," *Library* 2 (1890): 421–33.

214. See, e.g., Parkes, *Practical Hygiene*, 197; and Eassie, *Healthy Houses*, 179–80. See also Francis Jones, *The Air of Rooms: An Examination of the Effect Produced on the Air of Rooms by the Use of Gas, Coal, Electric Light etc., for Heating and Lighting Purposes* (Manchester: Taylor, Garnett, Evans, 1900).

215. This was particularly the case in France. See *Rapport de la commission sur le chauffage et la ventilation du Théâtre-Lyrique et du théâtre du Cirque Impérial* (Paris: Rapporteur le Général Morin, 1861); and Parkes, *Practical Hygiene*, 204.

216. "Gas and the Electric Light," *British Architect and Northern Engineer* 18 (October 13, 1882): 486.

217. Parkes, *Practical Hygiene*, 204.

218. Trotter, *Elements of Illuminating Engineering*, 31.

219. *Sanitary Engineer* 4 (December 1, 1880): 4; *Electrician* 19 (April 1, 1887): 453; Galton, *Healthy Hospitals*, 144, 177. Open arcs were particularly prone to releasing fumes. See Arthur P. Haslam, *Electricity in Factories and Workshops: Its Cost and Convenience: A Handy Book for Power Producers and Power Users* (London: Crosby, Lockwood, 1909), 300.

220. Iwan Rhys Morus, "The Measure of Man: Technologising the Victorian Body," *History of Science* 37, no. 3 (September 1999): 254–60, and *Frankenstein's Children*, 231–55; Ariès, *Hour of Our Death*, 389–90.

221. See, e.g., R. M. Simon, *Medical Electricity* (Birmingham: Hall & English, 1880); William Steavenson, *The Use of Electrolysis in Surgery* (London: J. & A. Churchill, 1890); and Herbert Tibbits, *Handbook of Medical Electricity* (Philadelphia: Lindsay & Blakiston, 1873).

222. *Electrician* 11 (September 15, 1883): 411.

223. C. W. Siemens, "On the Influence of Electric Light upon Vegetation, and on Certain Physical Principles Involved," *Electrician* 5 (March 13, 1880): 200–202. See also "Electro-Horticulture," *Science* 18, no. 451 (September 1891): 174; and Augustus Waller, "The Electrical Effect of Light upon Green Leaves," *Proceedings* (Royal Society of London) 67 (1900): 129–37. For detailed experiments on the impact of gas on plants, see William Crocker and Lee I. Knight, "Effect of Illuminating Gas and Ethylene upon Flowering Carnations," *Botanical Gazette* 46, no. 4 (1908): 259–76.

224. "Forcing Vegetables by Electricity," *Electrical Engineer* 10 (July 15, 1892): 61.

225. Hedges, *Electric Lighting*, 127.

226. Katherine Gamgee, *The Artificial Light Treatment of Children in Rickets, Anaemia and Malnutrition* (London: H. K. Lewis, 1927).

227. See, e.g., Morus, *Frankenstein's Children*; and Zajonc, *Catching the Light*, chap. 6.

228. M. Hertz, "The Identity of Light and Electricity," *Electrician* 14 (November 15, 1889): 32–33.

229. Morus, "Measure of Man," 274.

230. Alfred Smee, *Elements of Electro-Biology: or, The Voltaic Mechanism of Man; of Electro-Pathology, Especially of the Nervous System; and of Electro-Therapeutics* (London: Longman, Brown, Green & Longmans, 1849). For a detailed analysis of Smee's theories, see Morus, *Frankenstein's Children*, 147–50. See also Winter, *Mesmerized*, 119–20.

231. Preece, "Sanitary Aspects of Electric Lighting," 462, 464.

232. "Factories and Workshops Bill," *Electrician* 35 (May 17, 1895): 102; Stockman, *Practical Guide for Sanitary Inspectors*, 246.

233. W. Preece, "On the Relative Merit and Cost of Gas and Electricity for Lighting Purposes," *Proceedings* (Association of Municipal and Sanitary Engineers and Surveyors) 17 (June 27, 1891): 231.

CHAPTER 6

1. See Charles Ganton, "The Distribution of Electrical Energy by Secondary Generators," *Builder* 48 (April 18, 1885): 565–66.

2. "Electric Light Mains," *Electrical Review* 33 (November 24, 1893): 547.

3. For the legal dimension, see Hughes, *Networks of Power*. See also Leslie Hannah, *Electricity before Nationalisation: A Study of the Development of the Electricity Supply in Britain to 1948* (Baltimore: Johns Hopkins University Press, 1979). For the dynamo, see Percy Dunsheath, *A History of Electrical Engineering* (London: Faber & Faber, 1962), 99–122; and Charles Caesar Hawkins, *The Dynamo: Its Theory, Design and Manufacture*, 2nd ed. (London: Whittaker, 1896). On motors, see Dibdin, *Public Lighting*, 364–73. For the transformer, see J. Fleming, *The Alternate Transformer in Theory and Practice* (London: Benn, 1889). See also C. Mackechnie Jarvis, "The Generation of Electricity,"

in *A History of Technology*, ed. Charles Singer et al., 5 vols. (New York: Oxford University Press, 1954–59), 5:177–207, and "The Distribution and Utilization of Electricity," in ibid., 5:209–34.

4. This is obviously a simplified account. For more detail, see Hughes, *Networks of Power*.

5. "The Social Effects of Electric Power Supply," *Electrician* 29 (November 18, 1892): 63.

6. Gooday, *Morals of Measurement*, 213.

7. Scrutton, *Electricity*, 45–46.

8. "Electric Light Mains," 548.

9. Here, the word *mains* refers to all wire between plant and building: contemporaries often referred to *feeders* (for high-voltage electricity) and *distributors* (for low-voltage, higher-current electricity running from transformer stations into houses), for which *mains* was an umbrella term. I use the expression *internal wiring* for all wires within houses or public buildings, although terms like *mains* and *feeders* were sometimes used to describe arborescent formations of interior wiring.

10. H. Leaf, *The Internal Wiring of Buildings* (London: Archibald Constable, 1899), 135; H. R. Kempe, *The Electrical Engineer's Pocket-Book: Modern Rules, Formulae, Tables and Data*, 2nd ed. (London: Crosby, Lockwood, 1892), 79–80; B. C. Blake-Coleman, *Copper Wire and Electrical Conductors: The Shaping of a Technology* (Reading: Harwood, 1992); Robert Black, *The History of Electric Wires and Cables* (London: Peter Peregrinus, 1983), esp. 37–84.

11. Gutta-percha, a rubbery substance extracted from certain trees in Southeast Asia, was first used on a large scale by Werner von Siemens to insulate telegraph cables.

12. "The Art of Insulation," *Electrician* 24 (December 27, 1889): 192.

13. Fred Bathurst, "The Electric Wiring Question," *Electrician* 36 (December 6, 1895): 200.

14. David Pinkey, *Napoleon III and the Rebuilding of Paris* (Princeton, NJ: Princeton University Press), 1972, 133; Richardson, *Hygeia*, 27–28.

15. "Gas Mains in Subways," *Engineering* 1 (March 2, 1866): 137.

16. The first recorded scheme for metropolitan subways was that of the stationer John Williams (1817). See Newbigging and Fewtrell, eds., *King's Treatise*, 2:396–99.

17. Ibid., 405.

18. Charles Mason, "London Subways," *Engineering Record* 35 (May 1, 1897): 471.

19. "New York Subways," *Electrician* 24 (March 21, 1890): 503.

20. Trench and Hillman, *London under London*, 98.

21. Bathurst, "Electric Wiring Question," 201.

22. "Underground Systems of Distributing Electricity," *Electrician* 24 (February 28, 1890): 423.

23. Board of Trade regulations cited in Kempe, *Electrical Engineer's Pocket-Book*, 195.

24. "The Electrical Standardising Laboratory," *Electrician* 24 (May 30, 1890): 91. This article refers more specifically to routine inspection of instrumentation, but inspection was already essential outside the laboratory. The inspector's satchel and its contents were described in Kempe, *Electrical Engineer's Pocket-Book*, 208.

25. G. L. Addenbrooke, "Underground Mains," *Electrician* 24 (January 17, 1890): 261.

26. L. A. Ferguson, *Electrical Engineering in Modern Central Stations* (Madison: Engineering Bulletin of the University of Wisconsin, 1896), 244. See also G. L. Addenbrooke, "Underground Mains V," *Electrician* 24 (February 14, 1890): 368.

27. "Manhole Boxes," *Electrician* 33 (February 23, 1894): 441.

28. "Explosions in Manholes," *Electrical World* 23 (March 17, 1894): 377.

29. The same was true for turncocks, sewer gratings, and streetcar lines. See "Carriage-Way Pavements" (August 12), 188.

30. "Noiseless Manhole Covers," *Engineering Record* 34 (August 15, 1896): 196.

31. R. C. Quin, "The Localisation of Faults in Underground Mains," *Engineer* 80 (July 16, 1897): 52.

32. F. Charles Raphael, *The Localisation of Faults in Electric Light and Power Mains: With Chapters on Insulation Testing*, 2nd ed. (London: "The Electrician" Printing & Publishing Co., 1903), 135.

33. Raphael, *Localisation of Faults*, 68–69, 99–100 (quote).

34. For example, Kempe's *Electrical Engineer's Pocket-Book*.

35. *Builder* 32 (June 20, 1874): 521.

36. Trotter, *Illumination*, 262, 41 (quote).

37. For an explanation of this term, see n. 31 of chapter 1.

38. Bell, *Art of Illumination*, 257.

39. Tower lighting involved the fitting of intense arc lights to large towers. In Detroit, e.g., 122 towers were constructed. Those at the center of the city were 175 feet high, with six two-thousand-candlepower arc lamps each. See Jakle, *City Lights*, 48.

40. "The Hours of Lighting," *Electrician* 32 (January 19, 1894): 289.

41. Dibdin, *Public Lighting*, 300–301.

42. Firth, *Municipal London*, 307.

43. Newbigging and Fewtrell, eds., *King's Treatise*, 2:275.

44. For early modern streetlighting, see Ekirch, *At Day's Close*, 67–74. The expression *parish lantern* was used "in many parts of Britain" (ibid., 128).

45. Dibdin, *Public Lighting*, 301 (quote), app. ("Particulars with Respect to Public Street Lighting Kindly Supplied by Various Authorities").

46. C. R. Bellamy cited in ibid., 309.

47. *Electrician* 32 (March 2, 1894): 485.

48. "Competitive Lamps for the Thames Embankment," *Engineer* 29 (March 18, 1870): 154; Nicholas Taylor and David Watkin, "Lamp-Posts: Decline and

Fall in Cambridge," *Architectural Review* 129 (June 1961): 423–26; P. Varnon, "Street Furniture: Survey of Street Lighting," *Architectural Review* 110 (July 1951): 51.

49. "Lighting of Blackfriars Bridge," *Builder* 30 (November 30, 1872): 958.
50. "The Ownership of Street Lamps," *Sanitary Record* 7 (September 14, 1877): 177.
51. Hartman's paint was chosen for London's lampposts because of its "durability and appearance." See *Engineer* 77 (September 13, 1895): 266.
52. Dibdin, *Public Lighting*, 268–69.
53. Newbigging and Fewtrell, eds., *King's Treatise*, 3:62.
54. Barrows, *Electrical Illuminating Engineering*, 163. See also Bowers, *Lengthening the Day*, 145–46; and André Blondel, "Public Lighting by Arc Lamps," *Electrician* 36 (April 17, 1896): 820–22, and (April 24, 1896): 853–55.
55. André Blondel, "Street Lighting by Arc Lamps," *Electrician* 36 (November 8, 1895): 39–40.
56. Bell, *Art of Illumination*, 246.
57. *Electrician* 7 (August 13, 1881): 193–94.
58. Bryant and Hake, *Street Lighting*, 58.
59. Bell, *Art of Illumination*, 261.
60. Trotter, *Illumination*, 36.
61. Trotter, *Elements of Illuminating Engineering*, 97.
62. Newbigging and Fewtrell, eds., *King's Treatise*, 3:56.
63. Blondel, "Street Lighting" (November 8), 42.
64. The formula was candlepower $= EH^2/\cos^3\theta$, where $E =$ illumination and $H =$ height of lamp. Trotter, *Illumination*, 41.
65. This was another item on Firth's endless list of metropolitan anomalies. See *Municipal London*, 323.
66. Dibdin, *Public Lighting*, app. ("Particulars . . . "), n.p.
67. This reference is to Walter Benjamin, who wrote with great, and probably misleading, insight on Parisian streets as interiors. See, e.g., Benjamin, *Arcades Project*, 422. See also Schivelbusch, *Disenchanted Night*, 149.
68. "Street Lighting," *Electrician* 32 (February 23, 1894): 455.
69. Blondel, "Street Lighting" (November 8), 39.
70. Trotter, *Elements of Illuminating Engineering*, 12; Tyndall, *Light and Electricity*, 52.
71. Preece, "Electric Lighting in the City" (April 18), 481.
72. Trotter, *Illumination*, 28. Foster described Weber's law as follows: "The smallest change in the magnitude of a stimulus which we can appreciate through a change in our sensation always bears the same proportion to the whole magnitude of the stimulus. . . . This law holds good within certain limits only; it fails when the stimulus is either above or below a certain range of intensity." Michael Foster, *Textbook of Physiology*, 5th ed. (London: Macmillan, 1889–92), 4:1211, cited in Trotter, *Illumination*, 28–29.
73. Crary, *Suspensions of Perception*, 287.

74. Trotter, *Illumination*, 52.
75. The alternative to the tree system was the distribution board system. See Maycock, *Electric Wiring*, 237–40; and Noll, *How to Wire Buildings*, 34.
76. For example, Killingworth Hedges, *British Architect and Northern Engineer* 21 (February 22, 1884): 92.
77. Noll, *How to Wire Buildings*, 10.
78. P. Atkinson, *The Elements of Electric Lighting* (New York: D. Van Nostrand, 1897), 275.
79. Bathurst, "Electric Wiring Question," 200.
80. Maycock, *Electric Wiring*, 344.
81. Noll, *How to Wire Buildings*, 66.
82. Leaf, *Internal Wiring*, 144.
83. The switch was not solely an electrical technology. Pneumatic switches were devised for use with gaslight. See Grafton, *Practical Gas-Fitting*, 373; and Webber, *Town Gas*, 111.
84. Maycock, *Electric Wiring*, 145.
85. Hammond, *Electric Light in Our Homes*, 92.
86. *British Architect and Northern Engineer* 14 (September 18, 1885): 123.
87. Scrutton, *Electricity*, 24. On tactile practices during domestic noctambulations, see Ekirch, *At Day's Close*, 110.
88. Maycock, *Electric Wiring*, 265–66.
89. Kempe, *Electrical Engineer's Pocket-Book*, 192–93.
90. J. Gordon, "The Development of Electric Lighting" (paper read before the Society of Arts), *Electrician* 11 (June 23, 1883): 139–40.
91. Noll, *How to Wire Buildings*, 83. For gaslights in prison cells, see Henry Mayhew and John Binny, *The Criminal Prisons of London and Scenes of Prison Life* (New York: Augustus Kelley, 1968), 159–60, 542.
92. Scrutton, *Electricity*, 110–11.
93. Bowmaker, *Housing of the Working Classes*, 174.
94. E. J. Molera and J. C. Cebrian, "Practical Divisibility of the Electric Light," *Electrician* 3 (August 2, 1879): 128, 129.
95. "A Brief History of Light Guides," *International Association for Energy-Efficient Lighting Newsletter*, January 1997, http://www.iaeel.org/iaeel/newsl/1997/ett1997/LiTech_d_1_97.html.
96. "Division of Electric Light," *Mining and Scientific Press* 39 (September 27, 1879): 193. Fresnel's lens system was originally developed for lighthouses in the early nineteenth century. See Tyndall, *Sound*, 285.
97. *Electrician* 4 (November 29, 1879): 13, and (February 28, 1880): 169.
98. Harlan, *Eyesight*, 118.
99. Molera and Cebrian, "Practical Divisibility," 131 (emphasis added).
100. "Division of Electric Light," 193.
101. See, in particular, Siegfried Giedion, *Mechanization Takes Command: A Contribution to Anonymous History* (New York: Oxford University Press, 1948), 512–627.

102. Fred Schroeder, "More 'Small Things Forgotten': Domestic Electric Plugs and Receptacles, 1881–1931," *Technology and Culture* 27, no. 3 (1986): 525–43. See also Dillon, *Artificial Sunshine*, 171.

103. Maycock, *Electric Wiring*, 14–22.

104. Charles Vernon Boys, "On Meters for Power and Electricity," *Electrician* 11 (June 16, 1883): 110.

105. For a detailed analysis of this and debates over quantity vs. energy meters, see Gooday, *Morals of Measurement*, 247–52.

106. Gooday, *Morals of Measurement*, 232–39.

107. A comprehensive breakdown of metering systems was provided in Gerhardi, *Electricity Meters*. See also Henry Solomon, *Electricity Meters: A Treatise on the General Principles, Construction and Testing of Continuous Current and Alternating Current Meters* (London: Charles Griffin, 1906).

108. Gooday, *Morals of Measurement*, 244–45.

109. By *network*, I mean a material system that, along with its human operators, extends across large tracts of space. See Winner, *Autonomous Technology*, 11–12; and Andrew Feenberg, *Questioning Technology* (New York: Routledge, 1999), 118–19. On the networked society, see Manuel Castells, *The Rise of the Networked Society* (Cambridge, MA: Blackwell, 1996); Graham and Marvin, *Splintering Urbanism*; and Olivier Coutard, Richard E. Hanley, and Rae Zimmerman, eds., *Sustaining Urban Networks: The Social Diffusion of Large Technical Systems* (New York: Routledge, 2005). On risk, see Scott Lash, Bronislaw Szerszynski, and Brian Wynne, *Risk, Environment and Modernity: Towards a New Ecology* (London: Sage, 1996); and Ulrich Beck, *Risk Society: Towards a New Modernity*, trans. Mark Ritter (London: Sage, 1992).

110. "Electricity Supply Mains," *Electrician* (February 5, 1892): 351.

111. *Electrician* 29 (July 15, 1892): 269.

112. Hollingshead, *Underground London*, 2. On the underground and artistic and literary imagination, see Rosalind Williams, *Notes on the Underground: An Essay on Technology, Society and the Imagination* (Cambridge, MA: MIT Press, 1990); Flint, *Victorians and the Visual Imagination*, 139–66; and Pike, *Subterranean Cities*.

113. Girdlestone, *Unhealthy Condition of Dwellings*, 35–36.

114. Everard, *Gas Light and Coke Company*, 65.

115. Newbigging and Fewtrell, eds., *King's Treatise*, 2:331.

116. *Journal of Gas Lighting* 26 (November 2, 1880): 683.

117. *Electrician* 34 (February 15, 1895): 458.

118. Robert Hammond, "Municipal Electricity Works," *Builder* 65 (August 19, 1893): 138.

119. *Builder* 31 (December 6, 1873): 963.

120. Thorsheim, *Inventing Pollution*, 158.

121. On economical main laying, see Clegg, *Practical Treatise*, 161.

122. Merriman, *Gas-Burners*, 80.

123. Manchester City Council Watch Committee, *Minutes* 23 (June 4, 1896): 245.

124. Humphry Davy, "Extracts from the Minutes of Evidence, Printed by Order of the House of Commons, July 7, 1823," cited in Matthews, *Gas-Lighting*, 340.

125. "A Gas Leak Detector," *Engineering* 30 (December 31, 1880): 631; Sugg, *Domestic Uses of Coal Gas*, 60, 51–52.

126. Samuel Clegg, "Evidence to the Committee of the House of Commons, 1823," cited in Matthews, *Gas-Lighting*, 67–68. See also Clegg, *Practical Treatise*, 18.

127. Wood, *Gas-Lighting*, 7.

128. For example, *British Architect and Northern Engineer* 13 (February 6, 1880): 70.

129. On the brief, but eventful, history of London's first, gaslit, traffic light, see Miles Ogborne, "Traffic Lights," in *City A-Z*, ed. Steve Pile and Nigel Thrift (London: Routledge, 2000), 262.

130. "A Startling Occurrence," *Builder* 30 (September 21, 1872): 753.

131. *Journal of Gas Lighting* 36 (July 13, 1880): 63.

132. Thomas Bartlett Simpson, *Gas-Works: The Evils Inseparable from Their Existence in Populous Places, and the Necessity of Removing Them from the Metropolis* (London: William Freedman, 1866), 9 (first quote), 15 (on *The Times*), 20 (on the *Telegraph*), 24 (second quote). See also Nead, *Victorian Babylon*, 94–96.

133. Schivelbusch, *Disenchanted Night*, 112.

134. Everard, *Gas Light and Coke Company*, 177.

135. An explosion at Tradeston gasworks in Glasgow in 1883 was traced to Fenians, and a later explosion, at the city's Dawsholm gasworks in 1891, was immediately blamed on the same source. See *Engineer* 71 (February 6, 1891): 112.

136. For a review of gas blackouts in the United States and a comparison with electricity blackouts, see Peter C. Baldwin, "In the Heart of Darkness: Blackouts and the Social Geography of Lighting in the Gaslight Era," *Journal of Urban History* 30, no. 5 (July 2004): 749–68.

137. On Congreve and the 1823 committee, see Clifford, *Private Bill Legislation*, 1:213–16.

138. William Hyde Wollaston, "Extract . . . 1823," cited in Matthews, *Gas-Lighting*, 345.

139. Thomas Bolan, *Builder* 42 (May 13, 1882): 597.

140. *Builder* 58 (June 28, 1890): 475.

141. Hammond, *Electric Light in Our Homes*, 71.

142. *Engineer* 73 (February 17, 1892): 148.

143. *Lancet*, 1889, no. 1 (May 4): 907.

144. *Journal of Gas Lighting* 36 (October 26, 1880): 642.

145. *British Architect and Northern Engineer* 24 (November 13, 1885): 208.

146. *Electrician* 6 (May 14, 1881): 335.

147. *Engineer* 77 (April 12, 1895): 306.

148. "Electric Light Mains," 548.

149. Rankin Kennedy, *Electrical Installations of Electric Light, Power, Traction and Industrial Electrical Machinery*, 4 vols. (London: Blackwood, Le Bas, 1902), 4:231.

150. *Lancet*, 1887, no. 2 (July 2): 32. See also Benjamin Ward Richardson, "The Painless Extinction of Life," *Popular Science Monthly* 26 (1884–85): 641–52. For a more skeptical contemporary view of the efficacy of electric death, see "Painless Killing," *Lancet*, 1888, no. 2 (August 4): 219–20.

151. See, e.g., "Electric Currents and the Human Body," *Engineering* 50 (October 3, 1890): 387. See also Morus, "Measure of Man," 267–69.

152. Henry Lewis Jones, "The Lethal Effects of Electrical Currents," *British Medical Journal*, 1895, no. 1 (March 2, 1895): 469 (quote), 468. On D'Arsonval, see Thomas Oliver and Robert Bolam, *On the Cause of Death by Electric Shock* (London: British Medical Association, 1898), 5.

153. "The Peril of Electric Lamps," *Engineering* 30 (October 29, 1880): 378.

154. Hammond, *Electric Light in Our Homes*, 57; *Lancet*, 1885, no. 1 (March 7): 440.

155. Jones, "Lethal Effects," 469–70. See also *Builder* 66 (January 27, 1894): 65.

156. See Richard Moran, *Executioner's Current: Thomas Edison, George Westinghouse, and the Invention of the Electric Chair* (New York: Vintage, 2003); and Mark Essig, *Edison and the Electric Chair: A Story of Light and Death* (New York: Walker, 2003).

157. Testimony cited in *Electrician* 20 (February 3, 1888): 326.

158. "Execution by Electricity," *Engineering* 50 (August 8, 1890): 165, 166. See also Tim Armstrong, *Modernism, Technology and the Body: A Cultural History* (Cambridge: Cambridge University Press, 1998), 34; and Hughes, *Networks of Power*, 107–9.

159. *Engineering Magazine* 1 (1891): 262.

160. *Electrician* 24 (January 3, 1890): 226. See also Joseph P. Sullivan, "Fearing Electricity: Overhead Wire Panic in New York City," *IEEE Technology and Society Magazine* 14, no. 3 (1995): 8–16.

161. "Overhead Wires Again," *Electrician* 11 (November 3, 1883): 588.

162. Preece cited in *Report from the Select Committee of the House of Lords on the Electric Lighting Act (1882) Amendment Bill* (London: Hansard, 1886), 219.

163. "Overhead Wires," *Electrician* 12 (January 26, 1884): 253.

164. "Report of the Overhead Wires Committee" cited in *Electrician* 15 (May 15, 1885): 11.

165. Kempe, *Electrical Engineer's Pocket-Book*, 98–99.

166. Board of Trade Regulations for Securing the Safety of the Public cited in Dibdin, *Public Lighting*, 504.

167. See Owen, *Government of Victorian London*, 226–59.

168. "Street Illumination in the City," *Builder* 61 (August 8, 1891): 96, 97. See also De Beer, "London Street-Lighting."

169. Charles Booth, *On the City: Physical Pattern and Social Structure: Selected Writings*, ed. Harold Pfautz (Chicago: University of Chicago Press, 1967), 107.

170. "Electric Lighting in the City of London," *Electrician* 26 (February 13, 1891): 455.

171. Charles Welch, *Modern History of the City of London: A Record of Municipal and Social Progress from 1760 to the Present Day* (London: Blades, East & Blades, 1896), 319.

172. "Electric Lighting for the City of London," *Electrician* 6 (April 2, 1881): 244.

173. "Electric Lighting in the City," *Electrician* 10 (November 25, 1882): 29.

174. *Electrician*, August 22, 1890, 423.

175. "Electric Lighting in London," *Engineering* 52 (August 21, 1891): 221.

176. "Electric Lighting in London," *Engineering* 52 (September 21, 1891): 302.

177. "The City Lighting," *Electrician* 30 (November 25, 1892): 84.

178. "The City Lighting," *Electrical Review* 33 (December 8, 1893): 616.

179. Maycock, *Electric Wiring*, 418.

180. "The Electric Lighting of the City," *Electrical Review* 30 (January 29, 1892): 135. See also "The City of London Electric Lighting Company, Ltd.," *Electrical Review* 30 (January 22, 1892): 107.

181. "The City Lighting," *Electrician* 28 (March 11, 1892): 480.

182. C. E. Webber, "Some Notes on the Electric Lighting of the City of London," *Electrician* 32 (March 2, 1894): 482.

183. This has only intensified over the past century. The City of London has, at the time of writing, at least six optic fiber cable grids occupying sewers and tunnels beneath its square mile. Graham and Marvin refer to this as a "massive concentration of electronic infrastructure" (*Splintering Urbanism*, 319).

184. C. E. Webber, "Some Notes on the Electric Lighting of the City of London," *Electrician* 32 (February 23, 1894): 447. Webber was a founder member of the Society for Telegraph Engineers in 1871. See Bazerman, *Edison's Light*, 118.

185. Webber, "Some Notes on the Electric Lighting of the City of London," *Electrician* 32 (February 16, 1894): 424.

186. Webber, "Some Notes" (February 23), 449.

187. Ibid., 450.

188. "City of London Electric Lighting Co., Bankside, Southwark," *Proceedings* (Institution of Mechanical Engineers) 3–4 (1900): III–IV, 475–77.

189. Webber, "Some Notes" (February 23), 452.

190. Trotter, *Illumination*, 253.

191. "Lamp Posts for Oxford," *Electrician* 29 (October 28, 1892): 696.

192. "Electric Lighting of the City of London," *Engineering* 57 (February 16, 1894): 235.

193. "City Electric Light Standards," *Electrical Engineer* 10 (July 8, 1892): 34.

194. "The City Lighting," *Electrician* 30 (December 9, 1892): 148.

195. *Builder* 66 (January 27, 1894): 77.

196. *City Press* cited in "The City Lighting," *Electrician* 30 (March 24, 1893): 590.

197. "The Electric Light in the City," *Electrical Review* 33 (November 24, 1893): 562.

198. "Electric Lighting Notes," *Journal of Gas Lighting* 74 (December 5, 1899): 1387.
199. "The City Electric Lighting Breakdown," *Electrical Review* 35 (October 19, 1894): 458.
200. Everard, *Gas Light and Coke Company*, 283; "Electric Lighting Notes," *Journal of Gas Lighting* 74 (December 19, 1899): 1504.
201. Webber, *Town Gas*, 121.
202. Willy and Cardew cited in "The Cannon-Street Accident: Board of Trade Inquiry," *Electrician* 34 (November 23, 1894): 106–7, 107.
203. "Laboratory of the Electrical Inspector to the City Commission of Sewers," *Electrician* 35 (June 7, 1895): 195.
204. "Electric Lighting in the City," *Electrician* 27 (July 3, 1891): 242.
205. "The City Company's Mains," *Electrical Review* 35 (October 26, 1894): 508.
206. Hughes, *Networks of Power*, 227–61. A similar argument is made, implicitly, in Platt's *Shock Cities*.
207. This is a central argument of Latour's. See *Science in Action*, 132–44.
208. Arthur Guy, "Electric Light and Power," *Electrical Engineer* 12 (July 28, 1893): 78.
209. Hughes, *Networks of Power*, 261, 260.

CONCLUSION

1. There is a vast literature on these domains, particularly on art, photography, and cinema. For the history of aesthetic perception, see Clark, *Painting of Modern Life*; and Linda Nochlin, *The Politics of Vision: Essays on Nineteenth-Century Art and Society* (New York: HarperCollins, 1991). On photography, see Liz Wells, ed., *The Photography Reader* (London: Routledge, 2002); Susan Sontag, *On Photography* (New York: Picador, 2001); Roland Barthes, *Camera Lucida: Reflections on Photography*, trans. Richard Howard (London: Vintage, 1993); Lorraine Daston and Peter Galison, "The Image of Objectivity," *Representations* 40 (1992): 81–123; John Tagg, *The Burden of Representation: Essays on Photographies and Histories* (Amherst: University of Massachusetts Press, 1988); and Lalvani, *Photography*. On film, see Linda Williams, ed., *Viewing Positions: Ways of Seeing Film* (New Brunswick, NJ: Rutgers University Press, 1995); Laura Mulvey, *Visual and Other Pleasures* (Bloomington: Indiana University Press, 1989); and Teresa de Lauretis and Stephen Heath, ed., *The Cinematic Apparatus* (New York: St. Martin's, 1980). For the paranormal and extrabodily perception, see Roger Luckhurst, *The Invention of Telepathy, 1870–1901* (Oxford: Oxford University Press, 2002); Winter, *Mesmerized*; and Alex Owen, *The Darkened Room: Women, Power and Spiritualism in Late Victorian England* (Chicago: University of Chicago Press, 2004).
2. On gender and perception, see Rose, *Sexuality and the Field of Vision*; Pollock, *Vision and Difference*; and Stephen Kern, *Eyes of Love: The Gaze in English and French Paintings and Novels, 1840–1900* (London: Reaktion, 1996). For race,

see Pratt, *Imperial Eyes*; Richard Dyer, *White* (London: Routledge, 1997); and Fanon, *Black Skin, White Masks*.

3. These comparisons, obviously, would need to be pursued along multiple axes. The development of electricity systems in various national contexts (France, Germany, the United States, and Russia) is explored in Hughes, *Networks of Power*; Schivelbusch, *Disenchanted Night*; Nye, *Electrifying America*; Coopersmith, *Electrification of Russia*; and Andreas Killen, *Berlin Electropolis: Shock, Nerves and German Modernity* (Berkeley and Los Angeles: University of California Press, 2006). The constitution of networks within various parts of the British Empire is explored in Mitchell, *Colonising Egypt*; Goswami, *Producing India*; and Headrick, *Tentacles of Progress*. One would need to explore how these various national and colonial technological schemes intermeshed with political and government formations that were either nonliberal or liberal in a different way than in Britain.

4. For a discussion of the relation between invisibility, hygiene, and spectacle, see Alain Corbin, "The Blood of Paris: Reflections on the Genealogy of the Image of the Capital," in *Time, Desire and Horror*, 172–80.

5. Voyeurism is curiously understudied as a social and material phenomenon. Some suggestive insights are provided in Corbin's *Women for Hire* and Crook's "Power, Privacy and Pleasure," and there is a substantial literature on voyeurism in film.

6. An indicator was a screen, complete with holes and numbers, that showed the status of the library's books (in, out, overdue, etc.). By the early twentieth century, there were many models, and they were used in nearly all British libraries. For these and other systems of public signification used in libraries, see Burgoyne, *Library Construction*, 75–93.

7. Lefebvre, *Production of Space*, 313.

8. Leder, *Absent Body*, 150. See also chapter 1 above.

9. See n. 58 of the introduction.

10. This is a central theme of David Edgerton, *The Shock of the Old: Technology and Global History since 1900* (Oxford: Oxford University Press, 2007).

11. On fluorescent lighting, see Wiebe Bijker, *Of Bicycles, Bakelites and Bulbs: Towards a Theory of Sociotechnical Change* (Cambridge, MA: MIT Press, 1995), 199–269.

12. Pubs that are entirely or partially gaslit include the Pulteney Arms in Bath, Belfast's the Crown, Hale's Bar, Harrogate, the Globe, Leicester, Smithies Ale House in Edinburgh, and the New Beehive Inn, Bradford. I would like to thank Iain Loe, Michael Slaughter, and Geoff Brandwood of the Campaign for Real Ale for help on this point. For an excellent survey of the history of pub interiors, see Geoff Brandwood, Mike Slaughter, and Andrew Dawson, *Licensed to Sell: The History and Heritage of the Public House* (Swindon: English Heritage, 2004). Working gaslights can be seen at two National Trust houses at least (see Dillon, *Artificial Sunshine*, 19), while the Park Estate claims to have the largest system of gaslighting in Europe.

13. Dibdin, *Public Lighting*, 18; Armstrong, "Reign of the Engineer," 462.
14. The term *modern* or *modernity* has often been employed in quotation marks in this book. I do not think that the term is useless, but it very often has a nebulousness that blunts analysis. See Bernard Yack, *The Fetishism of Modernities: Epochal Self-Consciousness in Contemporary Society* (Notre Dame, IN: University of Notre Dame Press, 1997); Latour, *We Have Never Been Modern*; and Frederick Cooper, *Colonialism in Question: Theory, Knowledge, History* (Berkeley and Los Angeles: University of California Press, 2005), 113–49.
15. Nonetheless, actor-network theory has provided useful ways to think round the distinction. See, e.g., Callon, "Society in the Making"; and Latour, *Reassembling the Social*.
16. For "unintended consequences," see Winner, *Autonomous Technology*; Beniger, *Control Revolution*; and Edward Tenner, *Why Things Bite Back: Technology and the Revenge of Unintended Consequences* (New York: Knopf, 1996).
17. The literature here is voluminous. For a useful overview, see J. R. McNeil, *Something New under the Sun: An Environmental History of the Twentieth-Century World* (New York: Norton, 2000).
18. On the "superpanopticon," see Lyon, *Surveillance Society*.

Bibliography

This bibliography is not exhaustive. It lists only those books and articles that have been of greatest relevance to the construction of my argument. For more detail on sources, readers are directed to the notes.

PRIMARY SOURCES

Abady, Jacques. *Gas Analyst's Manual: Incorporating F. W. Hartley's "Gas Analyst's Manual" and "Gas Measurement."* London: E. & F. Spon, 1902.

Abney, William de Wiveleslie. *Colour Vision: Being the Tyndall Lectures Delivered in 1894, Etc.* London: Sampson Law, Marston, 1895.

Accum, Frederick. *Description of the Process of Manufacturing Coal Gas, for the Lighting of Streets, Houses and Public Buildings.* 2nd ed. London: Thomas Boys, 1820.

Addenbrooke, G. L. "Underground Mains." *Electrician* 24 (January 17, 1890).

———. "Underground Mains V." *Electrician* 24 (February 14, 1890).

Aitcheson, G. "Coloured Decorations." *Builder* 31 (March 29, 1873).

Allnutt, H. *Wood Pavement as Carried Out on Kensington High Road, Chelsea &c.* London: E. & F. Spon, 1880.

Arlidge, J. T. *The Hygienic Diseases and Mortality of Occupations.* London: Perceval & Co., 1892.

Armstrong, Henry. "The Reign of the Engineer." *Quarterly Review* 198, no. 396 (October 1903): 462–84.

"The Art of Insulation." *Electrician* 24 (December 27, 1889).

"Asphalte and Other Pavements, II." *Engineer* 33 (April 26, 1872).

"Assaulting a Sanitary Inspector." *Sanitary Record* 8 (June 14, 1878).

Atkinson, P. *The Elements of Electric Lighting*. New York: D. Van Nostrand, 1897.

Atkinson, William. "The Orientation of Buildings and of Streets in Relation to Sunlight." *Technology Quarterly and Proceedings of the Society of Arts* 18 (September 1, 1905): 204–27.

"The Atmosphere of Great Towns." *Engineer* 78 (May 15, 1896).

Ayling, R. Stephen. *Public Abattoirs: Their Planning, Design and Equipment*. London: E. & F. Spon, 1908.

Barr, James, and Charles Phillips. "The Brightness of Light: Its Nature and Measurement." *Electrician* 32 (March 9, 1894).

Barrows, William Edward. *Electrical Illuminating Engineering*. New York: McGraw, 1908.

Bathurst, Fred. "The Electric Wiring Question." *Electrician* 36 (December 6, 1895).

Baudelaire, Charles. "The Painter of Modern Life." In *The Painter of Modern Life and Other Essays*, trans. and ed. Jonathan Mayne. New York: Da Capo, 1986.

Bazalgette, J. W., and T. W. Keates. "The Report on the Electric Light on the Thames Embankment." *Electrician* 3 (June 7, 1879).

Beames, Thomas. *The Rookeries of London: Past, Present and Prospective*. London: T. Bosworth, 1852.

"The Beckton Works of the Chartered Gas Company (I)." *Engineer* 29 (February 4, 1870).

Bell, Louis. *The Art of Illumination*. New York: McGraw, 1902.

Bentham, Jeremy. *Panopticon; or, The Inspection-House: Containing the Idea of a New Principle of Construction Applicable to Any Sort of Establishment, in which Persons of Any Description Are to Be Kept under Inspection; and in Particular to Penitentiary-Houses, Prisons, Houses of Industry, Work-Houses, Poor-Houses, Manufactories, Mad-Houses, Lazarettos, Hospitals, and Schools*. In Jeremy Bentham, *The Panopticon Writings*, ed. Miran Božovič. London: Verso, 1995.

Besant, Walter. *London in the Nineteenth Century*. London: Adam & Charles Black, 1909.

"The Bethnal Green Museum." *Builder* 30 (May 18, 1872).

Bickerton, Thomas H. *Colour Blindness and Defective Eyesight in Officers and Sailors of the Mercantile Marine*. Edinburgh: James Thin, 1890.

Blanchard, A. H., and H. B. Drowne. *Text-Book on Highway Engineering*. New York: John Wiley & Sons, 1913.

Bloch, Leopold. *The Science of Illumination: An Outline of the Principles of Artificial Lighting*. Translated by W. C. Clinton. London: John Murray, 1912.

Blok, Arthur. *Elementary Principles of Illumination and Artificial Lighting*. London: Scott, Greenwood & Son, 1914.

Blondel, André. "Street Lighting by Arc Lamps." *Electrician* 36 (November 8, 1895; and November 15, 1895).

———. Public Lighting by Arc Lamps." *Electrician* 36 (April 3, 1896; April 17, 1896; and April 24, 1896).

Blyth, Alexander W. *A Dictionary of Hygiène and Public Health*. London: Charles Griffin & Co., 1876.

———. *A Manual of Public Health*. London: Macmillan, 1890.

Booth, Charles. *Life and Labour of the People of London: First Series: Poverty: Streets and Population Classified*. London: Macmillan, 1902.

Bowmaker, Edward. *The Housing of the Working Classes*. London: Methuen, 1895.

Boys, Charles Vernon. "On Meters for Power and Electricity." *Electrician* 11 (June 16, 1883).

Brewster, David. "The Sight and How to See." *North British Review* 26 (1856): 145–84.

Browne, Edgar. *How to Use the Ophthalmoscope: Being Elementary Instructions in Ophthalmoscopy*. Philadelphia: Henry O. Lea, 1877.

Browning, John. *How to Use Our Eyes and How to Preserve Them by the Aid of Spectacles*. London: Chatto & Windus, 1883.

Bryant, J. M., and H. C. Hake. *Street Lighting*. Engineering Experiment Station Bulletin no. 51. Urbana: University of Illinois, 1911.

Buckingham, James Silk. *National Evils and Practical Remedies, with the Plan of a Model Town*. London: Peter Jackson, 1849.

Burgoyne, Frank J. *Library Construction, Architecture, Fittings, and Furniture*. New ed. London: George Allen, 1905.

Byles, C. B. *The First Principles of Railway Signalling, Including an Account of the Legislation in the United Kingdom Affecting the Working of Railways and the Provision of Signalling and Safety Devices*. London: Railway Gazette, 1910.

Caminada, Jerome. *Twenty-five Years of Detective Life*. Manchester: John Heywood, 1895.

"The Cannon-Street Accident." *Electrician* 34 (November 23, 1894).

Carnelly, Thomas, J. S. Haldane, and A. M. Anderson. "The Carbonic Acid, Organic Matter and Micro-Organisms in Air, More Especially of Dwellings and Schools." *Philosophical Transactions of the Royal Society of London* 178 (1887): 61–111.

Carpenter, Alfred. "London Fogs." *Westminster Review*, n.s., 61 (January 1882): 136–57.

"Carriage-Way Pavements." *Engineering* 54 (July 22, 1892; July 29, 1892; and August 12, 1892).

Carter, B. "Colour Blind Engine Drivers." *Engineer* 69 (January 31, 1890).

Carter, Robert. *A Practical Treatise on Diseases of the Eye*. Philadelphia: Henry C. Lea, 1876.

———. *Eyesight: Good and Bad: A Treatise on the Exercise and Preservation of Vision*. 2nd ed. London: Macmillan, 1880.

Cash, C. *Our Slaughter-House System: A Plea for Reform*. London: George Bell, 1907.

Chadwick, Edwin. *Report on the Sanitary Condition of the Labouring Population of Great Britain*. Edited by M. W. Flinn. Edinburgh: Edinburgh University Press, 1965.

Chubb, Harry. "The Supply of Gas to the Metropolis." *Journal of the Statistical Society of London* 39, no. 2 (June 1876): 350–80.

"The City Company's Mains." *Electrical Review* 35 (October 26, 1894).

"The City Electric Lighting Breakdown." *Electrical Review* 35 (October 19, 1894).

"The City Lighting." *Electrician* 28 (March 11, 1892).

"The City Lighting." *Electrician* 30 (November 25, 1892).

"The City Lighting." *Electrician* 30 (December 9, 1892).

"The City Lighting." *Electrician* 30 (March 24, 1893).

"The City Lighting." *Electrical Review* 33 (December 8, 1893).

"City of London Electric Lighting Co., Bankside, Southwark." *Proceedings* (Institution of Mechanical Engineers) 3–4 (1890).

Clark, D. K. "Stone Pavements of Manchester." In *The Construction of Roads and Streets*, by Henry Law and D. Clark. London: Lockwood, 1890.

Clegg, Samuel, Jr. *A Practical Treatise on the Manufacture and Distribution of Coal-Gas; Its Introduction and Progressive Improvement, Illustrated by Engravings from Working Drawings.* London: John Weale, 1841.

Clewell, Clarence. *Factory Lighting.* London: McGraw-Hill, 1913.

"Competitive Lamps for the Thames Embankment." *Engineer* 29 (March 18, 1870).

Cooper, C. "Progress in Oil Lighting at Wimbledon." *Proceedings* (Association of Municipal and Sanitary Engineers and Surveyors) 15 (September 28, 1889).

Crimp, W. "An Experiment in Street Lighting." *British Architect and Northern Engineer* 24 (August 28, 1885).

Critchley, Harry G. *Hygiene in School: A Manual for Teachers.* London: Allman, 1906.

Darbyshire, A. "On Public Abattoirs." *Builder* 33 (February 6, 1875).

"Defects of Vision in Candidates for the Public Services." *Lancet*, 1886, no. 2 (August 28).

Dibdin, W. J. *Practical Photometry: A Guide to the Study of the Measurement of Light.* London: Walter King, 1889.

———. *Public Lighting by Gas and Electricity.* London: The Sanitary Publishing Co., 1902.

"Division of Electric Light." *Mining and Scientific Press* 39 (September 27, 1879).

Dobson, Benjamin A. "On the Artificial Lighting of Workshops." *Proceedings* (Institution of Mechanical Engineers) 3–4 (1893).

Donders, F. C. *On the Anomalies of Accommodation and Refraction of the Eye: With a Preliminary Essay on Physiological Dioptrics.* Translated by William Daniel Moore. London: New Sydenham Society, 1864.

Douglas, J. M. "Electric Lighting Applied to Lighthouse Illumination." *British Architect and Northern Engineer* 11 (March 21, 1879).

Dudgeon, R. E. *The Human Eye; Its Optical Construction Popularly Explained.* London: Hardwicke & Bogue, 1878.

Eassie, William. *Healthy Houses: A Handbook to the History, Defects and Remedies of Drainage, Ventilation, Warming, and Kindred Subjects: With Estimates for the Best Systems in Use, and Upward of Three Hundred Illustrations.* New York: D. Appleton, 1872.

Edwards, Percy. *History of London Street Improvements, 1855–1897.* London: P. S. King, 1898.

"The Electric 'Candle.'" *Builder* 35 (June 30, 1877).

"The Electric Light." *Electrician* 5 (November 13, 1880).

"The Electric Light from a Medical Standpoint." *Electrician* 7 (October 15, 1881).

"The Electric Light in the City." *Electrical Review* 33 (November 24, 1893).

"Electric Light Mains." *Electrical Review* 33 (November 24, 1893).

"Electric Lighting for the City of London." *Electrician* 6 (April 2, 1881).

"Electric Lighting in London." *Engineering* 52 (August 21, 1891; and September 21, 1891).

"Electric Lighting in the City." *Electrician* 10 (November 25, 1882).

"Electric Lighting in the City of London." *Electrician* 26 (February 13, 1891).

"Electric Lighting of the City of London." *Engineering* 57 (February 16, 1894).

"The Electric Search Light in Warfare." *Engineer* 73 (March 25, 1892).

"Electricity Supply Mains." *Electrician* 29 (February 5, 1892).

Ellice-Clark, E. B. "On the Construction and Maintenance of Public Highways." *Proceedings* (Association of Municipal and Sanitary Engineers and Surveyors) 3 (July 6, 1876).

———. "Asphalt and Its Application to Street Paving." *Proceedings* (Association of Municipal and Sanitary Engineers and Surveyors) 5 (July 31, 1879).

Engels, Friedrich. *The Condition of the Working Class in England.* Edited by David MacLellan. Oxford: Oxford University Press, 1993.

"Execution by Electricity." *Engineering* 50 (August 8, 1890).

"Explosions in Manholes." *Electrical World* 23 (March 17, 1894).

Fahie, J. A. "Electric Lighting from a Sanitary Point of View." *Electrician* 13 (October 18, 1884).

Fenwick, Edwin Hurry. *A Handbook of Clinical Electric-Light Cystoscopy.* London: J. & A. Churchill, 1904.

Firth, J. *Municipal London; or, London Government as It Is, and London under a Municipal Council.* London: Longmans, Green, 1876.

Fleming, J. A. *A Handbook for the Electrical Laboratory.* 2 vols. London: "The Electrician" Printing and Publishing Co., 1901.

Fletcher, Banister. *Light and Air: A Text-Book for Architects and Surveyors.* London: B. T. Batsford, 1879.

Foster, W. "Notes on Photometry." *Engineer* 32 (October 6, 1871).

Fox, Cornelius. *Sanitary Examinations of Water, Air and Food.* 3rd ed. London: J. & A. Churchill, 1878.

Galton, Douglas. "Some of the Sanitary Aspects of House Construction." *Builder* 32 (February 21, 1874).

———. *Healthy Hospitals: Observations on Some Points Connected with Hospital Construction.* Oxford: Clarendon, 1893.

———. *Observations on the Construction of Healthy Dwellings: Namely Houses, Hospitals, Barracks, Asylums, Etc.* 2nd ed. Oxford: Clarendon, 1896.

Gamgee, Katherine. *The Artificial Light Treatment of Children in Rickets, Anaemia and Malnutrition.* London: H. K. Lewis, 1927.

Ganton, Charles, "The Distribution of Electrical Energy by Secondary Generators." *Builder* 48 (April 18, 1885).

"Gas Companies and Electric Supply." *Electrician* 15 (August 1, 1890).

"Gas Consumption through Prepayment Meters." *Engineering Record* 36 (November 20, 1897).

"Gas in Libraries." *Electrician* 11 (September 23, 1883).

"A Gas Leak Detector." *Engineering* 30 (December 31, 1880): 631.

"Gas-Lighting by Incandescence." *Builder* 51 (September 18, 1886).

"Gas Mains in Subways." *Engineering* 1 (March 2, 1866).

"The Gas Supply of the City." *Chemical News and Journal of Physical Science* 4 (March 1869): 160.

Gerhardi, G. H. W. *Electricity Meters: Their Construction and Management.* London: Benn Bros., 1917.

Girdlestone, C. *Letters on the Unhealthy Condition of the Lower Class of Dwellings, Especially in Large Towns.* London: Longman, Brown, Green & Longmans, 1845.

Gordon, J. "The Development of Electric Lighting." *Electrician* 11 (July 23, 1883).

Grafton, Walter. *A Handbook of Practical Gas-Fitting: A Treatise on the Distribution of Gas in Service Pipes, the Use of Coal Gas, and the Best Means of Economizing Gas from Main to Burner.* 2nd ed. London: B. T. Batsford, 1907.

Greenaway, Henry. "A New Mode of Hospital Construction." *Builder* 30 (June 29, 1872).

Haab, O. *Atlas and Epitome of Ophthalmoloscopy and Ophthalmoscopic Diagnosis.* Edited by G. E. Schweintz. 2nd American ed. Philadelphia: W. B. Saunders, 1910.

Hammond, Robert. *The Electric Light in Our Homes: Popularly Explained and Illustrated.* London: Warne, 1884.

———. "Municipal Electricity Works." *Builder* 45 (August 19, 1893).

Harlan, George. *Eyesight, and How to Care for It.* New York: Julius King Optical Co., 1887.

Haslam, Arthur P. *Electricity in Factories and Workshops: Its Cost and Convenience: A Handy Book for Power Producers and Power Users.* London: Crosby, Lockwood, 1909.

Hassall, Arthur Hill. *Food: Its Adulterations, and the Methods for Their Detection.* London: Longmans, Green, 1876.

Haywood, John. "On Health and Comfort in House-Building." *Builder* 31 (August 23, 1873).

Hedges, Killingworth. *Useful Information on Electric Lighting.* London: E. & F. N. Spon, 1882.

Hepworth, T. *The Electric Light: Its Past History and Present Position.* London: Routledge, 1879.

Higgs, Paget. *The Electric Light in Its Practical Applications.* London: E. & F. Spon, 1879.

Hobhouse, L. T. *Liberalism.* New York: H. Holt, 1911.

Hope, Edward W., and Edgar A. Browne. *A Manual of School Hygiene: Written for the Guidance of Teachers in Day-Schools*. Cambridge: Cambridge University Press, 1901.

"The Hours of Lighting." *Electrician* 32 (January 19, 1894).

Houston, Edwin J., and A. E. Kennelly. *Electric Arc Lighting*. New York: W. J. Johnston, 1896.

Hughes, Samuel. *Gas Works: Their Construction and Arrangement; and the Manufacture and Distribution of Coal Gas*. Revised, rewritten, and much enlarged by William Richards. London: Crosby, Lockwood, 1885.

Hummel, F. H. "Ancient Lights." *Architect*, July 28, 1877.

Hunt, Charles. *The Construction of Gas-Works*. London: Institution of Civil Engineers, 1894.

———. *A History of the Introduction of Gas Lighting*. London: Walter King, 1907.

Huxley, Thomas. *Lessons in Elementary Physiology*. Enlarged and rev. ed. New York: Macmillan, 1917.

"Illumination and Ventilation." *Builder* 54 (April 14, 1888).

"Illumination in Warfare." *Engineer* 32 (December 29, 1871).

Jenkin, Fleeming. "On Sanitary Inspection." *Sanitary Record* 8 (April 19, 1878).

Jensen, Gerard. *Modern Drainage Inspection and Sanitary Surveys*. London: The Sanitary Publishing Co., 1899.

Kay, James Phillips. *The Moral and Physical Condition of the Working Classes in 1832*. London: Frank Cass, 1970.

Kempe, H. R. *The Electrical Engineer's Pocket-Book: Modern Rules, Formulae, Tables and Data*. 2nd ed. London: Crosby, Lockwood, 1892.

Kennedy, Rankin. *Electrical Installations of Electric Light, Power, Traction and Industrial Electrical Machinery*. 4 vols. London: Blackwood, Le Bas, 1902.

Kingzett, C. T. "Some Analyses of Asphalte Pavings." *Analyst* 8 (January 1883): 4–10.

Kohn, H. "Comparative Determinations of the Acuteness of Vision and the Perception of Colours by Day-Light, Gas-Light and Electric Light." *Archives of Ophthalmology* 9 1880.

"Laboratory of the Electrical Inspector to the City Commission of Sewers." *Electrician* 35 (June 7, 1895).

Lascelles, W. "Glass Roofs for London." *Builder* 34 (December 16, 1876).

Law, H., and D. Clark. *The Construction of Roads and Streets*. London: Lockwood, 1877.

Leaf, H. *The Internal Wiring of Buildings*. London: Archibald Constable, 1899.

Le Conte, Joseph. *Sight: An Exposition of the Principles of Monocular and Binocular Vision*. 2nd ed. New York: D. Appleton, 1897.

Leigh, John. *Coal-Smoke: Report to the Health and Nuisance Committees of the Corporation of Manchester*. Manchester: John Heywood, 1883.

Lemmoin-Cannon, Henry. *The Sanitary Inspector's Guide: A Practical Treatise on the Public Health Act, 1875, and the Public Health Act, 1890, So Far as They Affect the Inspector of Nuisances*. London: P. S. King, 1902.

Letheby, Henry. *On Noxious and Offensive Trades and Manufactures, with Special Reference to the Best Practicable Means of Abating the Several Nuisances Therefrom.* London: Statham, 1875.

"Light in London (III)." *Builder* 33 (November 13, 1875).

"Lighting in School-Rooms." *Builder* 30 (September 7, 1872).

Lincoln, D. F. "Sanitary School Construction." *Plumber and Sanitary Engineer* 2 (November 1, 1879).

Locke, John. *An Essay concerning Human Understanding.* Edited by Peter H. Nidditch. Oxford: Clarendon, 1975.

"London Gas." *Lancet*, 1877, no. 2 (October 20).

"London Noise." *Sanitary Record* 5 (October 28, 1876).

"London Private Slaughter-Houses." *Sanitary Record* 1 (December 19, 1874).

"London Roads and the Cost of Them." *Builder* 33 (July 24, 1875).

Long, Charles A. *Spectacles: When to Wear and How to Use Them.* London: Bland & Long, 1855.

MacKenzie, William. *A Practical Treatise on the Diseases of the Eye.* 4th ed., rev. and enlarged. London: Blanchard & Lea, 1855.

Maier, Julius. *Arc and Glow Lamps: A Practical Handbook on Electric Lighting.* London: Whittaker, 1886.

Maltbie, Milo Roy. "Gas Lighting in Great Britain." *Municipal Affairs* 4 (1900): 538–73.

Manchester City Council. *Proceedings.* 1866–96.

Manchester City Council Gas Committee. *Minutes.* 1880–98.

Manchester City Council Improvement and Buildings Committee. *Minutes.* 1863–86.

Manchester City Council Paving, Sewering and Highways Committee. *Minutes.* 1867–82.

"Manhole Boxes." *Electrician* 32 (February 23, 1894).

Marcet, William. "On Fogs." *Quarterly Journal of the Royal Meteorological Society* 15, no. 70 (April 1889): 59–72.

Marr, T. C. *Housing Conditions in Manchester and Salford.* Manchester: Sherratt & Hughes, 1904.

Mason, Charles. "London Subways." *Engineering Record* 35 (May 1, 1897).

Matthews, William. *An Historical Sketch of the Origin and Progress of Gas-Lighting.* 2nd ed. London: Simpkin & Marshall, 1832.

Maycock, William Perren. *Electric Wiring, Fittings, Switches and Lamps: A Practical Book for Electric Light Engineers, &c.* London: Whittaker, 1899.

Mearns, Andrew, et al. *The Bitter Cry of Outcast London.* Leicester: Leicester University Press, 1970.

Merriman, Owen. *Gas-Burners: Old and New.* London: Walter King, 1884.

"Microbes in Wood Pavement." *Engineering Record* 32 (September 5, 1896).

Mill, John Stuart. *Considerations on Representative Government.* New York: Harper & Bros., 1862.

———. "On Liberty." In *Essential Works of John Stuart Mill*, ed. Max Lerner. London: Bantam, 1961.

Millington, F. H. *The Housing of the Poor*. London: Cassell, 1891.

Minutes of Evidence and Appendix as to England and Wales. Vol. 2 of *Royal Commission on the Housing of the Working Classes*. London: Eyre & Spottiswoode, 1885.

Molera, E. J., and J. C. Cebrian. "Practical Divisibility of the Electric Light." *Electrician* 3 (August 2, 1879).

Mouat, Frederic, and H. Saxon Snell. *Hospital Construction and Management*. London: J. & A. Churchill, 1883.

Moyle, William. "Inspecting London." In *Living London: Its Work and Its Play, Its Humour and Its Pathos, Its Sights and Its Scenes*, ed. George Sims, vol. 3. London: Cassell, 1902.

Müller, Johannes. *Elements of Physiology*. Translated by William Baly. 2nd ed. London: Taylor & Walton, 1839.

Municipal Glasgow: Its Evolution and Enterprises. Glasgow: Glasgow Corporation, 1914.

Murdoch, William. "An Account of the Application of the Gas from Coal to Economical Purposes." *Philosophical Transactions of the Royal Society of London* 98 (1808): 124–32.

"Needless Noise." *Lancet*, 1876, no. 2 (September 23).

Nerz, F. *Searchlights: Their Theory, Construction and Applications*. Translated by Charles Rogers. London: Archibald Constable, 1907.

Nettleship, Edward. *Diseases of the Eye*. 5th ed. Philadelphia: Lea Bros., 1890.

Newbigging, T., and W. T. Fewtrell, eds. *King's Treatise on the Science and Practice of the Manufacture and Distribution of Coal Gas*. 3 vols. London: W. B. King, 1878–82.

Newsholme, Arthur. *School Hygiene: The Laws of Health in Relation to School Life*. London: Swan Sonnenschein, Lowrey, 1887.

Nightingale, Florence. *Notes on Hospitals*. 3rd ed. London: Longman, Green, Longman, Roberts & Green, 1863.

———. *Notes on Nursing: What It Is, and What It Is Not*. London: Churchill Livingstone, 1980.

"Noiseless Manhole Covers." *Engineering Record* 34 (August 15, 1896).

"Noiseless Pavements." *Lancet*, 1873, no. 2 (December 13).

"Noiseless Pavements." *Sanitary Record* 4 (February 6, 1875).

Noll, Augustus. *How to Wire Buildings: A Manual of the Art of Interior Wiring*. 4th ed. New York: D. Van Nostrand, 1906.

"Noxious Vapours." *Sanitary Record* 6 (June 8, 1877).

Oliver, Thomas, and Robert Bolam. *On the Cause of Death by Electric Shock*. London: British Medical Association, 1898.

"On Some Nuisances and Absurdities in London." *Builder* 33 (September 26, 1874).

"On the Technology of Glass." *Builder* 31 (January 25, 1873): 69–70.

"The Opening of the Beckton Gasworks." *Engineer* 31 (February 10, 1871).

"Optic Nerve Atrophy in Smokers." *British Medical Journal*, 1890, no. 1 (May 10).

"Overhead Wires." *Electrician* 12 (January 26, 1884).

"Overhead Wires Again." *Electrician* 11 (November 3, 1883).

"The Ownership of Street Lamps." *Sanitary Record* 7 (September 14, 1877).

"Painless Killing." *Lancet*, 1888, no. 2 (August 4): 219–20.

"The Panopticon of Science and Art." In *Year-Book of Facts in Science and Art*, ed. John Timbs, 9–11. London: David Bogue, 1855.

Papworth, John W., and Wyatt Papworth. *Museums, Libraries, and Picture Galleries, Public and Private; Their Establishment, Formation, Arrangement and Architectural Content*. London: Chapman & Hall, 1853.

Parkes, Edmund A. *A Manual of Practical Hygiene*. 7th ed. Philadelphia: P. Blakiston, 1887.

Peckston, Thomas. *A Practical Treatise on Gas-Lighting*. 3rd ed. London: Hebert, 1841.

"The Peril of Electric Lamps." *Engineering* 30 (October 29, 1880).

Pfeiffer, Carl. "Light: Its Sanitary Influence and Importance in Building." *Builder* 35 (July 21, 1877).

"Photometric Standards." *Sanitary Engineer* 5 (December 1, 1881).

"Photometric Units and Dimensions." *Electrician* 33 (September 28, 1894).

Pickering, W. "Concerning the Gas-Flame, Electric and Solar Spectra, and Their Effect on the Eye." *Nature* 26 (February 9, 1882).

Pillsbury, Walter. *Attention*. London: Swan, 1908.

"Plate Glass." *Engineering* 1 (January 12, 1866).

"A Plea for Sunshine." *Plumber and Sanitary Engineer* 2 (May 1879).

Powell, Harry J. *The Principles of Glass-Making*. London: George Bell, 1883.

Preece, William. "The Electric Light at the Paris Exhibition." *Electrician* 8 (December 24, 1881).

———. "On a New Standard of Illumination and the Measurement of Light." *Proceedings* (Royal Society of London) 36 (1884).

———. "Report on Electric Lighting in the City." *Electrician* 15 (April 18, 1885; and April 25, 1885).

———. "The Measure of Illumination." *Electrician* 23 (September 13, 1889).

———. "The Sanitary Aspects of Electric Lighting." *Electrician* 25 (August 29, 1890).

———. "On the Relative Merit and Cost of Gas and Electricity for Lighting Purposes." *Proceedings* (Association of Municipal and Sanitary Engineers and Surveyors) 17 (June 27, 1891).

Preston-Thomas, Herbert. *The Work and Play of a Government Inspector*. London: William Blackwood, 1909.

"Prestwich Union Workhouse." *Builder* 30 (August 17, 1872).

"Progress of the Gas Supply." *Engineer* 33 (February 28, 1872).

"Public Latrines in London." *Engineering News* 31 (May 31, 1894).

"Queer Pavements." *Engineering Record* 32 (September 19, 1896).

Quin, R. C. "The Localisation of Faults in Underground Mains." *Engineer* 80 (July 16, 1897).

Raphael, F. Charles. *The Localisation of Faults in Electric Light and Power Mains: With Chapters on Insulation Testing*. 2nd ed. London: "The Electrician" Printing & Publishing Co., 1903.

Redlich, Josef. *Local Government in England*. Edited with additions by Francis W. Hirst. 2 vols. London: Macmillan, 1903.

Reiniger, Erwin, Max Gebbert, and Karl Schall. *Electro-Medical Instruments and Their Management, and Illustrated Price List of Electro-Medical Apparatus*. Bristol: John Wright, 1893.

Report from the Select Committee of the House of Lords on the Electric Lighting Act (1882) Amendment Bill. London: Hansard, 1886.

"Report of the Overhead Wires Committee." *Electrician* 15 (May 15, 1885).

Richardson, Benjamin Ward. *Hygeia: A City of Health*. London: Macmillan, 1876.

———. "The Painless Extinction of Life." *Popular Science Monthly* 26 (1884–85): 641–52.

———. "Public Slaughter-Houses: A Suggestion for Farmers." *New Review* 8 (1893): 631–44.

Robins, E. C. "Buildings for Secondary Schools." *British Architect and Northern Engineer* 13 (April 23, 1880).

Robinson, Joseph. *The Sanitary Inspector's Practical Guide, with Inspection of Lodging-Houses (under the Sanitary Acts), and the Sale of Food and Drugs Amendment Act 1879*. 2nd ed. London: Shaw, 1884.

Rosenhain, Walter. *Glass Manufacture*. London: Archibald Constable, 1908.

Royle, H. "Street Pavements as Adopted in the City of Manchester." *British Architect and Northern Engineer* 6 (July 21, 1876).

"Rules to Be Observed in Planning and Fitting up Schools." *Builder* 30 (May 4, 1872).

Rumney, R. "On the Ashpit System of Manchester." *Engineer* 30 (September 30, 1870).

Russell, James Burn. *Public Health Administration in Glasgow: A Memorial Volume of the Writings of James Burn Russell*. Edited by A. K. Chalmers. Glasgow: James Maclehose, 1905.

Russell, R. *London Fogs*. London: Edward Stanford, 1880.

Rutter, J. *Gas-Lighting: Its Progress and Its Prospects; with Remarks on the Rating of Gas-Mains, and a Note on the Electric-Light*. London: John Parker, 1849.

Ryder, Thomas. "Electric Lighting of Schools." *Electrician* 22 (December 14, 1888).

Schultze, Max. *Zur Anatomie und Physiogie der Retina*. Bonn: Cohen, 1866.

Scrutton, Percy E. *Electricity in Town and Country Houses*. London: Archibald Constable, 1898.

Semon, Felix. "Electric Illumination of the Various Cavities of the Human Body." *Lancet*, 1885, no. 1 (March 21): 512–13.

Shoolbred, J. "Electric Lighting, and Its Application to Public Illumination by Municipal and Other Bodies." *Proceedings* (Association of Municipal and Sanitary Engineers and Surveyors) 6 (July 31, 1879).

———. "Illumination by Electricity." *Proceedings* (Association of Municipal and Sanitary Engineers and Surveyors) 8 (June 29, 1882).

Siemens, C. W. "On the Influence of Electric Light upon Vegetation, and on Certain Physical Principles Involved." *Electrician* 5 (March 13, 1880): 200–202.

"Silent Paving." *Lancet*, 1875, no. 2 (August 21).

Simon, E. D., and Marion Fitzgerald. *The Smokeless City*. London: Longmans, Green, 1922.

Simpson, Thomas Bartlett. *Gas-Works: The Evils Inseparable from Their Existence in Populous Places, and the Necessity of Removing Them from the Metropolis*. London: William Freedman, 1866.

Sims, George. *How the Poor Live; and, Horrible London*. London: Garland, 1984.

———, ed. *Living London: Its Work and Its Play, Its Humour and Its Pathos, Its Sights and Its Scenes*. 3 vols. London: Cassell, 1902.

Slater, J. "Electric Lighting Applied to Buildings." *Electrician* 6 (April 30, 1881).

———. "Artificial Illumination." *Builder* 22 (February 9, 1889).

"Slaughter-Houses in the Metropolis." *Builder* (August 17, 1872).

Smiles, Samuel. *Lives of the Engineers: Metcalfe-Telford: History of Roads*. London: John Murray, 1904.

———. *Lives of the Engineers: Vermuyden-Myddleton-Perry-James Brindley*. London: John Murray, 1904.

———. *Self-Help: With Illustrations of Character, Conduct and Perseverance*. Edited by Peter Sinnema. Oxford: Oxford University Press, 2002.

Smith, Adam. *An Enquiry into the Nature and Causes of the Wealth of Nations*. Edited by Edwin Cannan. Chicago: University of Chicago Press, 1976.

———. *Theory of Moral Sentiments*. Amherst, NY: Prometheus, 2000.

"Smoke Abatement." *Nature* 23 (January 13, 1881).

"Smoke Prevention." *Engineer* 78 (July 24, 1896).

"The Social Effects of Electric Power Supply." *Electrician* 29 (November 18, 1892).

Southard, W. F. *The Modern Eye: With an Analysis of 1300 Errors of Refraction*. San Francisco: W. A. Woodward, 1883.

"Staite's Improvements in Lighting, Etc." *Patent Journal and Inventor's Magazine* 4 (1848).

Steavenson, William. "Electricity, and Its Manner of Working in the Treatment of Disease." *Electrician* 14 (March 28, 1886).

Stevenson, Thomas. *Lighthouse Illumination: Being a Description of the Holophotal System and of Azimuthal Condensing and Other New Forms of Lighthouse Apparatus*. 2nd ed. Edinburgh: Adam & Charles Black, 1871.

Stewart, James Dugald, et al. *Open Access Libraries: Their Planning, Equipment and Organisation*. London: Grafton, 1915.

Stockman, Frank. *A Practical Guide for Sanitary Inspectors*. 3rd ed. London: Butterworth, 1915.

"The Strand: And Some Items Noteworthy in It." *Builder* 35 (May 19, 1877).

"Street Illumination in the City." *Builder* 61 (August 8, 1891).

"Street Paving and Hygiene." *Engineer* 78 (December 28, 1894): 573–74.

Sugg, William. *Gas as an Illuminating Agent Compared with Electricity*. London: Walter King, 1882.

———. *The Domestic Uses of Coal Gas, as Applied to Lighting, Cooking and Heating, Ventilation.* London: Walter King, 1884.

Sumpner, W. "The Diffusion of Light." *Electrician* 30 (February 3, 1893).

Swinburne, James. "Electrical Measuring Instruments." *Minutes of Proceedings* (Institute of Civil Engineers) 110 (April 26, 1892).

"Tar Pavements and Tar Varnish." *Engineering News* 26 (December 19, 1891): 598–99.

Taylor, Albert. *The Sanitary Inspector's Handbook.* 3rd ed. London: H. K. Lewis, 1901.

———. *The Sanitary Inspector's Handbook.* 5th ed. London: H. K. Lewis, 1914.

"Terra Metallic Pavings." *Builder* 35 (February 3, 1877).

Thwaite, B. "Hygiene Applied to Dwellings." *British Architect and Northern Engineer* 11 (January 10, 1879).

———. *Our Factories, Workshops, and Warehouses, Their Sanitary and Fire-Resisting Arrangements.* London: E. & F. N. Spon, 1882.

Tidy, Charles Meymott. *Handbook of Modern Chemistry: Inorganic and Organic; for the Use of Students.* London: J. & A. Churchill, 1878.

Titchener, Edward. *Lectures on the Elementary Psychology of Feeling and Attention.* New York: Macmillan, 1908.

"Toughened Glass." *Sanitary Record* 3 (December 11, 1875).

Triggs, Inigo, *Town Planning: Past, Present and Possible.* London: Methuen, 1909.

Trotter, A. M. "The Inspection of Meat and Milk in Glasgow." *Journal of Comparative Pathology and Therapeutics* 14 (1901): 84–90.

Trotter, Alexander Pelham. "The Distribution and Measurement of Illumination." *Minutes of Proceedings* (Institute of Civil Engineers) 110 (May 10, 1892).

———. *Illumination: Its Distribution and Measurement.* London: Macmillan, 1911.

———. *The Elements of Illuminating Engineering.* London: Sir Isaac Pitman, 1921.

Turnbull, George. "Lighting." In *London in the Nineteenth Century*, by Walter Besant, 316–26. London: Adam & Charles Black, 1909.

———. "Pavements." In *London in the Nineteenth Century*, by Walter Besant, 340–46. London: Adam & Charles Black, 1909.

Tyndall, John. *Light and Electricity: Notes on Two Courses of Lectures before the Royal Institution of Great Britain.* New York: D. Appleton, 1871.

———. *Sound.* 4th ed. London: Longmans, Green, 1883.

"Underground Systems of Distributing Electricity." *Electrician* 24 (February 28, 1890).

Ure, Andrew. *The Philosophy of Manufactures; or, An Exposition of the Scientific, Moral, and Commercial Economy of the Factory System of Great Britain.* London: Frank Cass, 1967.

Vacher, Francis. *A Healthy Home.* London: The Sanitary Record, [1894?].

Walker, Sydney F. *The Pocket Book of Electric Lighting and Heating.* New York: Norman W. Henley, 1907.

Walley, Thomas. *A Practical Guide to Meat Inspection*. Edinburgh: Young J. Pentland, 1890.

Walsh, John Henry. *A Manual of Domestic Economy: Suited to Familes Spending from £100 to £1000 a Year*. 2nd ed. London: G. Routledge, 1857.

Walton, H. Haynes. *A Treatise on Operative Ophthalmic Surgery*. 1st American ed., ed. S. Little. Philadelphia: Lindsay & Blakiston, 1853.

Wanklyn, James. "Need Gas-Making Be a Public Nuisance?" *British Architect and Northern Engineer* 24 (July 3, 1885).

"Want of Tact on the Part of Sanitary Inspectors." *Lancet*, 1893, no. 2 (December 30): 1643.

Webb, Aston. "Relation between Size of Windows and Rooms." *Plumber and Sanitary Engineer* 2 (March 1879).

Webber, C. E. "Some Notes on the Electric Lighting of the City of London." *Electrician* 32 (February 16, 1894; February 23, 1894; and March 2, 1894).

Webber, W. H. Y. *Town Gas and Its Uses for the Production of Light, Heat and Motive Power*. New York: D. Van Nostrand, 1907.

"The Wenham System of Lighting and Ventilating." *Builder* 58 (April 12, 1890).

White, A. "The Silent Electric Arc." *Electrician* 14 (November 29, 1884).

Winslow, Forbes. *Light: Its Influence on Life and Health*. London: Longmans, Green, Reader & Dyer, 1867.

Wood, A. *A Guide to Gas-Lighting*. Lewes: George P. Bacon, 1872.

"Wood and Asphalt Paving." *Engineer* 74 (May 20, 1892).

"Wood Pavement." *Lancet*, 1882, no. 2 (September 30).

Young, Desmond. "Lighting London." In *Living London: Its Work and Its Play, Its Humour and Its Pathos, Its Sights and Its Scenes*, ed. George Sims, vol. 2. London: Cassell, 1902.

SECONDARY SOURCES

Agar, Jon. "Bodies, Machines and Noise." In *Bodies/Machines*, ed. Iwan Rhys Morus. Oxford: Berg, 2002.

———. *The Government Machine: A Revolutionary History of the Computer*. Cambridge, MA: MIT Press, 2003.

Allen, Rick. "Observing London Street-Life: G. A. Sala and A. J. Munby." In *The Streets of London: From the Great Fire to the Great Stink*, ed. Tim Hitchcock and Heather Shore. London: Rivers Oram, 2003.

Appleyard, Rollo. *The History of the Institution of Electrical Engineers (1871–1931)*. London: Institution of Electrical Engineers, 1939.

Ariès, Philippe. *The Hour of Our Death*. Translated by Helen Weaver. Oxford: Oxford University Press, 1991.

Asendorf, Christoph. *Batteries of Life: On the History of Things and Their Perception in Modernity*. Translated by Don Reneau. London: University of Berkeley Press, 1993.

Ashby, E., and M. Anderson. *The Politics of Clean Air*. Oxford: Clarendon, 1981.

Ashworth, W. *The Genesis of Modern British Town Planning: A Study in Economic and Social History of the Nineteenth and Twentieth Centuries*. London: Routledge & Kegan Paul, 1954.

Awty, Brian G. "The Introduction of Gas-Lighting to Preston." *Transactions of the Historic Society of Lancashire and Cheshire* 125 (1975): 82–118.

Bailkin, Jordanna. "Colour Problems: Work, Pathology and Perception in Modern Britain." *International Labour and Working-Class History* 68 (2005): 93–111.

Baldwin, Peter. "In the Heart of Darkness: Blackouts and the Social Geography of Lighting in the Gaslight Era." *Journal of Urban History* 30, no. 5 (July 2004): 749–68.

Barry, Andrew. *Political Machines: Governing a Technological Society*. London: Athlone, 2001.

Bartrip, Peter. "British Government Inspection, 1832–1875: Some Observations." *Historical Journal* 25, no. 3 (1982): 605–26.

———. "How Green Was My Valance? Environmental Arsenic Poisoning and the Victorian Domestic Ideal." *English Historical Review* 111 (1994): 891–913.

Bazerman, Charles. *The Languages of Edison's Light*. Cambridge, MA: MIT Press, 1999.

Bellamy, Richard, ed. *Victorian Liberalism: Nineteenth-Century Political Thought and Practice*. London: Routledge, 1990.

———. "J. S. Mill, T. H. Green and Isaiah Berlin on the Nature of Liberty and Liberalism." In *Rethinking Liberalism*, ed. Richard Bellamy, 22–46. London: Pinter, 2000.

Beniger, James. *The Control Revolution: Technological and Economic Origins of the Information Society*. Cambridge, MA: Harvard University Press, 1986.

Benjamin, Walter. *Illuminations*. Edited by Hannah Arendt. Translated by Harry Zohn. London: Fontana, 1991.

———. *The Arcades Project*. Translated by Howard Eiland and Kevin McLaughlin. Cambridge, MA: Belknap, 2002.

Bennett, Tony. "The Exhibitionary Complex." *New Formations* 4 (1988): 73–100.

Berlin, Isaiah. "Two Concepts of Liberty." In *Four Essays on Liberalism*, 118–72. Oxford: Oxford University Press, 1969.

Black, Robert. *The History of Electric Wires and Cables*. London: Peter Peregrinus, 1983.

Blake-Coleman, B. C. *Copper Wire and Electrical Conductors: The Shaping of a Technology*. Reading: Harwood, 1992.

Borgmann, Albert. *Technology and the Character of Contemporary Life: A Philosophical Enquiry*. Chicago: University of Chicago Press, 1984.

Bowers, Brian. *Lengthening the Day: A History of Lighting Technology*. Oxford: Oxford University Press, 1998.

Boyne, Roy. "Post-Panopticism." *Economy and Society* 29, no. 2 (May 2000): 285–307.

Brand, Dana. *The Spectator and the City in Nineteenth-Century American Literature*. Cambridge: Cambridge University Press, 1991.

Brandwood, Geoff, Mike Slaughter, and Andrew Dawson. *Licensed to Sell: The History and Heritage of the Public House*. Swindon: English Heritage, 2004.

Brewer, John. *The Sinews of Power: War, Money and the English State, 1688–1783*. New York: Knopf, 1989.

Briggs, Asa. *Victorian Things*. London: Penguin, 1988.

Brunton, Deborah. "Evil Necessaries and Abominable Erections: Public Conveniences and Private Interests in the Scottish City, 1830–1870." *Social History of Medicine* 18, no. 2 (2005): 187–202.

Buchanan, R. A. "Engineers and Government in Nineteenth-Century Britain." In *Government and Expertise: Specialists, Administrators, and Professionals, 1860–1919*, ed. Roy MacLeod. Cambridge: Cambridge University Press, 1988.

Buck-Morss, Susan. *The Dialectics of Seeing: Walter Benjamin and the Arcades Project*. Cambridge, MA: MIT Press, 1989.

Burchell, Graham, Colin Gordon, and Peter Miller, eds. *The Foucault Effect: Studies in Governmentality*. London: Harvester, 1991.

Bynum, W. F. *Science and the Practice of Medicine in the Nineteenth Century*. Cambridge: Cambridge University Press, 1994.

Callon, Michel. "Society in the Making: The Study of Technology as a Tool for Sociological Analysis." In *The Social Construction of Technological Systems: New Directions in the Sociology and History of Technology*, ed. W. Bijker, Thomas Hughes, and Trevor Pinch. Cambridge, MA: MIT Press, 1987.

Campbell, John, and Matt Carlson. "Panopticon.com: Online Surveillance and the Commodification of Privacy." *Journal of Broadcasting and Electronic Media* 46, no. 4 (2002): 586–606.

Cantor, G. N., and M. J. S. Hodge, eds. *Conceptions of Ether: Studies in the History of Ether Theories, 1740–1900*. Cambridge: Cambridge University Press, 1981.

Carroll, Patrick. *Science, Culture, and Modern State Formation*. Berkeley and Los Angeles: University of California Press, 2006.

Chaloner, W. H. *The Social and Economic Development of Crewe, 1780–1923*. Manchester: Manchester University Press, 1950.

Clarke, E., and L. S. Jacyna. *Nineteenth-Century Origins of Neurological Concepts*. Berkeley and Los Angeles: University of California Press, 1987.

Clark, T. J. *The Painting of Modern Life: Paris in the Art of Manet and His Followers*. Princeton, NJ: Princeton University Press, 1984.

Clayson, Susan Hollis. "Outsiders: American Painters and Cosmopolitanism in the City of Light, 1871–1914." In *La France dans le regard des États-Unis/France as Seen by the United States*, ed. Frédéric Monneyron and Martine Xiberras, 57–71. Perpignan: Presses Universitaires de Perpignan; Montpellier: Publications de l'Université Paul Valéry—Montpellier III, 2006.

Cody, Lisa Forman. "Living and Dying in Georgian London's Lying-in Hospitals." *Bulletin of the History of Medicine* 78 (2004): 309–48.

Cooper, Frederick. *Colonialism in Question: Theory, Knowledge, History*. Berkeley and Los Angeles: University of California Press, 2005.

Coopersmith, Jonathan. *The Electrification of Russia, 1880–1926*. Ithaca, NY: Cornell University Press, 1992.

Corbin, Alain. *The Foul and the Fragrant: Odour and the French Social Imagination*. Translated by Miriam L. Kochan. Cambridge, MA: Harvard University Press, 1986.

———. "Backstage." In *A History of Private Life IV: From the Fires of Revolution to the Great War*, ed. Michelle Perrot. Cambridge, MA: Belknap, 1990.

———. *Time, Desire and Horror: Towards a History of the Senses*. Translated by Jean Birrell. Cambridge: Polity, 1995.

———. *Village Bells: Sound and Meaning in the Nineteenth-Century French Countryside*. Translated by Martin Thom. New York: Columbia University Press, 1998.

Corrigan, Philip, and Derek Sayer. *The Great Arch: English State Formation as Cultural Revolution*. Oxford: Blackwell, 1985.

Crampton, Jeremy. "Cartographic Rationality and the Politics of Geosurveillance and Security." *Cartography and Geographic Information Science* 30, no. 2 (2003): 135–48.

Crary, Jonathan. *Techniques of the Observer: On Vision and Modernity in the Nineteenth Century*. London: MIT Press, 1990.

———. *Suspensions of Perception: Attention, Spectacle and Modern Culture*. London: MIT Press, 1999.

Crook, Tom. "Norms, Forms and Bodies: Public Health, Liberalism and the Victorian City, 1830–1900." Ph.D. thesis, University of Manchester, 2004.

———. "Power, Privacy and Pleasure: Liberalism and the Modern Cubicle." *Cultural Studies* 21, nos. 4–5 (2007): 549–69.

Daunton, Martin. *House and Home in the Victorian City: Working-Class Housing, 1850–1914*. London: Edward Arnold, 1983.

Dean, Mitchell. *Governmentality: Power and Rule in Modern Society*. London: Sage, 1999.

Debord, Guy. *Society of the Spectacle*. Translated by Donald Nicholson-Smith. New York: Zone, 1994.

Delaporte, François. *Disease and Civilization: The Cholera in Paris, 1832*. Translated by Arthur Goldhammer. Cambridge, MA: MIT Press, 1986.

Dillon, Maureen. *Artificial Sunshine: A Social History of Domestic Lighting*. London: National Trust, 2002.

Dreyfus, Hubert, and Paul Rabinow. *Michel Foucault: Beyond Structuralism and Hermeneutics*. 2nd ed. Chicago: University of Chicago Press, 1983.

Dunsheath, Percy. *A History of Electrical Engineering*. London: Faber & Faber, 1962.

Durbach, Nadja. *Bodily Matters: The Anti-Vaccination Movement in England, 1853–1907*. Durham, NC: Duke University Press, 2005.

Edgerton, David. *The Shock of the Old: Technology and Global History since 1900*. Oxford: Oxford University Press, 2007.

Ekirch, A. Roger. *At Day's Close: Night in Times Past*. New York: Norton, 2005.

Elden, Stuart. "Plague, Panopticon, Police." *Surveillance and Society* 1, no. 3 (2003): 240–53.

Everard, Stirling. *History of the Gas Light and Coke Company, 1812–1949*. London: Benn, 1949.

Eyler, J. M. *Victorian Social Medicine: The Ideas and Methods of William Farr*. Baltimore: Johns Hopkins University Press, 1979.

Ezrahi, Yaron. "Technology and the Civil Epistemology of Democracy." In *Technology and the Politics of Knowledge*, ed. Andrew Feenberg and Alastair Hannay. Bloomington: Indiana University Press, 1995.

Falkus, M. E. "The British Gas Industry before 1850." *Economic History Review*, n.s., 20, no. 3 (1967): 494–508.

Figlio, Karl. "Theories of Perception and the Physiology of Mind in the Late Eighteenth Century." *History of Science* 13 (1975): 177–212.

Flick, Carlos. "The Movement for Smoke Abatement in Nineteenth-Century Britain." *Technology and Culture* 21, no. 1 (1980): 29–50.

Flint, Kate. *The Victorians and the Visual Imagination*. Cambridge: Cambridge University Press, 2000.

Forty, A. "The Modern Hospital in England and France: The Social and Medical Uses of Architecture." In *Buildings and Society: Essays on the Social Development of the Built Environment*, ed. Anthony King. London: Routledge, 1984.

Foucault, Michel. *The Order of Things: An Archaeology of the Human Sciences*. New York: Routledge, 1974.

———. "The Eye of Power." In *Power/Knowledge: Selected Interviews and Other Writings 1972–1977*, ed. Colin Gordon. New York: Pantheon, 1980.

———. "The Subject and Power." In *Michel Foucault: Beyond Structuralism and Hermeneutics* (2nd ed.), ed. Hubert Dreyfus and Paul Rabinow. Chicago: University of Chicago Press, 1983.

———. "Space, Knowledge and Power." In *The Foucault Reader*, ed. Paul Rabinow. New York: Pantheon, 1984.

———. *The History of Sexuality*. Vol. 1, *An Introduction*. Translated by Robert Hurley. New York: Vintage, 1990.

———. *Discipline and Punish: The Birth of the Prison*. Translated by Alan Sheridan. London: Penguin, 1991.

———. "Governmentality." In *The Foucault Effect: Studies in Governmentality*, ed. Graham Burchell, Colin Gordon, and Peter Miller. Chicago: University of Chicago Press, 1991.

———. *Dits et écrits*. Vol. 2. Paris: Gallimard, 1994.

Frankel, Oz. *States of Inquiry: Social Investigations and Print Culture in Nineteenth-Century Britain and the United States*. Baltimore: Johns Hopkins University Press, 2006.

Frazer, W. M. *Duncan of Liverpool: Being an Account of the Work of Dr. W. H. Duncan, Medical Officer of Health of Liverpool, 1847–63*. London: Hamish Hamilton Medical, 1947.

Gandy, Matthew. "Rethinking Urban Metabolism: Water, Space and the Modern City." *City* 8, no. 3 (December 2004): 371–87.

Gauchet, Marcel, and Gladys Swain. *Madness and Democracy: The Modern Psychiatric Universe.* Translated by Catherine Porter. Princeton, NJ: Princeton University Press, 1999.

Giedion, Siegfried. *Mechanization Takes Command: A Contribution to Anonymous History.* New York: Oxford University Press, 1948.

Gooday, Graeme. "Spot-Watching, Bodily Postures and the 'Practiced Eye': The Material Practice of Instrument Reading in Late Victorian Electrical Life." In *Bodies/Machines*, ed. Iwan Rhys Morus, 165–96. Oxford: Berg, 2002.

———. *The Morals of Measurement: Accuracy, Irony, and Trust in Late Victorian Electrical Practice.* Cambridge: Cambridge University Press, 2004.

Goodlad, Lauren M. E. *Victorian Literature and the Victorian State: Character and Governance in a Liberal Society.* Baltimore: Johns Hopkins University Press, 2003.

Goswami, Manu. *Producing India: From Colonial Economy to National Space.* Chicago: University of Chicago Press, 2004.

Graham, Stephen, and Simon Marvin. *Splintering Urbanism: Networked Infrastructures, Technological Mobilities and the Urban Condition.* London: Routledge, 2001.

Gunn, Simon. *The Public Culture of the Victorian Middle Class: Ritual and Authority and the English Industrial City.* Manchester: Manchester University Press, 2000.

Hacking, Ian. *Representing and Intervening: Introductory Topics in the Philosophy of Natural Science.* Cambridge: Cambridge University Press, 1983.

Hamlin, Christopher. *Public Health and Social Justice in the Age of Chadwick: Britain, 1800–1854.* Cambridge: Cambridge University Press, 1998.

———. "Public Sphere to Public Health: The Transformation of 'Nuisance.'" In *Medicine, Health and the Public Sphere in Britain, 1600–2000*, ed. Steve Sturdy, 189–204. London: Routledge, 2002.

———. "Sanitary Policing and the Local State, 1873–1874: A Statistical Study of English and Welsh Towns." *Social History of Medicine* 28, no. 1 (2005): 39–61.

Hannah, Leslie. *Electricity before Nationalisation: A Study of the Development of the Electricity Supply in Britain to 1948.* Baltimore: Johns Hopkins University Press, 1979.

Hardy, Anne. "Food, Hygiene, and the Laboratory: A Short History of Food Poisoning in Britain, *circa* 1850–1950." *Social History of Medicine* 12, no. 2 (1999): 293–311.

Harris, Stanley. *The Development of Gas Supply on North Merseyside, 1815–1949: A Historical Survey of the Former Gas Undertakings of Liverpool, Southport, Prescot, Ormskirk and Skelmersdale.* Liverpool: North Western Gas Board, 1956.

Hassan, John. *The Seaside, Health and Environment in England and Wales since 1800.* Aldershot: Ashgate, 2003.

Hennock, E. P. *Fit and Proper Persons: Ideal and Reality in Nineteenth-Century Urban Government.* Montreal: McGill-Queen's University Press, 1973.

Howes, David, ed. *The Varieties of Sensory Experience: A Sourcebook in the Anthropology of the Senses*. London: University of Toronto Press, 1991.

Hughes, Thomas. *Networks of Power: Electrification and Western Society, 1880–1930*. London: Johns Hopkins University Press, 1983.

Inglis, Simon. *The Football Grounds of England and Wales*. London: Willow, 1983.

Jacyna, L. S. "Immanence or Transcendance: Theories of Life and Organization in Britain, 1790–1835." *Isis* 74, no. 243 (September 1983): 311–29.

Jakle, John. *City Lights: Illuminating the American Night*. Baltimore: Johns Hopkins University Press, 2001.

Jay, Martin. *Downcast Eyes: The Denigration of Vision in Twentieth-Century French Thought*. London: University of California Press, 1994.

Johnston, Sean F. *A History of Light and Colour Measurement: Science in the Shadows*. London: Institute of Physics Publishing, 2001.

Jonas, Hans. *The Phenomenon of Life: Toward a Philosophical Biology*. New York: Harper & Row, 1966.

Joyce, Patrick. *The Rule of Freedom: Liberalism and the Modern City*. London: Verso, 2003.

Jütte, Robert. *A History of the Senses from Antiquity to Cyberspace*. Translated by James Lynn. Cambridge: Polity, 2005.

Kearns, Gerry. "Cholera, Nuisances and Environmental Management in Islington, 1830–55." In *Living and Dying in London*, ed. W. T. Bynum and Roy Porter, 94–125. London: Welcome Institute for the History of Medicine, 1991.

Koskela, Hille. "Cam Era—the Contemporary Urban Panopticon." *Surveillance and Society* 1, no. 3 (2003): 292–313.

Koven, Seth. *Slumming: Sexual and Social Politics in Victorian London*. Princeton, NJ: Princeton University Press, 2004.

Laporte, Dominique. *History of Shit*. Translated by Rodolphe el-Khoury and Nadia Benabid. London: MIT Press, 2000.

Laqueur, Thomas. *Solitary Sex: A Cultural History of Masturbation*. New York: Zone, 2003.

Latour, Bruno. *Science in Action: How to Follow Scientists and Engineers through Society*. Milton Keynes: Open University Press, 1987.

———. *We Have Never Been Modern*. Translated by Catherine Porter. London: Harvester, 1993.

———. *Reassembling the Social: An Introduction to Actor-Network Theory*. Oxford: Oxford University Press, 2005.

Latour, Bruno, and Emilie Hermant. *Paris ville invisible*. Paris: Institut Synthélabo pour le progrès de la connaissance, 1998.

Lawrence, Christopher. "The Nervous System and Society in the Scottish Enlightenment." In *Natural Order: Historical Studies of Scientific Culture*, ed. Barry Barnes and Steven Shapin. London: Sage, 1979.

Leder, Drew. *The Absent Body*. Chicago: University of Chicago Press, 1990.

Lefebvre, Henri. *The Production of Space*. Translated by Donald Nicholson-Smith. Oxford: Blackwell, 1991.

Levin, David, ed. *Modernity and the Hegemony of Vision*. London: University of California Press, 1993.

Lubenow, William C. *The Politics of Government Growth: Early Victorian Attitudes toward State Intervention, 1833–1848*. Newton Abbot: David & Charles Archon, 1971.

Luckin, Bill. "'The Heart and Home of Horror': The Great London Fogs of the Late Nineteenth Century." *Social History* 28, no. 1 (January 2003): 31–48.

MacDonagh, O. "The Nineteenth-Century Revolution in Government: A Reappraisal." *Historical Journal* 1, no. 1 (1958): 59–73.

———. *Early Victorian Government, 1830–1870*. New York: Holmes & Meier, 1977.

MacLeod, Roy. "Science and Government in Victorian England: Lighthouse Illumination and the Board of Trade, 1866–1886." *Isis* 60 (1969): 5–38.

———, ed. *Government and Expertise: Specialists, Administrators, and Professionals, 1860–1919*. Cambridge: Cambridge University Press, 1988.

McLaren, John. "Nuisance Law and the Industrial Revolution—Some Lessons from Social History." *Oxford Journal of Legal Studies* 3, no. 2 (1983): 155–221.

McQuire, Scott. "Dream Cities: The Uncanny Powers of Electric Light." *SCAN: Journal of Media Arts Culture* 1, no. 2 (2004), http://scan.net.au/scan/journal/display.php?journal_id=31.

Mandler, Peter, ed. *Liberty and Authority in Victorian Britain*. Oxford: Oxford University Press, 2006.

Marcus, Sharon. *Apartment Stories: City and Home in Nineteenth-Century Paris and London*. Berkeley and Los Angeles: University of California Press, 1999.

Markus, Thomas. *Buildings and Power: Freedom and Control in the Origin of Modern Building Types*. London: Routledge, 1993.

Marsden, Ben, and Crosbie Smith. *Engineering Empires: A Cultural History of Technology in Nineteenth-Century Britain*. Basingstoke: Palgrave Macmillan, 2005.

Matheisen, T. "The Viewer Society." *Theoretical Criminology* 1, no. 2 (1997): 215–34.

Merleau-Ponty, Maurice. *Phenomenology of Perception*. Translated by Colin Smith. London: Routledge & Kegan Paul, 1962.

Miller, Jacques-Alain. "La despotisme de l'utile: La machine panoptique de Jeremy Bentham." *Ornicar?* 3 (1975): 3–36.

Miller, William. *Anatomy of Disgust*. London: Harvard University Press, 1997.

Millward, Robert. "The Political Economy of Urban Utilities." In *The Cambridge Urban History of Britain*, vol. 3, *1840–1950*, ed. Martin Daunton, 315–50. Cambridge: Cambridge University Press, 2000.

Mitchell, Timothy. *Colonising Egypt*. Cambridge: Cambridge University Press, 1988.

Moran, Richard. *Executioner's Current: Thomas Edison, George Westinghouse, and the Invention of the Electric Chair*. New York: Vintage, 2003.

Mort, Frank. *Dangerous Sexualities: Medico-Moral Politics in England since 1830*. London: Routledge & Kegan Paul, 1987.

Morus, Iwan Rhys. *Frankenstein's Children: Electricity, Exhibition, and Experiment in Early-Nineteenth-Century London*. Princeton, NJ: Princeton University Press, 1998.

⸻. "The Measure of Man: Technologising the Victorian Body." *History of Science* 37, no. 3 (1999): 249–82.

⸻, ed. *Bodies/Machines*. Oxford: Berg, 2002.

Mosley, Stephen. *The Chimney of the World: A History of Smoke Pollution in Victorian and Edwardian Manchester*. Cambridge: White Horse, 2001.

Mumford, Lewis. *Technics and Civilization*. New ed. New York: Harbinger, 1963.

Musselman, Elizabeth Green. *Nervous Conditions: Science and the Body Politic in Early Industrial Britain*. Albany: State University of New York Press, 2006.

Nead, Lynda. *Victorian Babylon: People, Streets and Images in Nineteenth-Century London*. London: Yale University Press, 2000.

⸻. "Animating the Everyday: London on Camera circa 1900." *Journal of British Studies* 43 (2004): 65–90.

⸻. "Velocities of the Image, c. 1900." *Art History* 27, no. 5 (November 2004): 746–70.

Norris, Clive, and Gary Armstrong. *The Maximum Surveillance Society: The Rise of CCTV*. New York: Berg, 1999.

Nye, David. *Electrifying America: Social Meanings of a New Technology, 1880–1940*. London: MIT Press, 1990.

Ogborne, Miles. "Traffic Lights." In *City A-Z*, ed. Steve Pile and Nigel Thrift. London: Routledge, 2000.

Ong, Walter. "The Shifting Sensorium." In *Varieties of Sensory Experience: A Sourcebook in the Anthropology of the Senses*, ed. David Howes. London: University of Toronto Press, 1991.

Otis, Laura. *Networking: Communicating with Bodies and Machines in the Nineteenth Century*. Ann Arbor: University of Michigan Press, 2001.

Owen, David. *The Government of Victorian London, 1855–1889: The Metropolitan Board of Works, the Vestries, and the City Corporation*. Edited by Roy Macleod. Cambridge, MA: Belknap, 1982.

Park, David. *The Fire within the Eye: A Historical Essay on the Nature and Meaning of Light*. Princeton, NJ: Princeton University Press, 1997.

Parry, Jonathan. *The Rise and Fall of Liberal Government in Victorian Britain*. New Haven, CT: Yale University Press, 1993.

Penner, B. "A World of Unmentionable Suffering: Women's Public Conveniences in Victorian London." *Journal of Design History* 14, no. 1 (2001): 35–52.

Picker, John. "The Soundproof Study: Victorian Professionals, Work Space and Urban Noise." *Victorian Studies* 42, no. 3 (Spring 1999/2000): 427–48.

Pickering, Andrew. *The Mangle of Practice: Time, Agency and Science*. London: University of Chicago Press, 1995.

Platt, Harold. *Shock Cities: The Environmental Transformation and Reform of Manchester and Chicago*. Chicago: University of Chicago Press, 2005.

Poovey, Mary. *Making a Social Body: British Cultural Formation, 1830–1864.* Chicago: University of Chicago Press, 1995.

Porter, Theodore. *The Rise of Statistical Thinking.* Princeton, NJ: Princeton University Press, 1986.

Prest, John. *Liberty and Locality: Parliament, Permissive Legislation and Ratepayers' Democracies in the Nineteenth Century.* Oxford: Clarendon, 1990.

Rabinbach, Anson. *The Human Motor: Energy, Fatigue and the Origins of Modernity.* Berkeley and Los Angeles: University of California Press, 1992.

Rabinow, Paul. *French Modern: Norms and Forms of the Social Environment.* Cambridge, MA: MIT Press, 1989.

Redford, A. *The History of Local Government in Manchester.* 3 vols. London: Longmans, Green, 1939.

Richards, Thomas. *The Commodity Culture of Victorian Britain: Advertising and Spectacle, 1851–1914.* Stanford, CA: Stanford University Press, 1990.

Ripps, H., and R. A. Weale. "The Visual Photoreceptors." In *The Eye*, vol. 2A, *Visual Function in Man*, ed. Hugh Davson. New York: Academic, 1969.

Ritvo, Harriet. *The Animal Estate: The English and Other Creatures in the Victorian Age.* Cambridge, MA: Harvard University Press, 1987.

Roberts, Robert. *The Classic Slum: Salford Life in the First Quarter of the Century.* Harmondsworth: Penguin, 1974.

Rose, Nikolas. *Powers of Freedom: Reframing Political Thought.* Cambridge: Cambridge University Press, 1999.

Schapiro, Gary. *Archaelogies of Vision: Foucault and Nietzsche on Seeing and Saying.* Chicago: University of Chicago Press, 2003.

Schivelbusch, Wolfgang. *Disenchanted Night: The Industrialisation of Light in the Nineteenth Century.* Translated by Angela Davis. New York: Berg, 1988.

Schlör, Joachim. *Nights in the Big City: Paris, Berlin, London, 1840–1930.* Translated by Pierre Gottfried Imhof and Dafydd Rees Roberts. London: Reaktion, 1998.

Schroeder, Fred. "More 'Small Things Forgotten': Domestic Electric Plugs and Receptacles, 1881–1931." *Technology and Culture* 27, no. 3 (1986): 525–43.

Schwartz, Vanessa. *Spectacular Realities: Early Mass Culture in Fin-de-Siècle Paris.* Berkeley and Los Angeles: University of California Press, 1998.

Sennett, Richard. *The Fall of Public Man.* New York: Norton, 1976.

Shapin, Steven. *A Social History of Truth: Civility and Science in Seventeenth-Century England.* Chicago: Chicago University Press, 1994.

Shapin, Steven, and Simon Schaffer. *Leviathan and the Air-Pump: Hobbes, Boyle and the Experimental Life.* Princeton, NJ: Princeton University Press, 1985.

Simmel, Georg. "Sociology of the Senses: Visual Interaction." In *Introduction to the Science of Sociology.* ed. Robert E. Park and Ernest W. Burgess, 356–61. Chicago: University of Chicago Press, 1921.

Stallybrass, Peter, and Allon White. *The Politics and Poetics of Transgression.* Ithaca, NY: Cornell University Press, 1986.

Sternberger, Dolf. *Panorama of the Nineteenth Century*. Translated by Joachim Neu-
    groschel. New York: Urizen, 1977.
Stevens, F. L. *Under London: A Chronicle of London's Underground Life-Lines and
    Relics*. London: J. M. Dent & Sons, 1939.
Stocking, George W. *Victorian Anthropology*. New York: Free Press, 1987.
Sullivan, Joseph P. "Fearing Electricity: Overhead Wire Panic in New York City."
    IEEE Technology and Society Magazine 14, no. 3 (1995): 8–16.
Sykes, Alan. *The Rise and Fall of British Liberalism, 1776–1988*. New York: Long-
    man, 1997.
Taylor, Nicholas, and David Watkinn. "Lamp-Posts: Decline and Fall in Cam-
    bridge." *Architectural Review* 129 (June 1961): 423–26.
Tester, Keith, ed. *The Flâneur*. London: Routledge, 1994.
Thorsheim, Peter. *Inventing Pollution: Coal, Smoke, and Culture in Britain since 1800*.
    Athens: Ohio University Press, 2006.
Trench, Richard, and Ellis Hillman. *London under London: A Subterranean Guide*.
    New ed. London: J. Murray, 1993.
Varnon, P. "Street Furniture: Survey of Street Lighting." *Architectural Review* 110
    (July 1951): 51–55.
Vernon, James. "The Ethics of Hunger and the Assembly of Society: The Techno-
    Politics of the School Meal in Modern Britain." *American Historical Review*
    110, no. 3 (June 2005): 693–725.
Vigarello, Georges. *Concepts of Cleanliness: Changing Attitudes in France since the
    Middle Ages*. Translated by Jean Birrell. Cambridge: Cambridge University
    Press, 1988.
Waddington, Kier. "Unfit for Human Consumption: Tuberculosis and the Prob-
    lem of Infected Meat in Late-Victorian Britain." *Bulletin of the History of
    Medicine* 77, no. 3 (2003): 636–61.
Wade, Nicholas. *A Natural History of Vision*. Cambridge, MA: MIT Press, 1998.
Walkowitz, Judith. *City of Dreadful Delight: Narratives of Sexual Danger in Late-
    Victorian London*. Chicago: University of Chicago Press, 1992.
Williams, Rosalind. *Dream Worlds: Mass Consumption in Late Nineteenth-Century
    France*. Berkeley and Los Angeles: University of California Press, 1982.
Wilson, John. *Lighting the Town: A Study of Management in the North West Gas In-
    dustry, 1805–1880*. London: P. Chapman, 1991.
Winner, Langdon. *Autonomous Technology: Technics-out-of-Control as a Theme in
    Political Thought*. Cambridge, MA: MIT Press, 1978.
Winter, Alison. *Mesmerized: Powers of Mind in Victorian Britain*. Chicago: Univer-
    sity of Chicago Press, 1998.
Winter, James. *London's Teeming Streets: 1830–1914*. London: Routledge, 1993.
Wohl, Anthony. *Endangered Lives: Public Health in Victorian Britain*. London: Dent,
    1983.
———. *The Eternal Slum: Housing and Social Policy in Victorian London*. New
    Brunswick, NJ: Transaction, 2002.

Woodward, G. "Staite and Petrie: Pioneers of Electric Lighting." *Science, Measurement and Technology: IEE Proceedings A* 136, no. 6 (November 1989): 290–96.

Yar, Majid. "Panoptic Power and the Pathologisation of Vision: Critical Reflections on the Foucauldian Thesis." *Surveillance and Society* 1, no. 3 (2003): 254–71.

Young, R. *Mind, Brain and Adaptation in the Nineteenth Century: Cerebral Localization from Gall to Ferrier.* Oxford: Clarendon, 1970.

Zajonc, Arthur. *Catching the Light: The Entwined History of Light and Mind.* New York: Oxford University Press, 1993.

Index

Page numbers in italics refer to illustrations.

liberalism, 10–12; bodily capacity and, 18; economic dynamism and, 97; engineering and, 17; and infrastructure, 153; and inspection, 123–24, 134; and local government, 68, 132, 251; material history of Western, 2; negative liberty, 11; and noninvasive scrutiny, 196; party politics and, 12, 272n65; positive liberty, 11, 14; and privacy, 123–24, 133; and smoke, 86, 96; and technology, 19; unsystematic nature of, 12; in Victorian Britain, 11; and vision, 46–54, 96, 258–60. *See also* freedom; government; liberal subject

liberal subject: architecture and, 75; definition, 11; and desensitized subject, 56–57; engineering and, 17, 69; exclusive nature of, 11, 54, 260, 271n58; gas meters and, 148; gendered nature of, 58, 271n58; and illumination, 174, 193, 194, 205, 207, 222, 234; infrastructure and, 18, 262–63; inspection and, 113, 134; privacy and, 100, 257; technology and, 262–63; visual practice and, 19, 46–54, 56, 61, 63, 69, 132–33, 258, 258–60

libraries: effect of gaslight on books in, 208; illumination in, 202–3, 208, 324n172; multiple perceptual patterns in, 256, 259; oligoptic elements, 259; perception in, 19, 97, 255; visual form of, 76, *76*

Liddle, John, 55

light: ancient, law of, 70–72, 183; and broader environment, 17, 20; calculation of quantity, 70, *71,* 104; distribution of, in streets, 79, *80;* electric, nature of, 211; floods of, 8; industrialization of, 136; levels of, 162–65, 227; measurement of artificial, 136, 154–71; nature of, 155, 313n89; in nineteenth-century Britain, 1; and power, 4; right to, 70–72; standards of, 158–61; white, 185, 192. *See also* electric light; illumination; lighting; photometry

lighthouses, 169, 189–90, 191, *191,* 322n112

lighting: basement, 87, *88;* and crime, 99; in houses, 197; Mill on, 13; nuanced, in urban areas, 224–26; in prisons, 232; in rural areas, 223; Smith on, 13; schools,

90; streets, 69, 220–29, *229,* 244–50; for specific purposes, 173, 175; technology, 173–213; tower, 221–22, 270n48

lighting-up times, 222–23, *222*

lights, traffic, 112

Liverpool: cellar dwellings in, 69; gas mantles in, 176, 178; illumination in, 178, 223; inspection in, 101, 103, 108, 146; lodging houses in, 104; streets, 78

Livesey, George, 148

Local Government in England (Redlich), 99

Locke, John, 26

Lodge, Oliver, 211

lodging houses: definition of, 104, 300n38; illumination of, 232; inspection of, 101, 104–5, 106, 115, 121, 133, 260; privacy in, 127

Lombroso, Cesare, 52

London, City of: electric light in, 240, 243–50, *249,* 261; government of, 244; history of, 244; infrastructure in, 335n183; inspection in, 102; light levels in, 162, 164, 165, 314n120

London: cabs in, 192; electric plants in, 240; Embankment, illumination of, 181; fogs, 83, 84–85, 86, 106; food, 107; gas companies in, 153–54, 162, 312n82; gas examiners, 163, 315n130; gas explosions, 238; gas holders, 151, *152;* gas meters in, 146; gasworks in, 151–53, 239; government of, 67–68; illumination of, 137, *177, 180,* 193, 214, 220–21, 222–23, 227, 243–50; inspection in, 101, 110; light levels, 162; lodging houses, 104; markets, 184–85; photometry, 161; slaughterhouses, 110; smoke and prepayment meters, 149; spectacle in, 7; streets, 92, 93; subways, 216–18, *217,* 246–47, 328n16; technological "backwardness" of, 250–51. *See also* London, City of

London Building Act (1894), 90

London County Council, 68, 112

London Overhead Wires Act (1891), 243

London Slaughterhouse Act (1874), 129

Loudon, James, 87

Lubenow, William, 100

lumen, 314n106

lux, 170, 314n106

Lynde, John, 70

macadam, 92
MacDonagh, Oliver, 100
MacLeod, Roy, 190, 273n76
MacKenzie, William, 34, 35
MacNish, Robert, 43
macula lutea, 28, *31*
Magasins du Louvre, 184
Magendie, François, 27
magnifying glasses, 45
Maier, Julius, 187
Maine de Biran, 25
mains, electricity, 215, 216–20, *219*,
    245–47, *246*, 247, 328n9
mains, gas, 20, 138–39, 236, 237, 238, 239;
    distance covered, 151–53; expansion of
    network, 151; inspection of, 111; laying,
    139, *140*
mains, water, 111
Malebranche, Nicolas, 58
Manchester: cellar dwellings in, 69; court
    with shared facilities in, *65;* Deansgate
    improvement, 79–82; electric light in,
    178; fog in, 190–91; gaslight, 137, 261;
    gasworks, 153; government of, 68;
    inspection in, 101; lodging houses of,
    105; measurement, municipal, 155;
    Nuisance Committee, 106, 119;
    photometry, 164; physical
    disorganization of, in early nineteenth
    century, 67, 72; senses, and class in, 55,
    64; smoke in, 84, 117; streets, 78, 94, 95
Manchester and Salford Sanitary
    Association, 117
Manchester Association for the Prevention
    of Smoke, 84
Mandler, Peter, 11
manholes, 112, 218–20, 247, 253
Mansion House, 245, 248, *249*
Manual of School Hygiene (Hope and
    Browne), 43
maps, 53–54
Marcet, William, 84
Marcus, Sharon, 266n8, 292n52
Marylebone, 119, 162
masturbation, 126, 306n170
material agency, 16, 253, 274n87
materiality, 5, 16, 192, 253, 262
Matthews, William, 136, 151, 154, 156, 200
Maudsley, Henry, 32, 57
Maxwell, James Clerk, 211
measurement, 135, 136, 154–55, 312n87

meat, 108–9, 110, *111*, 120, 123, 184–85
medical illumination, 10, 90, 173, 195–96,
    195, 196
Merleau-Ponty, Maurice, 28, 46
Merriman, Owen, 237
meters: abusing, 144; accuracy of, 146;
    average system, 147–48; and class, 148;
    dry and wet, for gas, 144–45; electricity,
    20, 235, 332n107; gas, 113, 144–49, *144*,
    253; prepayment (penny-in-the-slot),
    148–49, *149*, 213, 263; reading, 144,
    146–47; repairing, *147*; standardizing,
    146; testing, 146; two-rate, 235
Metropolis Gas Act (1860), 154
Metropolis Management Act (1855), 67–68
Metropolitan Board of Works (MBW): and
    City of London, 244; establishment of,
    68; and illumination, 222–23;
    inspection and, 101; inspectors of, 117;
    and photometry, 161, 163; streets, 78,
    79
Metropolitan Building Act (1844), 124, 129
Metropolitan Gas Act (1860), 13, 162, 163
microscopy, 107, 123
military illumination, 10, 173, 186–87
milk, 109
Mill, John Stuart: ethology, 52; on the state,
    11, 132; on infrastructure, 13; on
    lighting, 13; on technology, 18; on
    disturbing spectacles, 129–31
Miller, William, 59
mills, perception in, 75, 183–84
mines, 118, 134, *233*
Miquel, Pierre, 95
Mitchell, Timothy, 268n21
model cities, 72–73
model clauses acts, 68
model housing, 110, 124, 127, 232
modernity, 338n14
Molera, E. J., 232
Molesworth, Reverend John, 84
moonlight, 192, 223
Moorfields eye hospital, 34, 110
Mordey, William, 174
Morris, William, 9
Morus, Iwan Rhys, 317n3
motor functions, separated from senses, 27
motorist, perception of, 257
motors, 215, 234, 327n3
Mouat, Frederick, 307n191
Moyle, William, 109

Plumber and Sanitary Engineer, 67
policing, 64, 101, 105, 108, 193–95
pollution, 106–7, 148–49, 236–37
polyopticon, 5, 268n26
population, government of, 14
portability, and inspection, 113
Porterfield, William, 33
Porter, Roy, 278n18
postal service, 114, 119
power, 1, 3, 4, 5, 258–60
Pray's astigmatism test, *39*
Preece, William: on adaptation of eye, 31;
    on color of illuminants, 185; on dioptric
    reflectors, 228; on electric cables, 242;
    on electric light, 17, 211–12, 240; on
    electric switches, 231; on illumination
    of the City of London, 244; on
    photometry, 165, 168, 169, 170
presbyopia, 32, 33, 38
Prescription Act (1832), 70
Preston-Thomas, Herbert, 133
prisons, illumination of, 232, 331n91
privacy: in bylaw housing, 75; constitution
    of, 63, 73, 123–31, 257; in hospitals,
    127; in Hygeia, 73; illumination and, 9,
    222; inspection and, 20, 100, 103,
    120–21, 255; liberalism and, 123–24,
    133, 257; lodging houses and, 105;
    oligoptic space and, 74, 255; suspicion
    of, 72; and toilets, 124–26
private space, 124–31
privies. *See* toilets
Production of Space, The (Lefebvre), 24;
proprioception, 277n9
prostitutes, 58, 115, 287n181
proxemics, 60
public analysts, 108
public conveniences, 126
Public Health Act (1875), 101, 104, 118,
    127, 128
Public Health Act (1890), 243
Public Health (London) Act (1891), 101,
    305n140
Public Health (Scotland) Act (1897), 101
public space, 50, 72–73, 75, 84–85, 96–97
public sphere, 50
pubs, 337n12
pulmonary disorders, 84
pupil, 9, 30
pupil, artificial, 35
Purkinje effect, 29, 168

Rabinow, Paul, 73
race, and perception, 57–58, 254, 336n2
radiometry, 155, 313n91
railway carriages, illumination of, 170, 198
railway signals, illumination of, 188–89
reading: in bed, 10; by candlelight, 204,
    205; correct practice, 41, *42;*
    illumination and, 201–3; illumination
    levels and, 170; incorrect practice, 41,
    42; lamps, 198, *199*, 256
Rebuilding of Manchester, The (Simon and
    Inman), 96
Redlich, Josef, 99, 100, 133
refraction: of the eye, 33–34; of spectacle
    lenses, 38
Regent Street, 78
Reid, George, 117
Reid, James, 128
Reid, Thomas, 27
Rendle, William, 116
respiratory disorders, 84
retina: artificial illuminants, effect of upon,
    9, 183, 192, 204, 206; cross section of,
    29; electric nature of, 211; rods and
    cones of, 28–29, *30;* searchlights and,
    187; social, Tarde on, 50; Thomas Reid
    on, 27. *See also* eye; perception; vision
Richardson, Benjamin, 62–63, 77, 241
rickets, 65
Ritchie, D. G., 18
Roberts, David, 100
Roberts, Robert, 56, 307n185
Rose, Nikolas, 50
Rosherville Zoological Gardens, 131
Royal Zoological Gardens, 131
Rumford, Count, photometer of, 135, 154,
    155, 156
Russell, James, 55, 61, 104–5
Russell, Rollo, 83, 84

saccadic motion, 32
Sala, George Augustus, 6
Sale of Food and Drugs Act (1875), 108, 164
Sale of Gas Act (1859), 13, 145–46, 311n43
San Francisco, 232, 234
Sanitary Act (1866), 104
Sanitary Evolution of London, The (Jephson),
    59, 104
Sanitary Inspector's Handbook (Taylor), 101
Sanitary Record, 73
sanitation, illumination and, 208–10